ROBUST INFERENCE

STATISTICS: Textbooks and Monographs

A SERIES EDITED BY

D. B. OWEN, Coordinating Editor

Department of Statistics
Southern Methodist University
Dallas, Texas

- Vol. 1: The Generalized Jackknife Statistic, *H. L. Gray and W. R. Schucany*
- Vol. 2: Multivariate Analysis, *Anant M. Kshirsagar*
- Vol. 3: Statistics and Society, *Walter T. Federer*
- Vol. 4: Multivariate Analysis: A Selected and Abstracted Bibliography, 1957-1972, *Kocherlakota Subrahmaniam and Kathleen Subrahmaniam* (out of print)
- Vol. 5: Design of Experiments: A Realistic Approach, *Virgil L. Anderson and Robert A. McLean*
- Vol. 6: Statistical and Mathematical Aspects of Pollution Problems, *John W. Pratt*
- Vol. 7: Introduction to Probability and Statistics (in two parts), Part I: Probability; Part II: Statistics, *Narayan C. Giri*
- Vol. 8: Statistical Theory of the Analysis of Experimental Designs, *J. Ogawa*
- Vol. 9: Statistical Techniques in Simulation (in two parts), *Jack P. C. Kleijnen*
- Vol. 10: Data Quality Control and Editing, *Joseph I. Naus* (out of print)
- Vol. 11: Cost of Living Index Numbers: Practice, Precision, and Theory, *Kali S. Banerjee*
- Vol. 12: Weighing Designs: For Chemistry, Medicine, Economics, Operations Research, Statistics, *Kali S. Banerjee*
- Vol. 13: The Search for Oil: Some Statistical Methods and Techniques, *edited by D. B. Owen*
- Vol. 14: Sample Size Choice: Charts for Experiments with Linear Models, *Robert E. Odeh and Martin Fox*
- Vol. 15: Statistical Methods for Engineers and Scientists, *Robert M. Bethea, Benjamin S. Duran, and Thomas L. Boullion*
- Vol. 16: Statistical Quality Control Methods, *Irving W. Burr*
- Vol. 17: On the History of Statistics and Probability, *edited by D. B. Owen*
- Vol. 18: Econometrics, *Peter Schmidt*
- Vol. 19: Sufficient Statistics: Selected Contributions, *Vasant S. Huzurbazar (edited by Anant M. Kshirsagar)*
- Vol. 20: Handbook of Statistical Distributions, *Jagdish K. Patel, C. H. Kapadia, and D. B. Owen*
- Vol. 21: Case Studies in Sample Design, *A. C. Rosander*
- Vol. 22: Pocket Book of Statistical Tables, *compiled by R. E. Odeh, D. B. Owen, Z. W. Birnbaum, and L. Fisher*
- Vol. 23: The Information in Contingency Tables, *D. V. Gokhale and Solomon Kullback*
- Vol. 24: Statistical Analysis of Reliability and Life-Testing Models: Theory and Methods, *Lee J. Bain*
- Vol. 25: Elementary Statistical Quality Control, *Irving W. Burr*
- Vol. 26: An Introduction to Probability and Statistics Using BASIC, *Richard A. Groeneveld*
- Vol. 27: Basic Applied Statistics, *B. L. Raktoe and J. J. Hubert*
- Vol. 28: A Primer in Probability, *Kathleen Subrahmaniam*
- Vol. 29: Random Processes: A First Look, *R. Syski*

Vol. 30: Regression Methods: A Tool for Data Analysis, *Rudolf J. Freund and Paul D. Minton*
Vol. 31: Randomization Tests, *Eugene S. Edgington*
Vol. 32: Tables for Normal Tolerance Limits, Sampling Plans, and Screening, *Robert E. Odeh and D. B. Owen*
Vol. 33: Statistical Computing, *William J. Kennedy, Jr. and James E. Gentle*
Vol. 34: Regression Analysis and Its Application: A Data-Oriented Approach, *Richard F. Gunst and Robert L. Mason*
Vol. 35: Scientific Strategies to Save Your Life, *I. D. J. Bross*
Vol. 36: Statistics in the Pharmaceutical Industry, *edited by C. Ralph Buncher and Jia-Yeong Tsay*
Vol. 37: Sampling from a Finite Population, *J. Hajek*
Vol. 38: Statistical Modeling Techniques, *S. S. Shapiro*
Vol. 39: Statistical Theory and Inference in Research, *T. A. Bancroft and C.-P. Han*
Vol. 40: Handbook of the Normal Distribution, *Jagdish K. Patel and Campbell B. Read*
Vol. 41: Recent Advances in Regression Methods, *Hrishikesh D. Vinod and Aman Ullah*
Vol. 42: Acceptance Sampling in Quality Control, *Edward G. Schilling*
Vol. 43: The Randomized Clinical Trial and Therapeutic Decisions, *edited by Niels Tygstrup, John M. Lachin, and Erik Juhl*
Vol. 44: Regression Analysis of Survival Data in Cancer Chemotherapy, *Walter H. Carter, Jr., Galen L. Wampler, and Donald M. Stablein*
Vol. 45: A Course in Linear Models, *Anant M. Kshirsagar*
Vol. 46: Clinical Trials: Issues and Approaches, *edited by Stanley H. Shapiro and Thomas H. Louis*
Vol. 47: Statistical Analysis of DNA Sequence Data, *edited by B. S. Weir*
Vol. 48: Nonlinear Regression Modeling: A Unified Practical Approach, *David A. Ratkowsky*
Vol. 49: Attribute Sampling Plans, Tables of Tests and Confidence Limits for Proportions, *Robert E. Odeh and D. B. Owen*
Vol. 50: Experimental Design, Statistical Models, and Genetic Statistics, *edited by Klaus Hinkelmann*
Vol. 51: Statistical Methods for Cancer Studies, *edited by Richard G. Cornell*
Vol. 52: Practical Statistical Sampling for Auditors, *Arthur J. Wilburn*
Vol. 53: Statistical Signal Processing, *edited by Edward J. Wegman and James G. Smith*
Vol. 54: Self-Organizing Methods in Modeling: GMDH Type Algorithms, *edited by Stanley J. Farlow*
Vol. 55: Applied Factorial and Fractional Designs, *Robert A. McLean and Virgil L. Anderson*
Vol. 56: Design of Experiments: Ranking and Selection, *edited by Thomas J. Santner and Ajit C. Tamhane*
Vol. 57: Statistical Methods for Engineers and Scientists. Second Edition, Revised and Expanded, *Robert M. Bethea, Benjamin S. Duran, and Thomas L. Boullion*
Vol. 58: Ensemble Modeling: Inference from Small-Scale Properties to Large-Scale Systems, *Alan E. Gelfand and Crayton C. Walker*
Vol. 59: Computer Modeling for Business and Industry, *Bruce L. Bowerman and Richard T. O'Connell*
Vol. 60: Bayesian Analysis of Linear Models, *Lyle D. Broemeling*
Vol. 61: Methodological Issues for Health Care Surveys, *Brenda Cox and Steven Cohen*
Vol. 62: Applied Regression Analysis and Experimental Design, *Richard J. Brook and Gregory C. Arnold*

Vol. 63: Statpal: A Statistical Package for Microcomputers – PC-DOS Version for the IBM PC and Compatibles, *Bruce J. Chalmer and David G. Whitmore*

Vol. 64: Statpal: A Statistical Package for Microcomputers – Apple Version for the II, II+, and IIe, *David G. Whitmore and Bruce J. Chalmer*

Vol. 65: Nonparametric Statistical Inference, Second Edition, Revised and Expanded, *Jean Dickinson Gibbons*

Vol. 66: Design and Analysis of Experiments, *Roger G. Petersen*

Vol. 67: Statistical Methods for Pharmaceutical Research Planning, *Sten W. Bergman and John C. Gittins*

Vol. 68: Goodness-of-Fit Techniques, *edited by Ralph B. D'Agostino and Michael A. Stephen Stephens*

Vol. 69: Statistical Methods in Discrimination Litigation, *edited by D. H. Kaye and Mikel Aickin*

Vol. 70: Truncated and Censored Samples from Normal Populations, *Helmut Schneider*

Vol. 71: Robust Inference, *M. L. Tiku, W. Y. Tan, and N. Balakrishnan*

Vol. 72: Statistical Image Processing and Graphics, *edited by Edward J. Wegman and Douglas J. DePriest*

OTHER VOLUMES IN PREPARATION

ROBUST INFERENCE

M. L. Tiku
McMaster University
Hamilton, Ontario, Canada

W. Y. Tan
Memphis State University
Memphis, Tennessee

N. Balakrishnan
McMaster University
Hamilton, Ontario, Canada

MARCEL DEKKER, INC. New York and Basel

Dedicated to

Professor D. J. Finney, D.Sc., F.R.S., F.R.S.E.

and

Professor D. N. Lal, Ph.D., F.N.I.

Library of Congress Cataloging-in-Publication Data

Tiku, Moti Lal.
Robust inference.

(Statistics, textbooks and monographs ; vol. 71)
Bibliography: p.
Includes index.
1. Robust statistics. 2. Nonparametric statistics.
3. Estimation theory. I. Tan, W. Y. [date]
II. Balakrishnan, N. [date] III. Title.
IV. Series: Statistics, textbooks and monographs ;
v. 71
QA276.T52 1986 519.5 86-13447
ISBN 0-8247-7532-5

MARCEL DEKKER, INC.
270 Madison Avenue, New York, New York 10016

Current printing (last digit):
10 9 8 7 6 5 4 3 2 1

PRINTED IN THE UNITED STATES OF AMERICA

Preface

Most classical statistical procedures are based on two assumptions, that the sample observations are independently and identically distributed, and that the underlying distribution is normal. While the first assumption may not be unrealistic in certain situations, it is the second assumption that is rather unrealistic from a practical point of view. To quote R. C. Geary (Biometrika, 1947): "Normality is a myth; there never was, and never will be, a normal distribution." This might be an overstatement, but the fact is that nonnormal distributions are more prevalent in practice, and to assume normality instead might lead to erroneous statistical inferences. On the other hand, to assume that nothing is known about the underlying distribution is to deny the information that such procedures as graph-plotting techniques and goodness-of-fit tests can provide about the shape of the underlying distribution. It is true that for moderate sample sizes, these procedures are not very successful in discriminating between normal and moderately nonnormal distributions, but they are very successful in distinguishing normal from extremely nonnormal symmetric distributions and extremely skew distributions (Cauchy, exponential, and the like). For a sample of size 20, for example, the Shapiro-Wilk omnibus goodness-of-fit test of normality has, at the 10% significance level, power values of 0.99 and 0.92 against exponential and Cauchy distributions, respectively; against a logistic distribution, however, this test (and other available tests of normality) has a very low power value of 0.16 (Tiku, 1974a, Table I). Thus in situations where one finds it expedient to assume normality, what should perhaps be assumed instead is that the underlying distribution is one of a reasonably wide class of distributions (consisting particularly of those distributions that constitute plausible alternatives to normality, e.g., short-tailed and long-tailed symmetric distributions and moderately skew distributions, all having finite mean and variance); which one exactly is not known. The normal distribution is, of course, a member of this class, but extremely nonnormal symmetric distributions and extremely skew distributions need not necessarily be included, since they are distinguishable from a normal and may be considered separately. The objective of this work is to develop robust estimation and hypothesis-testing procedures; a robust (to deviations from normality) procedure is defined to be one that is nearly as efficient as the classical procedure (based on the sample mean $\bar{x}$ and variance s^2) for a normal distribution but is considerably more efficient overall for nonnormal distributions, particularly distributions that constitute plausible alternatives to normality. This kind of robustness, we believe, is of real value from a practical point of view (Box and Tiao, 1964b). Of course, there are other perceptions and definitions of robustness, which in application seem to be limited to long-tailed symmetric distributions, and for these the reader is referred to the authoritative works of Hoaglin et al. (1983) and Huber (1981). What we have essentially developed in this book are robust estimation and hypothesis-testing procedures for means and variances. They are nearly as efficient as the classical procedures (based on the sample mean $\bar{x}$ and variance s^2) for a normal distribution but are considerably more efficient overall for plausible alternatives to normality. They also have the feature of adaptability; for example, simple variants of them are also highly efficient for extremely nonnormal symmetric distributions and extremely skew distributions. These new procedures are based on the so-called modified maximum likelihood estimators of location and scale parameters. The hypothesis-testing procedures developed in this book have robustness of validity as well as robustness of efficiency. Robustness of validity is the phenomenon that the type I error of a test procedure is stable from distribution to distribution; at any rate the type I error for plausible alternatives to normality is never too large compared to its normal-theory value. Robustness of efficiency is the phenomenon that the power function of a test procedure is sensitive to underlying distributions and the test is almost as powerful as the classical test for a normal distribution, but is more powerful overall for plausible alternatives to normality. The robust procedures developed in this book have another endearing feature: their mathematics and computations are completely analogous to those of the classical procedures and are therefore easy to understand. It must, however, be pointed out that these procedures, like most other robust procedures, might not have good robustness properties in the pathological situations where discontinuities occur toward the middle of the underlying distribution; such situations are very rare, though, and need not cause too much concern.

In Chapter 1, the robustness of the classical Student t and ANOVA tests are studied; it is shown that these tests do not have robustness of efficiency. It is also shown that the classical F test for testing the equality of two population variances does not even have robustness of validity. In Chapter 2, the phenomenon of reducing nonnormality through censoring of extreme observations is explained. It is indicated how censoring of extreme observations can lead to robust procedures. A few well-known methods (BLUE, maximum likelihood, etc.) of estimation are presented. The method of modified maximum likelihood (MML) estimation is described in detail and the properties of the MML estimators of location and scale parameters, based on type II censored normal samples, are studied. Their efficiencies for normal and nonnormal distributions are evaluated and compared in particular with the

Huber M estimators. The MML estimators are shown to have remarkably high efficiencies besides being easy to compute. In Chapter 3, the MML estimators (based on type II censored samples) are worked out for a few other important distributions (e.g., log-normal, logistic, exponential, and Rayleigh distributions). The efficiencies of these estimators are evaluated and shown to be remarkably high compared to the BLUE and the maximum likelihood estimators. These MML estimators may be used in constructing robust (to deviations from the respective distribution) hypothesis-testing procedures, exactly on the same lines as in Chapter 4. In Chapter 4, the MML10 estimators (MML estimators based on normal samples with about 10% smallest and largest observations censored) are used in constructing robust (to deviations from normality) hypothesis-testing procedures. For testing the population means, statistics analogous to Student's t are defined. Their distributional and power properties are studied and the tests are shown to have robustness of validity as well as robustness of efficiency. Tests based on simple variants of the MML10 estimators (e.g., MML30 estimators and MML estimators based on samples censored only on one side) are developed. These tests are shown to be more powerful than the tests based on the MML10 estimators for extremely nonnormal symmetric distributions and extremely skew distributions, respectively. The tests based on the MML10 and MML30 estimators are compared with the Huber and Sprott procedures in particular and shown to perform remarkably well, if not better. Robust tests for testing equality of population variances are also given and are shown to be considerably more powerful than other available robust tests, including some of the popular nonparametric tests. In Chapter 5 the specialized topics of regression, analysis of variance, classification, and sample survey are considered, and the role of the MML estimators in deriving robust procedures in these areas is explained. The regression and analysis-of-variance procedures based on the MML estimators are shown to have robustness of validity as well as robustness of efficiency. A robust classification procedure based on the MML estimators is developed and is shown to be superior to the classical Fisher-Anderson procedure. In the context of sample survey, the MML estimator of a population mean is shown to be considerably more efficient overall than the traditional estimator, the sample mean. In Chapter 6, the important topic of goodness of fit is discussed. Various approaches to testing an assumed distribution, from complete as well as censored samples, are presented. A goodness-of-fit test can be of great value in choosing an appropriate robust procedure (MML10, MML30, etc.). A goodness-of-fit test can also explain the differences in results of the classical and robust procedures. This is illustrated through numerical examples. In Chapter 7, tests for detecting outliers are given. Since the classical procedures based on sample mean and variance are highly inefficient if the sample contains outliers, it is advisable not to use them if a test detects outliers. Instead, it is advantageous to use robust procedures if outliers are suspected. Finally, in a special appendix, Q-Q plots are described. Their use in capturing a departure from normality is outlined.

This book is intended as a reference for theoreticians and practitioners of statistics and can also be used as an auxiliary text for a senior-level course on robustness. The book is also intended to illustrate the appropriateness of various robust methods for solving numerous statistical inference problems in both one-sample and multisample situations. This monograph has provided the senior author an opportunity to consolidate and revise a series of his papers on robust methods and on censored sampling which have appeared over the past two decades in various statistical journals. A basic knowledge of sampling distributions, statistical inference, and order statistics is assumed. Numerical illustrations are given for most procedures, which, we hope, make them easy to understand. The book contains numerous examples dealing with real data which explain the usefulness of robust methods. For readers interested in research on robustness, fairly extensive references to recent works are given.

This book was written by invitation of Dr. Maurits Dekker to the senior author. We are grateful for this invitation, as it provided an impetus for writing the book. We are also very grateful to the reviewers for their comments and suggestions, which led to a substantial improvement of the original draft.

M. L. Tiku
W. Y. Tan
N. Balakrishnan

Contents

1
Robustness of Some Classical Tests

1.1 INTRODUCTION

Statistical methods have been used in almost every applied field to analyze experimental data. But unfortunately, abuses and misuses of statistical methods are very common. For example, Gore et al. (1977) noted that 32 of the 62 reports that appeared in the British Medical Journal from January to March 1976 contained errors in the use of statistics. Gibbons and Freund (1980) noted that in 28 of the 30 papers in the March/April 1978 issue of the Agronomy Journal (Vol. 70, No. 22), the statistical methodology was poorly documented or incomplete and also occasionally inappropriate. One of the major reasons for such misuses is that many practitioners are unaware of the fact that statistical methods are derived under certain sets of assumptions; yet, in practice, many of these assumptions do not hold. For example, many of the statistical procedures, such as the important classical t test and variance-ratio tests, are derived under the assumption of normality. However, as noted by Geary (1947) and Scheffé (1959), the assumption of normality very often is not a realistic one. A crucial question then is: What are the effects of nonnormality on the normal theory procedures? A statistical test has traditionally been called robust if both its type I error (significance level) and its power are not affected drastically by departures from normality. Box and Tiao (1964a,b), Tiao and Lund (1970), and Tiao and Ali (1971) have in fact advocated that when the underlying distribution is not necessarily normal, a statistical procedure should be derived under a reasonably wide class of distributions consisting of the normal and plausible alternatives to the normal distribution, thus enhancing the scope of application of this procedure. It is, therefore, advantageous and desirable to develop tests that have (1) type I error robustness and (2) high power (preferably maximum power) under normal and a reasonably wide class of nonnormal distributions. The purpose of this book is to illustrate how to develop such tests by using modified maximum likelihood (MML) estimators and other so-called robust estimators. To set the groundwork, we study first in this chapter the effects of nonnormality on the type I error and the power function of the important classical t and variance-ratio tests.

1.2 STUDENT'S t TEST

Let $x_1, \ldots, x_n$ be a random sample from a normal distribution $N(\mu, \sigma)$ with mean μ and standard deviation σ (i.e., variance σ^2). One wants to test the null hypothesis $H_0: \mu = 0$ against the alternative hypothesis $H_1: \mu \neq 0$. The uniformly most powerful and unbiased (UMPU; see Lehmann, 1959, p. 167) test is, under the assumption of normality, given by the well-known Student's t statistic,

$$t = \sqrt{n}\,\frac{\bar{x}}{s} \tag{1.2.1}$$

where $\bar{x} = \Sigma_{i=1}^{n} x_i/n$ and $s^2 = \Sigma_{i=1}^{n} (x_i - \bar{x})^2/(n - 1)$ are the sample mean and sample variance, respectively. Small or large values of t lead to the rejection of H_0 in favor of H_1. The null (H_0 true) distribution of t is, under the assumption of normality, the central t distribution with $\nu = n - 1$ degrees of freedom and the nonnull (H_1 true) distribution of t is the noncentral t distribution with $\nu = n - 1$ degrees of freedom and noncentrality parameter $\Delta^2 = n\mu^2/\sigma^2$.

Thus one calculates the statistic t from the given sample and rejects H_0 in favor of H_1 at significance level α (usually $\alpha = 0.01$ or 0.05) if t is either smaller than $-t_{\alpha/2}(\nu)$ or greater than $t_{\alpha/2}(\nu)$, where $t_{\alpha/2}(\nu)$ is the $100(1-\alpha/2)\%$ point of the Student's t distribution with ν degrees of freedom.

If $H_1: \mu > 0$ is true, then the UMP test exists (under normality) and is given by: Reject H_0 in favor of H_1 at level α if $t > t_\alpha(\nu)$; similarly, the UMP test for H_0 versus $H_1: \mu < 0$ is given by: Reject H_0 in favor of H_1 at level α if $t < -t_\alpha(\nu)$. For further details, one might refer to Lehmann (1959, p. 97).

1.3 ASYMPTOTIC ROBUSTNESS OF THE t TEST

In practice, the normality assumption is not a realistic one. To quote Geary (1947): "Normality is a myth; there never was, and never will be, a normal distribution." This might be an overstatement, but the fact is that nonnormal distributions are more prevalent in practice. For assessing the effects of nonnormality on the t test, we therefore let the population distribution be a member of a sufficiently wide class of distributions with mean μ, variance σ^2, and finite higher cumulants κ_r, $r \geq 3$. Then, since s^2 is the second Fisher κ statistic, $\mathrm{Var}(s^2) = (\kappa_4/n) + 2\sigma^4/(n - 1)$ (see Kendall and Stuart, 1969, p. 286), which converges to 0 as $n \to \infty$, and s^2/σ^2 converges in probability to 1 as $n \to \infty$. Further, by the central limit theorem, the distribution of $\sqrt{n}(\bar{x} - \mu)/\sigma$ converges to normal $N(0,1)$ as $n \to \infty$. Thus, by Slutsky's theorem (see Rao, 1975, Chap. 2), the distribution of t converges to normal $N(0,1)$ as $n \to \infty$, which is also the asymptotic distribution of t under the assumption of normality. This shows that asymptotically the t test is robust with respect to departures from normality. One may note that to test H_0

against H_1: $\mu > 0$, the asymptotic power function of the t test above is given by $P[t \geq t_\alpha(\nu) \mid H_1] \simeq P(Z \geq Z_\alpha - \Delta)$, where $\Delta = \sqrt{n}\mu/\sigma$; Z is the standard normal variate N(0, 1) and Z_α is the $100(1-\alpha)\%$ point of N(0, 1). Thus the asymptotic power function of the t test is also insensitive to departures from normality. This insensitivity may also be explained intuitively by the fact that $\text{var}(\bar{x}) = \sigma^2/n$ and $E(s) = \sigma$ (asymptotically), always irrespective of the nature of the underlying distribution and, consequently, $\Delta = \sqrt{n}\mu/\sigma$ never changes.

Although the t test above is asymptotically robust with respect to departures from normality, it is not inference robust in the sense of Box and Tiao (1964b) [e.g., in terms of having high power (see also Box and Tiao, 1973; Sprott, 1982)]. In fact, as will be seen later, the t test above has low power compared to certain other proposals, particularly for long-tailed symmetric distributions (Chapter 4).

1.4 VARIANCE-RATIO TEST FOR COMPARING SEVERAL MEANS

As an extension of the one-sample t test, we now consider the problem of comparing k means ($k \geq 2$). Let $x_{i1}, \ldots, x_{in_i}$ be a random sample from the ith normal distribution $N(\mu_i, \sigma)$ and one wants to test the hypothesis H_0: $\mu_1 = \cdots = \mu_k$ against the alternative hypotheses H_1: $\mu_i \neq \mu_j$ (for some i and j). The UMP similar test whose power is a function of $\Delta^2 = \Sigma_{i=1}^k n_i(\mu_i - \bar{\mu})^2/\sigma^2$, with $\bar{\mu} = \Sigma_{i=1}^k n_i\mu_i/N$ and $N = \Sigma_{i=1}^k n_i$, is based on the F statistic (see Lehmann, 1959, pp. 268, 313)

$$F = \frac{S_g^2}{S_e^2} \qquad (1.4.1)$$

where

$$S_g^2 = \frac{\Sigma_{i=1}^k n_i(\bar{x}_i - \bar{x})^2}{k - 1}$$

$$S_e^2 = \frac{\Sigma_{i=1}^k (n_i - 1)s_i^2}{N - k}$$

$$s_i^2 = \frac{\Sigma_{j=1}^{n_i} (x_{ij} - \bar{x}_i)^2}{n_i - 1}$$

$$\bar{x}_i = \frac{\Sigma_{j=1}^{n_i} x_{ij}}{n_i}$$

$$\bar{x} = \frac{\Sigma_{i=1}^{k} n_i \bar{x}_i}{N}$$

Large values of F lead to the rejection of H_0 in favor of H_1. The null distribution of F is the central F distribution with degrees of freedom $\nu_1 = k - 1$ and $\nu_2 = N - k$. The nonnull distribution of F is the noncentral F distribution with degrees of freedom ν_1 and ν_2 and noncentrality parameter Δ^2. Thus one computes the F statistic and rejects H_0 in favor of H_1 at level α if $F > F_\alpha(\nu_1, \nu_2)$, where $F_\alpha(\nu_1, \nu_2)$ is the $100(1 - \alpha)\%$ point of a central F distribution with degrees of freedom ν_1 and ν_2.

1.5 ASYMPTOTIC ROBUSTNESS OF THE VARIANCE-RATIO TEST

To assess the asymptotic robustness of the F test above, we assume, as in Section 1.4, a wider class of distributions for x_{ij} with mean μ_i, variance σ^2, and finite higher cumulants κ_j, $j = 3, 4, \ldots$. Then, since s_i^2 is the second Fisher κ statistic from the ith population, s_i^2/σ^2 converges to 1 in probability as $n_i \to \infty$ for all $i = 1, 2, \ldots, k$. Further, by the central limit theorem, $\bar{x}_i$ is distributed as $N(\mu_i, \sigma^2/n_i)$ asymptotically. By Slutsky's theorem, therefore, F converges in distribution to $\chi^2_{k-1}(\Delta^2)$ as $n_i \to \infty$, $i = 1, 2, \ldots, k$, where $\chi^2_{k-1}(\Delta^2)$ is the noncentral chi-square variate with degrees of freedom $\nu_1 = k - 1$ and noncentrality parameter Δ^2, irrespective of the underlying distribution in the class. Since this is precisely the limiting distribution of F under the assumption of normality, the F statistic is asymptotically robust with respect to departures from normality. The asymptotic power of F is, with β being the probability of type II error,

$$1 - \beta = \int_{\chi^2_{\alpha(\nu_1)}}^{\infty} p(\chi^2 \mid H_1)\, d\chi^2 \tag{1.5.1}$$

where $\chi^2_{\alpha(\nu_1)}$ is the $100(1 - \alpha)\%$ point of a central chi-square distribution with degrees of freedom $\nu_1 = k - 1$ and $p(\chi^2 \mid H_1)$ is the density function of a noncentral chi-square distribution, usually denoted by $\chi^2_{k-1}(\Delta^2)$.

Alternatively, Tan (1982) has justified the asymptotic robustness of the F test by comparing the first four cumulants $\kappa_j(F)$ of F under nonnormal distributions with the corresponding cumulants of F under the normal distribution. It is shown by Tan (1982) that under H_0, the first four cumulants of F under nonnormal distributions are closely approximated by the corresponding cumulants of F under the normal distribution. This also establishes the asymptotic type I robustness of the F test. Note that the asymptotic value of Δ^2 does not change if there is a departure from normality, and therefore one concludes that the asymptotic power function of F is also robust.

As in the case of the t test, we note that while the F test above is asymptotically robust, it is not inference robust in the sense of Box and Tiao (1964b); in fact, we shall show in Chapter 4 that it has low power compared to certain other proposals.

1.6 SMALL-SAMPLE DISTRIBUTIONS OF t AND F UNDER NONNORMAL DISTRIBUTIONS AND ROBUSTNESS

We have shown above that both the t test and the F test are asymptotically robust, but in practice, one seldom has very large samples. Therefore, the asymptotic robustness of the above tests is hardly a consolation. To study the type I robustness and the power of the foregoing tests for small samples, several investigations have been carried out; four excellent review papers by Tiku (1975a), Cressie (1980a), Ito (1980), and Tan (1982) describe the major works on this topic. The method based on Laguerre expansion, due to Tiku (1964), is known to have wide applicability (Tan and Wong, 1977, 1980; Tan, 1982); therefore, we adopt this approach here. Since the type I robustness and the power properties of the t and F tests are similar for nonnormal populations (Gayen, 1949, 1950; Tiku, 1964, 1971; Tan, 1982), we will only discuss in detail the type I robustness and the power of the F test. It may be noted that the square of the Student two-sample t, $t^2 = (\bar{x}_1 - \bar{x}_2)^2/s^2[(1/n_1) + (1/n_2)]$, $s^2 = [(n_1 - 1)s_1^2 + (n_2 - 1)s_2^2]/(n_1 + n_2 - 2)$, is the F statistic in (1.4.1) with $\nu_1 = 1$ and $\nu_2 = n_1 + n_2 - 2$; $\bar{x}_1$ and $\bar{x}_2$ and s_1^2 and s_2^2 are the means and variances of two independent random samples, respectively. In order not to interrupt the development of the subject matter, some of the details which are more mathematically involved are deferred to Appendix A. Here we only give the main steps in deriving approximations to the small-sample distributions of the t and F statistics.

1.7 SMALL-SAMPLE NULL DISTRIBUTION OF F AND TYPE I ROBUSTNESS

For assessing type I robustness of the F test for small samples, we proceed to derive approximations to the null distribution of F. For simplicity, we assume that $n_i = n$ and that the ith sample comes from a distribution with finite cumulants κ_{1i}, κ_r, $r = 2, 3, \ldots$, although our method is quite general (Tiku, 1964) with respect to n_i being unequal and κ_{ri} being different. The assumption that $\kappa_{ri} = \kappa_r$, $r = 2, 3, \ldots$, implies that the k samples $x_{i1}, \ldots, x_{in_i}$, $i = 1, 2, \ldots, k$, come from distributions with common variance and common third- and higher-order standard cumulants $\lambda_r = \kappa_r/\kappa_2^{r/2}$, $r = 3, 4, \ldots$; note that for the normal distribution, $\lambda_r = 0$ for all $r = 3, 4, \ldots$.

Put $X_1 = (k - 1)S_g^2/(2\sigma^2)$ and $X_2 = (N - k)S_e^2/(2\sigma^2)$ with $N = nk$ and let $g(x_1, x_2)$ be the joint density of (X_1, X_2). Following up on the work of Tiku (1964, 1965b), it was shown by Davis (1976) and Tan (1982) that $g(x_1, x_2)$ can be formally expanded as a convergent series involving the product of Laguerre polynomials $L_i^{(m_1)}(x_1)$ and $L_j^{(m_2)}(x_2)$, where $m_1 = (k - 1)/2$ and $m_2 = (N - k)/2$. Using the results given in Appendix A, $g(x_1, x_2)$ can be approximated quite accurately by a finite Laguerre polynomial series involving only a few terms (see Tiku, 1964, 1975a; Tan and Wong, 1977; Tan, 1982). In fact, we have

$$\begin{aligned} g(x_1, x_2) &\simeq \left[\sum_{0 \le i+j \le 4} \beta_{ij} L_i^{(m_1)}(x_1) L_j^{(m_2)}(x_2)\right] p_{m_1}(x_1) p_{m_2}(x_2) \\ &= \sum_{0 \le i+j \le 4} d_{ij} \sum_{u=0}^{i} \sum_{v=0}^{j} \binom{i}{u}\binom{j}{v} (-1)^{u+v} p_{m_1+u}(x_1) p_{m_2+v}(x_2) \end{aligned} \tag{1.7.1}$$

where d_{ij} and β_{ij} are given in (A.4) and (A.5), with κ_{rs} being the (r, s)th mixed cumulants of (X_1, X_2) and $p_m(x) = e^{-x}x^{m-1}/\Gamma(m)$, $0 \le x < \infty$. Now, for $0 \le r + s \le 4$, the (r, s)th mixed cumulants of (X_1, X_2) are available from David and Johnson (1951) and Tan and Cheng (1981). The coefficients β_{ij} (and hence $d_{ij} = c_i^{(m_1)} c_j^{(m_2)} \beta_{ij}$) in (1.7.1) can, therefore, be determined exactly and are given by (Tiku, 1964)

$$\beta_{00} = 1 \qquad \beta_{10} = \beta_{01} = 0$$

$$\beta_{20} = \frac{\nu_1}{N(\nu_1 + 2)}\lambda_4 \qquad \beta_{11} = \frac{1}{N}\lambda_4 \qquad \beta_{02} = \frac{\nu_2}{N(\nu_2 + 2)}\lambda_4$$

$$\beta_{30} = -\frac{1}{N(\nu_1 + 2)(\nu_1 + 4)}\left[\frac{\nu_1^2}{N}\lambda_6 + 4(\nu_1 - 1)\lambda_3^2\right] \tag{1.7.2}$$

$$\beta_{12} = -\frac{1}{N(\nu_2 + 2)}\left(\frac{\nu_2}{N}\lambda_6 + 4\lambda_3^2\right) \qquad \beta_{21} = -\frac{\nu_1}{N^2(\nu_1 + 1)}\lambda_6$$

$$\beta_{03} = -\frac{1}{N(\nu_2 + 2)(\nu_2 + 4)}\left[\frac{\nu_2^2}{N}\lambda_6 + 4(\nu_2 - \nu_1 - 1)\lambda_3^2\right]$$

and so on.

It should be noted that β_{ij}, $i + j \le 4$, are all the coefficients to order N^{-3} (David and Johnson, 1951). We remark here that the second-order coefficients in (1.7.2) involve only the fourth-order standard cumulants (i.e., λ_4); third-order coefficients involve only the sixth-order standard cumulants (i.e., λ_6 and λ_3^2); and so on.

Alternatively, Gayen (1950) was able to obtain the exact distribution $g(x_1, x_2)$ if the underlying distribution is completely represented by the first four terms of the Edgeworth series, that is, if the probability density of $Y = (X - \mu)/\sigma$ is given by

$$f(y) = \left[1 + \frac{\lambda_3}{6} H_3(y) + \frac{\lambda_4}{24} H_4(y) + \frac{\lambda_3^2}{72} H_6(y)\right] \phi(y) \tag{1.7.3}$$

where $\phi(y) = (2\pi)^{-\frac{1}{2}} \exp(-y^2/2)$ and $H_r(y)$ is the Hermite polynomial of degree r (see Szegö, 1959). We note, however, that by neglecting cumulants of order greater than 4, (1.7.1) agrees perfectly with Gayen's (1950) exact results. However, Barton and Dennis (1952) have shown that (1.7.3) is unimodal and has nonnegative ordinates only if $0 \leq \lambda_4 \leq 2.4$ and $\lambda_3^2 \leq 0.2$, so that Gayen's (1950) result is of limited applicability; on the other hand, Tan and Wong (1977) have studied distributions like (1.7.1) and concluded that they represent plausible distributions for a much wider range of λ_3 and λ_4 (and higher cumulants) than that anticipated by Gayen (1950).

Small-Sample Null Distribution of the Variance Ratio

Since the underlying distribution is not necessarily normal, we will write w for the variance ratio $F = S_g^2/S_e^2 = \nu_2 X_1/\nu_1 X_2$. Submitting (1.7.1) to the transformation $(\nu_1/\nu_2)w = X_1/X_2$ and integrating over X_2 from zero to infinity, we get the probability density function of w as

$$p(w \mid H_0) \simeq p_0(w) + \sum_{2 \leq i+j \leq 4} d_{ij} P_{ij}(w) \tag{1.7.4}$$

where $p_0(w)$ is the density of the central F distribution with degrees of freedom $[\nu_1 = k - 1,\ \nu_2 = k(n - 1)]$ and the $P_{ij}(w)$'s are given by

$$P_{ij}(w) = \sum_{u=0}^{i} \sum_{v=0}^{j} \binom{i}{u}\binom{j}{v} (-1)^{u+v} p(w;\ m_1 + u,\ m_2 + v)$$

$$p(w; m_1 + u,\ m_2 + v) = \frac{\nu_1/\nu_2}{B(m_1 + u,\ m_2 + v)} \left(\frac{\nu_1}{\nu_2} w\right)^{m_1+u-1} \left(1 + \frac{\nu_1}{\nu_2} w\right)^{-(m_1+m_2+u+v)}$$

$$0 < w < \infty$$

with $m_1 = \nu_1/2$ and $m_2 = \nu_2/2$.

In (1.7.4), $p_0(w)$ is the null distribution of F under the normality assumption; therefore, all other terms may be called nonnormality corrective functions.

Mean and Variance of w. It is easy to see that (Tiku, 1964)

$$\int_0^\infty w^a p(w;\ u, v)\, dw = \left(\frac{\nu_1}{\nu_2}\right)^{-a} \frac{B(u + a,\ v - a)}{B(u, v)}$$

for any $a \geq 0$. Applying this result and using (1.7.4), one can compute moments of w. In particular, the mean and variance of w are given, respectively, by

$$E(w) \simeq 1 + \frac{2}{N} + O(N^{-2})$$

and

$$V(w) \simeq \frac{2}{\nu_1}\left[1 + \frac{\nu_1 + 6}{N} - \frac{1}{N}\lambda_4 + O(N^{-2})\right]$$

It is clear that to order $O(N^{-2})$, λ_4 enters into V(w) but not in E(w); hence for moderate N, E(w) is not affected by nonnormality, although the population kurtosis $\lambda_4 = \beta_2 - 3$ ($\beta_2 = \mu_4/\mu_2^2 = 3$ for the normal) has some effect on the variance of w.

Probability Integral. Let w_0 be the $100(1 - \alpha)\%$ point of the central F distribution with degrees of freedom $\nu_1 = k - 1$ and $\nu_2 = N - k = k(n - 1)$. We are interested in the value of the integral

$$\begin{aligned} P(w_0) &= \int_{w_0}^{\infty} p(w \mid H_0)\, dw \\ &\simeq P_0(w_0) + \sum_{2 \le i+j \le 4}\sum d_{ij} I_{ij}(x_0) \end{aligned} \tag{1.7.5}$$

where $P_0(w_0)$ is the normal-theory level [i.e., $P_0(w_0) = \alpha$] and $x_0 = 1/[1 + (\nu_1/\nu_2)w_0]$,

$$I_{ij}(x_0) = \sum_{u=0}^{i}\sum_{v=0}^{j}\binom{i}{u}\binom{j}{v}(-1)^{u+v} I_{x_0}(m_2 + v,\ m_1 + u)$$

Here $I_x(a, b) = \int_0^x t^{a-1}(1 - t)^{b-1}\, dt/B(a, b)$ is Karl Pearson's (1934) incomplete beta function.

Retaining only the terms of order $O(N^{-2})$ in (1.7.2), formula (1.7.5) can be written explicitly as

$$P(w_0) \simeq P_0(w_0) - \lambda_4 P_{\lambda_4}(w_0) + \lambda_3^2 P_{\lambda_3^2}(w_0) + \lambda_6 P_{\lambda_6}(w_0) - \lambda_4^2 P_{\lambda_4^2}(w_0) \tag{1.7.6}$$

where $P_0(w_0) = \alpha$ is the normal-theory level and $P_{\lambda_4}(w_0)$, $P_{\lambda_3^2}(w_0)$, ... are free of the λ_r's and hence are the nonnormality corrective functions [explicit formulas for these functions are given by Tiku (1964)]. To get an idea about the relative magnitude of these corrective functions, we tabulate their values for a few representative values of ν_1 and ν_2 in Table 1.7.1. It is clear from the values given in Table 1.7.1 and (1.7.6) that the effects of λ_4 (kurtosis) and λ_3^2 (skewness) have a tendency to cancel each other.

Table 1.7.1 Values of the Nonnormality Corrective Functions, at 5% Normal-Theory Significance Level [i.e., $P_0(w_0) = 0.05$]

ν_2	$10^{+5}P_{\lambda_4}$	$10^{+5}P_{\lambda_3^2}$	$10^{+5}P_{\lambda_6}$	$10^{+5}P_{\lambda_4^2}$	$10^{+5}P_{\lambda_4}$	$10^{+5}P_{\lambda_3^2}$	$10^{+5}P_{\lambda_6}$	$10^{+5}P_{\lambda_4^2}$
	$\nu_1 = 1$				$\nu_1 = 4$			
2	480	840	76	194	322	904	31	-51
4	481	545	68	212	444	1024	49	61
8	296	147	26	76	414	591	42	112
20	110	5	3	9	240	97	15	40
60	31	-2	-0	1	91	-18	2	5
120	20	-1	-0	0	50	-17	1	1
	$\nu_1 = 8$				$\nu_1 = 12$			
2	210	866	13	-138	156	844	7	-164
4	312	1134	23	-89	238	1161	13	-152
8	336	877	26	27	273	1012	17	-35
20	238	281	14	40	216	425	11	30
60	104	18	3	8	105	66	3	9
120	57	-7	1	2	60	13	1	3

EXAMPLE 1.7.1 Suppose that the underlying distribution is one of the three distributions described by E. S. Pearson (1963, pp. 101-102) under his "case B," namely, the Pearson type IV curve,

$$\text{constant} \times \left(1 + \frac{y^2}{a^2}\right)^{-q} \exp\left(-d \tan^{-1} \frac{y}{a}\right) \qquad -\infty \leq y \leq \infty$$

the noncentral t distribution,

$$\text{constant} \times \left(1 + \frac{t^2}{v}\right)^{-(v+1)/2} \exp\left(-\frac{\frac{1}{2}v\tau^2}{v+t^2}\right) Hh_v\left(-\frac{t\tau}{\sqrt{v + t^2}}\right)$$

$$Hh_v(y) = \frac{1}{\Gamma(v + 1)} \int_0^\infty u^v e^{-(u+y)^2/2}\, du \qquad -\infty \leq t \leq \infty$$

and the Johnson S_U curve (see Johnson, 1949, 1965)

$$\text{constant} \times (\cosh^{-1} y) \exp\left[-\tfrac{1}{2}(\gamma_0 + \gamma_1 \sinh^{-1} y)^2\right] \quad -\infty \leq y \leq \infty \qquad (1.7.7)$$

for which $\lambda_3^2 = 0.58$, $\lambda_4 = 1.23$, have very similar distributions within the range of significance frequency and all have values of λ_5 and λ_6 not far from 3.0 and 11.0, respectively. Here

$$P(w_0) \simeq 0.05 - 1.23(0.00296) + 0.58(0.00147) + 11.0(0.00026) - 1.5129(0.00076)$$

$$= 0.0489 \quad \text{for } \nu_1 = 1 \text{ and } \nu_2 = 8$$

$$P(w_0) \simeq 0.05 - 1.23(0.00240) + 0.58(0.00097) + 11.0(0.00015) - 1.5129(0.00040)$$

$$= 0.0487 \quad \text{for } \nu_1 = 4 \text{ and } \nu_2 = 20$$

It is quite clear that the effect of the foregoing nonnormality is negligible, even for values of ν_2 as small as 8.

EXAMPLE 1.7.2 Another important situation from a practical point of view is when the underlying distribution is a mixture of two normals, for example

$0.90N(\mu, \sigma) + 0.10N(\mu, 2\sigma)$ (scale-contamination model)

This is a symmetric distribution with mean μ, variance $\kappa_2 = 1.3\sigma^2$, and

$$\lambda_3^2 = 0 \quad \lambda_4 = 1.4379 \quad \lambda_6 = 13.2726$$

Here

$$P(w_0) \simeq 0.05 - 1.4379(0.00296) + 13.2726(0.00026) - 2.0676(0.00076)$$

$$= 0.0476 \quad \text{for } \nu_1 = 1 \text{ and } \nu_2 = 8$$

$$P(w_0) \simeq 0.05 - 1.4379(0.00240) + 13.2726(0.00015) - 2.0676(0.00040)$$

$$= 0.0477 \quad \text{for } \nu_1 = 4 \text{ and } \nu_2 = 20$$

Again the effect of this nonnormality on the type I error α is negligible.

Subrahmaniam et al. (1975) evaluated the significance level when the underlying distribution is $\pi N(\mu, \sigma) + (1 - \pi)N(\mu + \delta\sigma, \sigma)$ (location-contamination model), and found them not very different from the normal-theory level α.

EXAMPLE 1.7.3 In some situations, the underlying distribution is symmetric and short-tailed ($\lambda_4 < 0$). Consider, for example, the uniform distribution $f(x) = 1/(\theta_2 - \theta_1)$, $\theta_1 \leq x \leq \theta_2$. For this distribution

$$\lambda_3^2 = 0 \quad \lambda_4 = -1.2 \quad \lambda_6 = 6.8567$$

Here

$$P(w_0) \simeq 0.05 + 1.2(0.00296) + 6.8567(0.00026) - 1.44(0.00076)$$

$$= 0.0542 \quad \text{for } \nu_1 = 1 \text{ and } \nu_2 = 8$$

Table 1.7.2 Type I Error of the F Test, Normal-Theory Value; $\alpha = 0.05$

		$\nu_1 = 3$ (k = 4)		
n	ν_2	Normal	Exponential	Log-normal
4	12	0.048	0.041	0.035
8	28	0.047	0.040	0.034
16	60	0.048	0.047	0.037

$$P(w_0) \simeq 0.05 + 1.2(0.00240) + 6.8567(0.00015) - 1.44(0.00040)$$
$$= 0.0533 \quad \text{for } \nu_1 = 4 \text{ and } \nu_2 = 20$$

The effect of short-tailed symmetric distributions is in general to make the level α^* larger than the normal-theory level α, although not by appreciable amounts.

There are situations when one is interested in testing hypotheses about location parameters but the underlying distributions are extremely nonnormal symmetric distributions or extremely skew distributions (Cauchy, exponential, etc.). The distributions of the former type might not even have a finite mean and a finite variance; Cauchy distribution, for example, does not even have a finite mean. In such situations, formula (1.7.6) cannot be expected to give close approximations since it ignores higher cumulants and, in spite of the fact that the correction terms due to these cumulants are of order $O(N^{-2})$, their contributions could be quite appreciable. However, Donaldson (1968) carried out a Monte Carlo investigation to study the effect of such departures on the type I error of the foregoing F test. His results are as shown in Table 1.7.2. The effect of such extreme departures is in general to make the level α^* smaller than its normal-theory value (conservativeness of the F test) and that might not be an undesirable feature from a practical point of view [see also Benjamin (1983), Posten (1978, 1982), Posten et al. (1982), and Rasch* (1980), who study in detail the robustness of the Student t test]. It will, however, be shown in Chapter 4 that the power of these classical tests is much lower than certain other proposals based on the so-called robust estimators.

*We also refer the reader to the pioneering work on robustness of the GDR group headed by D. Rasch [see Rasch (1983) and the references cited there].

1.8 SMALL-SAMPLE POWER OF THE t TEST AND THE F TEST

For nonnormal parent distributions with finite cumulants, the exact nonnull distribution $g(x_1, x_2)$ can again be expanded as a convergent infinite series involving products of Laguerre polynomials (Tiku, 1971; Davis, 1976; Tan, 1982). Thus, from results given in Appendix A, one can again develop a close approximation to $g(x_1, x_2)$ by using the first few terms of a Laguerre polynomial expansion with order up to 4, on exactly the same lines as in Section 1.7. This has been done by Tiku for the F test (Tiku, 1972a) and for the t test (Tiku, 1971). Alternatively, by assuming an Edgeworth series distribution with four terms, Srivastava (1959) has obtained the exact distribution of F. Numerical calculations have been done by Srivastava (1959), Donaldson (1968), and Tiku (1972a) for the F test and by Tiku (1971) for the t test. Since the method is exactly the same as that given in Section 1.7, although the algebra is somewhat more involved, we will not give theoretical details but present only conclusions based on numerous numerical calculations; for mathematical details and derivations, the reader is referred to the works by Srivastava (1959), Tiku (1971, 1972a,b), and Tan (1982).

1. For the F test, if the departure from normality is moderate and if $n_1 = \cdots = n_k = n$, the power function is quite insensitive. However, for severe departures such as exponential and log-normal distributions, the effects are quite considerable even if $n_1 = \cdots = n_k = n$. In general, the effects of departures from normality are more pronounced for unbalanced cases ($n_1 \neq \cdots \neq n_k$) than for the balanced cases.
2. For the t test, if the underlying distribution is symmetric but the departure is moderate, the power function is quite insensitive to nonnormality; for skew distributions with skewness λ_3 satisfying $|\lambda_3| > 0.5$, however, the effects are usually quite considerable, often resulting in lower power (and higher type I error) than under normality.

From the results noted above and various other studies (Bradley, 1980; Geary, 1947; Lee and D'Agostino, 1976; Lee and Gurland, 1977; Pearson, 1932; Pearson and Adyanthaya, 1929; Posten 1978, 1982; Sansing and Owen, 1974; Tan, 1977), it is therefore clear that the t and F tests make good test procedures (i.e., they have type I robustness as well as high power) only if the parent distribution departs only moderately from the normal. One would, therefore, prefer other test procedures which have type I robustness over a wider class of distributions, and have substantially the same power (if not more) for populations close to the normal but have higher power for other, more nonnormal populations. Several such tests are presented in Chapter 4.

1.9 TESTING EQUALITY OF TWO VARIANCES

Another important problem from a practical point of view is that of testing the equality of two population variances. Let $x_1, x_2, \ldots, x_{n_1}$ and

$y_1, y_2, \ldots, y_{n_2}$ be independent random samples from two populations specified by their finite standard cumulants $(\lambda_{r(1)}, \lambda_{r(2)})$, $r = 3, 4, \ldots$; in particular, $\lambda_{r(1)}$ could be equal to $\lambda_{r(2)}$. If the underlying populations are normal $N(\mu_1, \sigma_1)$ and $N(\mu_2, \sigma_2)$ and one wants to test the null hypothesis H_0: $\sigma_1^2 = \sigma_2^2$, the test that is usually employed is based on the statistic

$$W = \frac{f_2 S_1}{f_1 S_2} \tag{1.9.1}$$

where $S_1 = \Sigma_{i=1}^{n_1} (x_i - \bar{x})^2$ and $S_2 = \Sigma_{i=1}^{n_2} (y_i - \bar{y})^2$; $f_1 = n_1 - 1$ and $f_2 = n_2 - 1$. The null distribution of W is the central F distribution having f_1 and f_2 degrees of freedom. If the alternative hypotheses are $H_1: \sigma_1^2 > \sigma_2^2$, one rejects H_0 in favor of H_1 at an $\alpha\%$ significance level if the calculated value of W is greater than $F_\alpha(f_1, f_2)$. Under the assumption of normality, the test above is uniformly most powerful. To obtain the probability density function of the statistic W in a nonnormal situation, we write

$$X_i = \frac{\frac{1}{2} S_i}{\sigma_i^2} \qquad i = 1, 2 \qquad (\sigma_1^2 = \sigma_2^2 = \sigma^2 \text{ under } H_0)$$

and let $h(x_1, x_2)$ and $h_i(x_i)$ denote the density functions of (X_1, X_2) and X_i, respectively. Then, because of the independence of X_1 and X_2, we have

$$h(x_1, x_2) = h_1(x_1) h_2(x_2) \tag{1.9.2}$$

It is shown by Tiku (1964) and Davis (1976) that the distribution of X_i can be expanded as a convergent infinite series involving Laguerre polynomials. From results given in Appendix A, therefore, $h_i(x_i)$ can then be approximated by a finite Laguerre series expansion up to order 4 (Roy and Tiku, 1962; Tan and Wong, 1977), for example

$$h_i(x_i) \simeq \left[1 + \sum_{j=2}^{4} \alpha_{j(i)} L_j^{(m_i)}(x_i)\right] p_{m_i}(x_i)$$

$$= p_{m_i}(x_i) + \sum_{j=2}^{4} d_{j(i)} \left[\sum_{u=0}^{j} \binom{j}{u} (-1)^u p_{m_i+u}(x_i)\right] \tag{1.9.3}$$

where $m_i = f_i/2$, $d_{j(i)} = c_j^{(m_i)} \alpha_{j(i)} = E[L_j^{(m_i)}(X_i)]$, and $p_f(x) = (x^{f-1} e^{-x})/\Gamma(f)$, $x \geq 0$.

For $0 \leq r \leq 4$, the nonzero $\alpha_{r(i)}$ are given by Roy and Tiku (1962) as

$$\alpha_{0(i)} = 1 \qquad \alpha_{2(i)} = \frac{f_i}{(f_i + 1)(f_i + 2)} \lambda_{4(i)}$$

$$\alpha_{3(i)} = -\frac{1}{(f_i+1)(f_i+2)(f_i+4)}\left[\frac{f_i^2}{f_i+1}\lambda_{6(i)} + 4(f_i-1)\lambda_{3(i)}^2\right]$$

$$\alpha_{4(i)} = \frac{1}{(f_i+1)^2(f_i+2)(f_i+4)(f_i+6)}\left[\frac{f_i^3}{f_i+1}\lambda_{8(i)} + 32f_i(f_i-1)\lambda_{5(i)}\lambda_{3(i)} + (3f_i^3+32f_i^2-8f_i+8)\lambda_{4(i)}^2\right] \tag{1.9.4}$$

The coefficients $\alpha_{r(i)}$ for $r > 4$ are of order n^{-3} (see Fisher, 1930, p. 210; Kendall, 1945, p. 268); $n = \min(n_1, n_2)$.

Submitting (1.9.2) to the transformation $(f_1/f_2)W = X_1/X_2$ and integrating over X_2 from zero to infinity, the probability density function of W is obtained as

$$p(W) \simeq p_0(W) + \sum_{2 \le r+s \le 4} d_{rs}P_{rs}(W) \tag{1.9.5}$$

where $d_{rs} = d_{r(1)}d_{s(2)}$, $p_0(W)$ is the central F distribution with f_1 and f_2 degrees of freedom, and the $P_{rs}(W)$'s are the functions in (1.7.4) but with w, ν_1, and ν_2 now replaced by W, f_1, and f_2, respectively.

Mean and Variance of W

From (1.9.5) it is easy to obtain the moments of W. In particular, the mean and variance of W are given, respectively, by (Tiku, 1964)

$$\mu_1'(W) = 1 + \frac{2}{f_2}\left(1+\frac{2}{f_2}\right) + \frac{\lambda_{4(2)}}{f_2}\left(1-\frac{1}{f_2}\right) - \frac{4\lambda_{3(2)}^2}{f_2^2} - \frac{\lambda_{6(2)}}{f_2^2} + \frac{3\lambda_{4(2)}^2}{f_2^2} + O(n^{-3}) \tag{1.9.6}$$

and

$$\mu_2(W) = \frac{2}{f_1} + \frac{2}{f_2}\left(1+\frac{6}{f_1}\right) + \frac{\lambda_{4(1)}}{f_1}\left(1-\frac{1}{f_1}\right) + \frac{\lambda_{4(2)}}{f_2}\left(1-\frac{5}{f_2}\right) - \frac{8\lambda_{3(2)}^2}{f_2^2} - \frac{2\lambda_{6(2)}}{f_2^2} + \frac{2\lambda_{4(2)}^2}{f_2^2} + \frac{3\lambda_{4(1)}\lambda_{4(2)}}{f_1f_2} + O(n^{-3}) \tag{1.9.7}$$

$n = \min(n_1, n_2)$. The effect of population kurtosis $\lambda_{4(1)}$ and $\lambda_{4(2)}$ on the mean and variance of W is now substantial. It may also be noted that in (1.9.6) and

(1.9.7), the order of the terms in the sixth cumulant and the square of the third cumulants is the same as of the square of the fourth cumulants; the same is true with the coefficients in (1.9.4). Note that the exact distributions obtained by Gayen (1949, 1950) and Srivastava (1958, 1959), by assuming an Edgeworth series distribution with four terms, include only the fourth cumulant and square of third cumulants and does not contain the square of fourth cumulants and sixth cumulants; clearly, the latter results are not applicable in situations where the underlying distribution is not the Edgeworth series (1.7.3).

Probability Integral

Retaining only the terms of order $O(n^{-2})$ in (1.9.4), the type I error $P(W_0) = P(W \geq W_0 \mid H_0) = \int_{W_0}^{\infty} p(W \mid H_0)\, dW$, with $W_0 = F_\alpha(\frac{1}{2}f_1, \frac{1}{2}f_2)$, is given by

$$\begin{aligned} P(W_0) \simeq\; & P_0(W_0) + \lambda_{4(1)} P_{\lambda_{4(1)}}(W_0) + \lambda_{4(2)} P_{\lambda_{4(2)}}(W_0) - \lambda_{3(1)}^2 P_{\lambda_{3(1)}^2}(W_0) \\ & - \lambda_{3(2)}^2 P_{\lambda_{3(2)}^2}(W_0) - \lambda_{6(1)} P_{\lambda_{6(1)}}(W_0) - \lambda_{6(2)} P_{\lambda_{6(2)}}(W_0) \\ & + \lambda_{4(1)}^2 P_{\lambda_{4(1)}^2}(W_0) + \lambda_{4(2)}^2 P_{\lambda_{4(2)}^2}(W_0) - \lambda_{4(1)}\lambda_{4(2)} P_{\lambda_{4(1)}\lambda_{4(2)}}(W_0) \end{aligned} \tag{1.9.8}$$

$P_0(W_0) = \alpha$ is the normal-theory type I error α and the other terms in (1.9.8) are the nonnormality corrective functions which are free of the λ_r's. Explicit formulas for these functions are available in Tiku (1975a) but being similar to the ones in (1.9.4), are not reproduced here. For f_2 tending to infinity, (1.9.8) reduces to equation (2.12) of Roy and Tiku (1962) (see also Tan and Wong, 1977). Note that unless $\lambda_{r(1)} = \lambda_{r(2)} = 0$ $(r = 3, 4, \ldots)$, $P(W_0)$ does not converge to the normal-theory value $P_0(W_0)$, since $P_{\lambda_{4(1)}}(W_0)$, $P_{\lambda_{4(2)}}(W_0)$, and so on, do not diminish to zero as f_2 tends to infinity.

As an example, we assume that the underlying distribution is one of (1.7.7) (Pearson's distribution), with $\lambda_{3(1)}^2 = \lambda_{3(2)}^2 = 0.58$, $\lambda_{4(1)} = \lambda_{4(2)} = 1.23$, and $\lambda_{6(1)} = \lambda_{6(2)} = 11.0$. The values of $P(W_0)$ calculated from (1.9.8) are as shown in Table 1.9.1; $\alpha = P_0(W_0) = 0.05$. These values typically illustrate the very nonrobust character of the variance-ratio test for testing the equality of two variances. The use of this test cannot, therefore, be recommended [see also Box (1953), Box and Tiao (1964a), Brown and Forsythe (1974),

Table 1.9.1 Type I Error for the Distribution (1.7.7)

	f_1				
f_2	4	6	8	12	24
4	0.058	0.059	0.059	0.059	0.059
8	0.062	0.065	0.066	0.067	0.067
12	0.063	0.067	0.070	0.072	0.073
24	0.062	0.068	0.073	0.078	0.082
60	0.059	0.066	0.071	0.083	0.088

Conover et al. (1981), Gartside (1972), Ito (1980), and Tiao and Ali (1971), who express reservations about the usefulness of this test].

1.10 CONCLUSION AND SUMMARY

In this chapter the robustness with respect to departures from normality of the one-sample t test and the variance-ratio tests for comparing k means (under equal variances) and for comparing two variances are investigated, leading us to the following conclusions:

1. The one-sample t test and the variance-ratio test for comparing k means under equal variances are asymptotically robust with respect to departures from normality. However, the variance ratio test for comparing two variances does not have this asymptotic robustness; in fact, the null distribution here depends heavily on the kurtosis of the underlying population.
2. For small sample sizes, the numerical calculations based on Laguerre series approximations clearly indicate that:
   a. If the underlying population is symmetric and if the departure from normality is moderate, the effects on the one-sample t test are negligible. If the underlying population is skew, the effect on the test is considerable unless, of course, the skewness is slight (e.g., $|\lambda_3| \leq 0.5$). If the underlying population is symmetric but is extremely nonnormal (e.g., λ_4 is very large), the effect on the test is considerable.
   b. If the departure from normality is moderate with respect to skewness and/or kurtosis, the variance-ratio test for comparing k means (under equal population variances) is quite robust. For extreme departures, particularly with respect to the kurtosis of the underlying population, the effect is considerable.

c. The variance-ratio test for comparing two population variances is not robust even to small departures from normality.

EXERCISES

1.1 The distribution of $X = \frac{1}{2}\chi'^2$, where χ'^2 is a noncentral chi-square variate with ν degrees of freedom and noncentrality parameter λ, is given by

$$p_m^*(X) = e^{-\lambda/2} \sum_{r=0}^{\infty} \frac{1}{r!\Gamma(m+r)} \left(\frac{1}{2}\lambda\right)^r e^{-X} X^{m+r-1} \qquad 0 < X < \infty$$

$$\left(m = \frac{1}{2}\nu\right)$$

(a) Show that the moment generating function $M_\theta(X) = E(e^{\theta X})$ of X is given by

$$M_\theta(X) = (1-\theta)^{-m} \exp\left(\frac{1}{2}\frac{\lambda\theta}{1-\theta}\right) \qquad \text{(Patnaik, 1949)}$$

(b) Derive the moments of X from $M_\theta(X)$.

(c) Use the moments of X to develop a Laguerre series expansion of the distribution of X and show that

$$p_m^*(X) = \left[\sum_{r=0}^{\infty} \alpha_r L_r^{(m)}(X)\right] p_m(X) \qquad \text{(Tiku, 1965b)}$$

where

$$\alpha_r = \frac{\Gamma(m)}{\Gamma(m+r)}\left(-\frac{1}{2}\lambda\right)^r \quad \text{and} \quad p_m(X) = \frac{1}{\Gamma(m)} e^{-X} X^{r-1}$$

1.2 Consider the mixture of two normal distributions

$$0.95N(0, \sigma) + 0.05N(0, \sigma\sqrt{10})$$

(a) Show that this distribution is symmetric.

(b) Calculate its standard cumulants λ_4 and λ_6.

(c) Substitute these values of λ_4 and λ_6 in the expression (1.7.6) and compute the values of $P(w_0)$ for $\nu_1 = 1$ and $\nu_2 = 4$, 8, 20, and 120. Comment on the effect of this departure on the normal-theory value $P_0(w_0)$ of the Student t^2 test.

1.3 Consider the mixture

$$0.90N(0, \sigma) + 0.10N(3\sigma, \sigma)$$

(a) Show that this distribution is positively skew.

(b) Calculate its standard cumulants λ_r, r = 3, 4, 5, and 6. Substitute these values in the expression (1.7.6) and evaluate the effect of this departure on the normal-theory type I error of the F test.

APPENDIX A: Laguerre Polynomial Expansion of Nonnegative Random Variables

Let $m > 0$. The Laguerre polynomial $L_r^{(m)}(x)$ in x with degree r and parameter m is defined as (Szegö, 1959)

$$L_r^{(m)}(x) = \frac{1}{r!} \sum_{j=0}^{r} \binom{r}{j} (-x)^j \frac{\Gamma(m+r)}{\Gamma(m+j)} \qquad r = 0, 1, 2, \ldots$$

In particular,

$$\begin{aligned}
L_0^{(m)}(x) &= 1 \\
L_1^{(m)}(x) &= m - x \\
L_2^{(m)}(x) &= \frac{1}{2!} m(m+1) - (m+1)x + \frac{x^2}{2!} \\
L_3^{(m)}(x) &= \frac{1}{3!} m(m+1)(m+2) - \frac{1}{2!}(m+1)(m+2)x + \frac{x^2}{2!} - \frac{x^3}{3!} \\
L_4^{(m)}(x) &= \frac{1}{4!} m(m+1)(m+2)(m+3) - \frac{1}{3!}(m+1)(m+2)(m+3)x \\
&\quad + \frac{1}{2!}(m+2)(m+3)\frac{x^2}{2!} - (m+3)\frac{x^3}{3!} + \frac{x^4}{4!}
\end{aligned} \tag{A.1}$$

and so on. The Laguerre polynomials satisfy the following orthogonality property:

$$\int_0^{\infty} L_r^{(m)}(x) L_j^{(m)}(x) p_m(x)\, dx = \begin{cases} 0 & \text{if } r \neq j \\ c_r^{(m)} & \text{if } r = j \end{cases} \tag{A.2}$$

where

$$c_r^{(m)} = \frac{m(m+1)\cdots(m+r-1)}{r!} = \binom{m+r-1}{r} \quad \text{and} \quad p_m(x) = \frac{e^{-x} x^{m-1}}{\Gamma(m)}$$

$$0 < x < \infty$$

Let $X_1 \geq 0$ and $X_2 \geq 0$ be random variables with finite (r, s)th mixed cumulant κ_{rs}. Let $E(X_j) = m_j > 0$ and let $f(x_1, x_2)$ denote the joint density of (X_1, X_2). Suppose that $f(x_1, x_2)/[p_{m_1}(x_1) p_{m_2}(x_2)]$ can be formally expanded as a convergent series involving products of Laguerre polynomials, that is,

$$f(x_1, x_2) = \left[\sum_{i=0}^{\infty} \sum_{j=0}^{\infty} \beta_{ij} L_i^{(m_1)}(x_1) L_j^{(m_2)}(x_2) \right] p_{m_1}(x_1) p_{m_2}(x_2)$$

$$= \left[\sum_{i=0}^{\infty} \sum_{j=0}^{\infty} d_{ij} \sum_{u=0}^{i} \sum_{v=0}^{j} \binom{i}{u}\binom{j}{v} (-1)^{u+v} p_{m_1+u}(x_1) p_{m_2+v}(x_2) \right] \qquad (A.3)$$

noting that

$$L_i^{(m_1)}(x_1) p_{m_1}(x_1) = c_i^{(m_1)} \sum_{u=0}^{i} \binom{i}{u} (-1)^u p_{m_1+u}(x_1)$$

Then, by the orthogonality property (A.2),

$$d_{ij} = \beta_{ij} c_i^{(m_1)} c_j^{(m_2)} = E[L_i^{(m_1)}(X_1) L_j^{(m_2)}(X_2)] \qquad (A.4)$$

Using (A.4) one may readily compute β_{ij} if the mixed cumulants κ_{rs} or mixed moments of (X_1, X_2) are available. For computational convenience Tan (1982) and Tan and Cheng (1984) have expressed $d_{ij} = \beta_{ij} c_i^{(m_1)} c_j^{(m_2)}$ in terms of κ_{rs} for $0 \le r + s \le 6$. One has, in fact,

$$\begin{aligned}
(2!)\ d_{20} &= \kappa_{20} - \kappa_{10} \\
d_{11} &= \kappa_{11} \\
(3!)\ d_{30} &= \kappa_{30} - 6\kappa_{20} + 4\kappa_{10} \\
(2!)\ d_{21} &= \kappa_{21} - 2\kappa_{11} \\
(4!)\ d_{40} &= \kappa_{40} - 12\kappa_{30} + 3\kappa_{20}^2 + 6\kappa_{20}(6 - \kappa_{10}) + 3\kappa_{10}(\kappa_{10} - 6) \\
(3!)\ d_{31} &= \kappa_{31} - 6\kappa_{21} + 3\kappa_{20}\kappa_{11} - 3\kappa_{11}(\kappa_{10} - 2) \\
4\,d_{22} &= \kappa_{22} - 2(\kappa_{21} + \kappa_{12}) + \kappa_{20}\kappa_{02} + 2\kappa_{11}^2 - (\kappa_{10}\kappa_{02} + \kappa_{20}\kappa_{01}) \\
&\quad + 4\kappa_{11} + \kappa_{10}\kappa_{01} \quad \text{etc.}
\end{aligned} \qquad (A.5)$$

(d_{ij} is obtained from d_{ji} by replacing κ_{st} by κ_{ts} if $j > i$).

From (A.3), one also has for the distribution $f_i(x_i)$ of X_i:

$$f_i(x_i) = \left[\sum_{j=0}^{\infty} \beta_j L_j^{(m_i)}(x_i) \right] p_{m_i}(x_i)$$

$$= \left[\sum_{j=0}^{\infty} d_{j(i)} \sum_{u=0}^{j} \binom{j}{u} (-1)^u p_{m_i+u}(x_i) \right] \qquad (A.6)$$

where $d_{j(i)} = \beta_j c_j^{(m_i)} = E[L_j^{(m_i)}(X_i)]$.

From (A.3) and (A.6), one immediately has the following respective approximations for $f(x_1, x_2)$ and $f_i(x_i)$:

$$f(x_1, x_2) \simeq \left[\sum_{0 \le i+j \le k} \beta_{ij} L_i^{(m_1)}(x_1) L_j^{(m_2)}(x_2)\right] p_{m_1}(x_1) p_{m_2}(x_2)$$

$$= \sum_{0 \le i+j \le k} d_{ij} \left[\sum_{u=0}^{i} \sum_{v=0}^{j} \binom{i}{u}\binom{j}{v} (-1)^{u+v} p_{m_1+u}(x_1) p_{m_2+v}(x_2)\right] \qquad (A.7)$$

and

$$f_i(x_i) \simeq \sum_{j=0}^{k} d_{j(i)} \left[\sum_{u=0}^{j} \binom{j}{u} (-1)^u p_{m_i+u}(x_i)\right] \qquad (A.8)$$

Several justifications for the approximations (A.7) and (A.8) have been given by Tan (1982). Specifically, we mention the following three:

1. Since the joint distribution $f(x_1, x_2)$ of (X_1, X_2) can be formally expanded in infinite series involving product of Laguerre polynomials, it is logical to use (A.7) and (A.8) to approximate $f(x_1, x_2)$ and $f_i(x_i)$, respectively.
2. It is shown by Tan and Wong (1977, 1978) that the mixed moments up to order k equal the respective exact mixed moments up to order k if one approximates the exact distribution by a finite Laguerre polynomial expansion up to order k.
3. It has been shown by Pearson and Tiku (1970) and Tiku (1975a) that if the two Pearson coefficients β_1 and β_2 of X_1 and X_2 lie close to the type III line $\beta_2 = 3 + 1.5\beta_1$, then expansions in terms of Laguerre polynomials and gamma density functions provide close approximations to the joint density of (X_1, X_2). This condition is satisfied by the sums of squares such as S_g^2/σ^2 and S_e^2/σ^2.

2

Estimation Procedures for Censored Samples

2.1 INTRODUCTION

In this chapter we introduce type II censoring and illustrate why type II censoring is tantamount to reducing and bringing the standard cumulants λ_r, in particular λ_3 and λ_4, of a long-tailed symmetric distribution or a skew distribution close to those of a normal distribution. This phenomenon motivates us to develop inference procedures based on type II censored normal samples and regard them as genuine robust procedures; the robustness features of such procedures is discussed in detail in Chapter 3. The first important problem that arises, however, is that of estimation of location and scale parameters from type II censored samples. Various methods of estimation are available in the literature and these are presented in this chapter. The method of modified maximum likelihood (MML) estimation, which is relatively new, is given special attention and the optimality properties of the resulting MML estimators are discussed in detail. The efficiency properties of the MML estimators (based on type II censored normal samples) and various other "robust" estimators are then discussed and it is shown that the MML estimators are jointly at least as efficient as some of the celebrated "robust" estimators which have received a great deal of attention in recent times.

2.2 TYPE II CENSORING

Let $x_1, x_2, \ldots, x_n$ be a random sample of size n from a distribution. Arrange these n observations in ascending order of magnitude and censor (trim) the r_1 smallest and the r_2 largest observations. The remaining observations

$$x_{(r_1+1)}, x_{(r_1+2)}, \ldots, x_{(n-r_2)} \tag{2.2.1}$$

constitute a type II censored sample of size $n - r_1 - r_2$. We note in passing that type II censored samples arise naturally in certain experimental situations, which is illustrated by the following typical example.

EXAMPLE 2.2.1 The following data show the days on which the first 7 of a sample of 10 tested mice died after being inoculated with a uniform culture of human tuberculosis:

Days after inoculation:	41	44	46	54	55	58	60
Log-days after inoculation:	1.613	1.644	1.663	1.732	1.740	1.763	1.778

Here $n = 10$, $r_1 = 0$, and $r_2 = 3$, and the available observations constitute a type II censored sample of size 7; censoring occurs here only on the right-hand side. The distribution of log-days is usually assumed to be normal $N(\mu, \sigma)$ (see Gupta, 1952; Cohen, 1957).

A different type of censoring which censors observations beyond a given point(s) is called type I censoring. Thus type I censoring could be called fixed-point censoring and type II censoring could be called fixed-number censoring. The former is discussed in detail in Chapter 3.

To illustrate how type II censoring dampens the effects of the third and/or fourth standard cumulants, we consider two truncated distributions and work out their standard cumulants as follows.

2.3 TRUNCATED DISTRIBUTIONS AND THEIR CUMULANTS

Consider the following two families of distributions, both representing location-scale distributions:

$$f(x;p) = \frac{1}{\sigma\sqrt{k}\,B(\frac{1}{2}, p - \frac{1}{2})}\left[1 + \frac{(x-\mu)^2}{k\sigma^2}\right]^{-p} \qquad -\infty \le x \le \infty \tag{2.3.1}$$

$k = 2p - 3$ $(p \ge 2)$, and

$$g(x;a) = \frac{1}{\delta^a\Gamma(a)}\, e^{-(x-\theta)/\delta}(x - \theta)^{a-1} \qquad \theta \le x \le \infty \qquad (a > 0) \tag{2.3.2}$$

The distributions in the family (2.3.1) are symmetric with mean $E(x) = \mu$ and variance $V(x) = \sigma^2$ and have long tails, longer than the normal $N(\mu, \sigma)$ distribution, on both sides; note that $t = \sqrt{\nu}\,(x - \mu)/\sigma\sqrt{k}$ has Student's t distribution with $\nu = 2p - 1$ degrees of freedom. The distributions in the family (2.3.2) are positively skew and have long tails on the right-hand side; also, note that $\chi^2 = 2(x - \theta)/\delta$ has a chi-square distribution with 2a degrees of freedom and $V(x) = a\delta^2$.

Truncated Distribution of (2.3.1)

Consider the truncated distribution $[z = (x - \mu)/\sigma]$ of (2.3.1),

Table 2.3.1 Values of $\sqrt{\mu_2}$ and β_2^* for the Distribution (2.3.3)

p = 5/2			p = 7/2			p = ∞ (normal)		
z_0	$\sqrt{\mu_2}$	β_2^*	z_0	$\sqrt{\mu_2}$	β_2^*	z_0	$\sqrt{\mu_2}$	β_2^*
∞	1	∞	∞	1	6	∞	1	3
2.650	0.837	3.32	2.566	0.886	3.04	2.326	0.935	2.54
1.508	0.657	2.46	1.586	0.716	2.36	1.645	0.789	2.19
1.084	0.531	2.16	1.178	0.587	2.12	1.282	0.662	2.03

$$f_T(z;p) = \text{constant} \times \left(1 + \frac{z^2}{k}\right)^{-p} \qquad -z_0 \le z \le z_0 \tag{2.3.3}$$

which is symmetric with mean E(z) = 0. Its standard deviation $\sqrt{\mu_2}$ and its kurtosis $\beta_2^* = \mu_4/\mu_2^2$ ($\lambda_4 = \beta_2^* - 3$), are given in Table 2.3.1 for values of z_0 for which $\int_{z_0}^{\infty} f(z;p)\,dz = 0.0,\ 0.01,\ 0.05$, and 0.10, respectively. These values were calculated from the equations ($k = 2p - 3$)

$$\mu_2 = \frac{k\int_0^{1-u_0} u^{1/2}(1 - u)^{p-5/2}\,du}{\int_0^{1-u_0} u^{-1/2}(1 - u)^{p-3/2}\,du} \qquad u_0 = \frac{1}{1 + z_0^2/k} \tag{2.3.4}$$

and

$$\beta_2^* = \frac{\mu_4}{\mu_2^2} = \frac{\left[\int_0^{1-u_0} u^{-1/2}(1 - u)^{p-3/2}\,du\right]\left[\int_0^{1-u_0} u^{3/2}(1 - u)^{p-7/2}\,du\right]}{\left[\int_0^{1-u_0} u^{1/2}(1 - u)^{p-5/2}\,du\right]^2} \tag{2.3.5}$$

For $k > 2$, (2.3.4) and (2.3.5) can be expressed in terms of the incomplete beta function $I_u(a,b) = \int_0^u x^{a-1}(1 - x)^{b-1}\,dx/B(a,b)$, $0 < u < 1$ ($a > 0$, $b > 0$), that is,

$$\mu_2 = \frac{I_{1-u_0}(1.5,\ p - 1.5)}{I_{1-u_0}(0.5,\ p - 0.5)} \tag{2.3.6}$$

and

$$\beta_2^* = \frac{3k[I_{1-u_0}(0.5, p - 0.5) I_{1-u_0}(2.5, p - 2.5)]}{(k - 2)[I_{1-u_0}(1.5, p - 1.5)]^2} \tag{2.3.7}$$

For the normal ($p = \infty$), the equations are

$$\mu_2 = \frac{\Gamma_{(1/2)z_0^2}(1.5)}{\Gamma_{(1/2)z_0^2}(0.5)} \tag{2.3.8}$$

and

$$\beta_2^* = \frac{3[\Gamma_{(1/2)z_0^2}(0.5)\Gamma_{(1/2)z_0^2}(2.5)]}{[\Gamma_{(1/2)z_0^2}(1.5)]^2} \tag{2.3.9}$$

where $\Gamma_z(m) = \int_0^z e^{-u}u^{m-1}\,du/\Gamma(m)$ is the incomplete gamma function.

Truncated Distribution of (2.3.2)

Consider the truncated distribution $[w = (x - \theta)/\delta]$ of (2.3.2),

$$g_T(w;a) = \text{constant} \times e^{-w}w^{a-1} \qquad 0 \le w \le w_0 \tag{2.3.10}$$

The jth moment $\mu_j' = E(w^j)$ $(j = 1, 2, \ldots)$ is given by

$$\mu_j' = \frac{a(a+1)\cdots(a+j-1)\Gamma_{w_0}(a+j)}{\Gamma_{w_0}(a)} \tag{2.3.11}$$

We thus have the values shown in Table 2.3.2 of the mean μ_1', the standard deviation $\sqrt{\mu_2}$, and the coefficients of skewness and kurtosis $\beta_1^* = \mu_3^2/\mu_2^3 = \lambda_3$ and $\beta_2^* = \mu_4/\mu_2^2$, respectively, for values of w_0 for which $[1/\Gamma(a)] \int_{w_0}^{\infty} e^{-w}w^{a-1}\,dw = 0.0, 0.01, 0.05$, and 0.10. Note that if w_1 and w_2 are independent and both have the distribution (2.3.10), then $z = w_1 - w_2$ has a symmetric distribution with mean zero, variance $V(z) = 2V(w)$, and $\beta_2^*(z) = 1.5 + 0.5\beta_2^*(w)$; therefore, the distribution of z is much closer to a normal than the distribution of w.

From the results above, it is clear that one way of reducing the effect of long tails (nonnormality) on the tests mentioned in Chapter 1 is to truncate moderately the tail(s) of a distribution, if it has long tail(s), and sample this truncated distribution. Since the location and scale parameters [e.g., μ (or θ)

Table 2.3.2 Values of μ_1', $\sqrt{\mu_2}$, β_1^*, and β_2^* for the Distribution (2.3.10)

w_0	μ_1'	$\sqrt{\mu_2}$	β_1^*	β_2^*	w_0	μ_1'	$\sqrt{\mu_2}$	β_1^*	β_2^*
		a = 1					a = 2		
∞	1	1	4	9	∞	2	1.414	2	6
4.605	0.954	0.885	2.01	4.84	6.638	1.942	1.292	0.99	3.68
2.996	0.842	0.709	0.98	3.22	4.744	1.794	1.089	0.40	2.64
2.303	0.744	0.588	0.61	2.66	3.890	1.656	0.943	0.19	2.29
		a = 4					a = 8		
∞	4	2	1	4.5	∞	8	2.828	0.50	3.75
10.05	3.926	1.863	0.47	3.10	16.00	7.903	2.668	0.20	2.84
7.754	3.728	1.622	0.14	2.43	13.15	7.636	2.373	0.03	2.39
6.681	3.537	1.443	0.03	2.22	11.77	7.372	2.147	0.00	2.28

and σ] are not known, one cannot determine the point(s) of truncation. Instead, one may sample the entire distribution and censor (trim) the extreme observations, which clearly represent the tails, leading to type II censored samples. This illustrates the role of type II censoring in the search for robust procedures.

2.4 ESTIMATION OF PARAMETERS FROM TYPE II CENSORED SAMPLES

Suppose that the underlying distribution is of the type $(1/\sigma)f((x - \mu)/\sigma)$ and that one wants to estimate the location and scale parameters μ and σ from the type II censored sample as given in (2.2.1). Two methods of estimation are widely used, the method of weighted least squares gives the linear unbiased and minimum variance estimators (BLUEs) of μ and σ among all linear unbiased estimators. However, since the least-squares method makes use of the distribution only through its expected values and variances and covariances of order statistics and does not use moments of higher order, some information may be lost in using this method. On the other hand, the method of maximum likelihood uses the entire distribution, and therefore one would expect this method to produce better estimators than the BLUEs. This is in fact true for most of the distributions. Since, however, the least-squares method (based on ordered observations) has been used by a number of statisticians in solving numerous practical problems (see Hamouda and

Leone, 1974; Lloyd, 1952; Leone and Hamouda, 1973; Tiao and Lund, 1970), we shall give a brief description of this method before we introduce and discuss the method of maximum likelihood. Comparisons between the two methods will be made at various stages of development of the subject matter.

2.5 LEAST-SQUARES ESTIMATORS (BLUEs)

Write $z_{(i)} = (x_{(i)} - \mu)/\sigma$, $i = r_1 + 1, \ldots, n - r_2$, and let $E(z_{(i)}) = \mu_{i:n}$ be the expected values and $\underset{\sim}{V} = (\sigma_{i,j:n})$ be the variance-covariance matrix of $z_{(i)}$, $i = r_1 + 1, \ldots, n - r_2$; $\sigma_{i,j:n} = \text{cov}(z_{(i)}, z_{(j)})$. Expressions for $\mu_{i:n}$ and $\sigma_{i,j:n}$ are given in Appendix B. Now

$$E(x_{(i)}) = \mu + \sigma\mu_{i:n}$$

and (2.5.1)

$$\text{cov}(x_{(i)}, x_{(j)}) = \sigma^2\sigma_{i,j:n}$$

Let $\underset{\sim}{X}' = (x_{(r_1+1)}, \ldots, x_{(n-r_2)})$, $\underset{\sim}{\theta}' = (\mu, \sigma)$, $\underset{\sim}{E}' = (\mu_{r_1+1:n}, \ldots, \mu_{n-r_2:n})$, and $\underset{\sim}{W} = (\underset{\sim}{1}_{n-r_1-r_2}, \underset{\sim}{E})$, where $\underset{\sim}{1}_r$ is a $r \times 1$ column of 1's. In matrix notation,

$$E(\underset{\sim}{X}) = \underset{\sim}{W}\underset{\sim}{\theta}$$

and the variance-covariance matrix of $\underset{\sim}{X}$ is

$$V(\underset{\sim}{X}) = \sigma^2\underset{\sim}{V} \tag{2.5.2}$$

From (2.5.2), the weighted least-squares solution $\underset{\sim}{\theta}^{*\prime} = (\mu^*, \sigma^*)$ is

$$\underset{\sim}{\theta}^* = (\underset{\sim}{W}'\underset{\sim}{V}^{-1}\underset{\sim}{W})^{-1}\underset{\sim}{W}'\underset{\sim}{V}^{-1}\underset{\sim}{X}$$

which is the BLUE of $\underset{\sim}{\theta}$ based on the type II censored sample. From (2.5.2), one has $E(\underset{\sim}{\theta}^*) = \underset{\sim}{\theta}$ and the variance-covariance matrix of $\underset{\sim}{\theta}^*$ is $V(\underset{\sim}{\theta}^*) = \sigma^2(\underset{\sim}{W}'\underset{\sim}{V}^{-1}\underset{\sim}{W})^{-1} = (v_{ij})$. Let $\Delta_W = |\underset{\sim}{W}'\underset{\sim}{V}^{-1}\underset{\sim}{W}|$, the determinant of $\underset{\sim}{W}'\underset{\sim}{V}^{-1}\underset{\sim}{W}$, and put $\Gamma = (1/\Delta_W)\underset{\sim}{V}^{-1}(\underset{\sim}{1}_{n-r_1-r_2}\underset{\sim}{E}' - \underset{\sim}{E}\underset{\sim}{1}'_{n-r_1-r_2})\underset{\sim}{V}^{-1}$. Then

$$\mu^* = -\underset{\sim}{E}'\Gamma\underset{\sim}{X} = \sum_{i=r_1+1}^{n-r_2} a_i x_{(i)}$$

and (2.5.3)

$$\sigma^* = \underset{\sim}{1}'_{n-r_1-r_2}\Gamma\underset{\sim}{X} = \sum_{i=r_1+1}^{n-r_2} b_i x_{(i)}$$

$$v_{11} = \frac{\sigma^2(\underset{\approx}{E}'\underset{\sim}{V}^{-1}\underset{\approx}{E})}{\Delta_W}$$

$$v_{22} = \frac{\sigma^2(\underset{\sim}{1}'_{n-r_1-r_2}\underset{\sim}{V}^{-1}\underset{\sim}{1}_{n-r_1-r_2})}{\Delta_W} \tag{2.5.4}$$

and

$$v_{12} = v_{21} = \frac{-\sigma^2(\underset{\sim}{1}'_{n-r_1-r_2}\underset{\sim}{V}^{-1}\underset{\approx}{E})}{\Delta_W}$$

Remark: If the underlying distribution is symmetric, then $\mu_{i:n} = -\mu_{n-i+1:n}$ and $\sigma_{i,j:n} = \sigma_{n-j+1,n-i+1:n}$ $(i < j)$. Further, if $r_1 = r_2 = r$ (symmetric type II censoring), then clearly $\underset{\sim}{1}'\underset{\sim}{V}^{-1}\underset{\approx}{E} = \underset{\approx}{E}'\underset{\sim}{V}^{-1}\underset{\sim}{1} = 0$; in that case, $v_{12} = \operatorname{cov}(\mu^*, \sigma^*) = 0$ and

$$\mu^* = \frac{\underset{\sim}{1}'\underset{\sim}{V}^{-1}\underset{\approx}{X}}{\underset{\sim}{1}'\underset{\sim}{V}^{-1}\underset{\sim}{1}} \quad \text{and} \quad \sigma^* = \frac{\underset{\approx}{E}'\underset{\sim}{V}^{-1}\underset{\approx}{X}}{\underset{\approx}{E}'\underset{\sim}{V}^{-1}\underset{\approx}{E}} \tag{2.5.5}$$

which are symmetric linear functions of $x_{(i)}$ $(i = r + 1, \ldots, n - r)$, for example, $a_i = a_{n-i+1}$ and $b_i = -b_{n-i+1}$. Further,

$$V(\mu^*) = \frac{\sigma^2}{\underset{\sim}{1}'\underset{\sim}{V}^{-1}\underset{\sim}{1}} \quad \text{and} \quad V(\sigma^*) = \frac{\sigma^2}{\underset{\approx}{E}'\underset{\sim}{V}^{-1}\underset{\approx}{E}} \tag{2.5.6}$$

The least-squares estimators μ^* and σ^* (and their variances and covariance) can thus be calculated from the expected values and the variances and covariances of order statistics of random samples from a distribution of the type $(1/\sigma)f((x - \mu)/\sigma)$. The means, variances, and covariances of order statistics are available for numerous distributions; the normal (Tietjen et al., 1977); the outlier models $(n - 1)N(0, 1)$ and $1N(\lambda, 1)$ and $(n - 1)N(0, 1)$ and $1N(0, \tau^2)$ (David et al., 1977); the Cauchy distribution (Barnett, 1966); the double exponential (Govindarajulu, 1966); the logistic distribution (Gupta et al., 1967); the Student's t distributions (Tiku and Kumra, 1981); the gamma distributions (Prescott, 1974); the half logistic distribution (Balakrishnan, 1984); the double Weibull distributions (Balakrishnan and Kocherlakota, 1985a); and the log-logistic distributions (Balakrishnan and Malik, 1985a).

For the normal distribution, Sarhan and Greenberg (1962, pp. 218-251) tabulate the values of the coefficients a_i and b_i needed to calculate μ^* and σ^* from (2.5.3). For the data of Example 2.2.1, we have $n = 10$, $r_1 = 0$, $r_2 = 3$, and

i:	1	2	3	4	5	6	7
a_i	0.0244	0.0636	0.0818	0.0962	0.1089	0.1207	0.5045
b_i	-0.3252	-0.1758	-0.1058	-0.0502	-0.0006	0.0469	0.6107

The observed order statistics are

1.613 1.644 1.663 1.732 1.740 1.763 1.778

Therefore,

$$\mu^* = \sum_{i=1}^{8} a_i x_{(i)} = 1.746 \quad \text{and} \quad \sigma^* = \sum_{i=1}^{8} b_i x_{(i)} = 0.091$$

are the least-squares estimates of μ and σ of the assumed normal distribution. The variances and the covariance of these estimates are given by (Sarhan and Greenberg, 1962, p. 253)

$$(\underset{\sim}{W}'\underset{\sim}{V}^{-1}\underset{\sim}{W})^{-1}\sigma^2 = \begin{pmatrix} 0.1167 & 0.0260 \\ 0.0260 & 0.0989 \end{pmatrix} \sigma^2$$

The standard errors of μ^* and σ^* are, therefore, $\pm\sqrt{(0.1167)(0.091)^2} = \pm 0.031$ and $\pm\sqrt{(0.0989)(0.091)^2} = \pm 0.029$, respectively.

For the normal distribution $N(\mu, \sigma)$, the expected values $\mu_{i:n}$ and the variances and the covariances $\sigma_{i,j:n}$ of order statistics are available for samples of size $n \leq 50$ (Tietjen et al., 1977); for $n > 50$, however, the asymptotic equations (B.27)-(B.30) of Appendix B provide close approximations to $\mu_{i:n}$ and $\sigma_{i,j:n}$.

Complete Samples

For complete samples, $r_1 = r_2 = 0$. If the underlying distribution is normal, then $\underset{\sim}{1}'\underset{\sim}{V} = \underset{\sim}{1}'$ [see (B.26)], so that $\underset{\sim}{1}' = \underset{\sim}{1}'\underset{\sim}{V}\underset{\sim}{V}^{-1} = \underset{\sim}{1}'\underset{\sim}{V}^{-1}$. Substituting this value in (2.5.5), we get

$$\mu^* = \frac{1}{n}\sum_{i=1}^{n} x_{(i)} = \frac{1}{n}\sum_{i=1}^{n} x_i = \bar{x} \text{ (the sample mean)} \qquad (2.5.7)$$

$\bar{x}$ is the most efficient estimator (unbiased and minimum variance) since its variance is equal to the minimum variance bound $1/[-E(d^2 \log L/d\mu^2)] = \sigma^2/n$.

For nonnormal populations, $V(\bar{x}) = \sigma^2/n$ always, but it can be shown that $V(\mu^*)$ is strictly smaller than σ^2/n, except when $\underset{\sim}{1}'\underset{\sim}{V} = \underset{\sim}{1}'$ [for details, see Lloyd (1952), Downton (1953), and Govindarajulu (1968)]. A stronger result is given by Bondesson (1976), who proves that the sample mean is BLUE for the population mean if and only if the underlying distribution is either the normal or the gamma distribution.

For the normal $N(\mu, \sigma)$, the maximum likelihood estimator (which will be introduced in Section 2.6) $\hat{\hat{\sigma}}$ corrected for bias is given by

$$\hat{\hat{\sigma}} = \frac{\Gamma[(n-1)/2]}{\sqrt{2}\,\Gamma(n/2)} \left[\sum_{i=1}^{n} (x_i - \bar{x})^2 \right]^{\frac{1}{2}}$$

The efficiency of the BLUE σ^* relative to $\hat{\hat{\sigma}}$, for the normal distribution $N(\mu, \sigma)$ is quite high, as is clear from the following values:

n:	4	5	6	7	8	9	10
$V(\hat{\hat{\sigma}})/\sigma^2$	0.1781	0.1318	0.1045	.0865	0.0738	0.0653	0.0570
$V(\sigma^*)/\sigma^2$	0.1801	0.1333	0.1057	0.0875	0.0746	0.0650	0.0576

Simplified Linear Estimators

The BLUEs μ^* and σ^* work out in terms of the expected values and variances and covariances of order statistics. For most distributions, however, the covariances are particularly difficult to compute and, therefore, μ^* and σ^* are not readily available. Hence Gupta (1952) proposed the use of the unweighted least-squares method, which utilizes the expected values of order statistics but not their variances and covariances. The unweighted least-squares estimators μ^{**} and σ^{**} are obtained from (2.5.3) by replacing $\underset{\sim}{V}$ by the identity matrix and are given by

$$\mu^{**} = \sum_{i=r_1+1}^{n-r_2} c_i x_{(i)} \quad \text{and} \quad \sigma^{**} = \sum_{i=r_1+1}^{n-r_2} d_i x_{(i)}$$

where

$$c_i = \frac{\Sigma_{i=r_1+1}^{n-r_2} \mu_{i:n}^2 - \mu_{i:n} \Sigma_{i=r_1+1}^{n-r_2} \mu_{i:n}}{D}$$

$$d_i = \frac{(n - r_1 - r_2)\mu_{i:n} - \Sigma_{i=r_1+1}^{n-r_2} \mu_{i:n}}{D}$$

and

$$D = (n - r_1 - r_2) \sum_{i=r_1+1}^{n-r_2} \mu_{i:n}^2 - \left(\sum_{i=r_1+1}^{n-r_2} \mu_{i:n} \right)^2$$

Obviously, both μ^{**} and σ^{**} are unbiased and

$$V(\mu^{**}) = (\underset{\sim}{c}'\underset{\sim}{V}\underset{\sim}{c})\sigma^2 \qquad V(\sigma^{**}) = (\underset{\sim}{d}'\underset{\sim}{V}\underset{\sim}{d})\sigma^2$$

and

$$\text{cov}(\mu^{**}, \sigma^{**}) = (\underset{\sim}{c}'\underset{\sim}{V}\underset{\sim}{d})\sigma^2$$

where $\underset{\sim}{c}' = (c_{r_1+1}, c_{r_1+2}, \ldots, c_{n-r_2})$ and $\underset{\sim}{d}' = (d_{r_1+1}, d_{r_1+2}, \ldots, d_{n-r_2})$. Thus the variances and covariances of order statistics are needed to calculate the variances and the covariance of the simplified estimators μ^{**} and σ^{**}. Note that if the underlying distribution is symmetric and if $r_1 = r_2 = r$, then $\Sigma_{i=r+1}^{n-r} \mu_{i:n} = 0$ and hence $\mu^{**} = \Sigma_{i=r+1}^{n-r} x_{(i)}/(n - 2r)$, the trimmed sample mean.

To have an idea about some of the applications of the method of weighted least squares, we consider two important distributions here, namely, the exponential distribution $E(\theta, \sigma)$ and the uniform distribution $U(\theta_1, \theta_2)$. For these two distributions, the expected values and the variances and covariances of order statistics are available in closed form.

Exponential Distribution

For the exponential distribution $E(\theta, \sigma)$, the probability density function is given by

$$f(x) = \frac{1}{\sigma} \exp\left(-\frac{x - \theta}{\sigma}\right) \qquad \theta < x < \infty$$

The expected values and the variances and covariances of $z_{(i)} = (x_{(i)} - \theta)/\sigma$ are given, respectively, by [(B.15) and (B.16)]

$$E(z_{(i)}) = \mu_{i:n} = \sum_{j=1}^{i} \frac{1}{n - j + 1} \qquad \sigma_{i,i:n} = \sigma_{i,j:n} = \sum_{u=1}^{i} \frac{1}{(n - u + 1)^2} \qquad i < j \tag{2.5.8}$$

On substituting these results in (2.5.3), we obtain

$$\theta^* = x_{(r_1+1)} - \mu_{r_1+1:n}\sigma^*$$

and

$$\sigma^* = \frac{\Sigma_{i=r_1+1}^{n-r_2} x_{(i)} + r_2 x_{(n-r_2)} - (n - r_1)x_{(r_1+1)}}{n - r_1 - r_2 - 1}$$

[for details, see Sarhan and Greenberg (1955, 1957)]. Using (2.5.4), the variances of θ^* and σ^* and their covariance are obtained as

$$\mathrm{var}(\theta^*) = \left[\sigma_{r_1+1, r_1+1:n} + \frac{\mu^2_{r_1+1:n}}{n - r_1 - r_2 - 1}\right]\sigma^2$$

$$\mathrm{var}(\sigma^*) = \frac{\sigma^2}{n - r_1 - r_2 - 1} \qquad (2.5.9)$$

and

$$\mathrm{cov}(\theta^*, \sigma^*) = \frac{-\mu_{r_1+1:n}\sigma^2}{n - r_1 - r_2 - 1}$$

For complete samples, $r_1 = r_2 = 0$, in which case

$$\theta^* = x_{(1)} - \frac{1}{n}\sigma^* \quad \text{and} \quad \sigma^* = \frac{\Sigma_{i=1}^{n} x_{(i)} - nx_{(1)}}{n - 1}$$

$$V(\theta^*) = \frac{\sigma^2}{n(n - 1)} \quad \text{and} \quad V(\sigma^*) = \frac{\sigma^2}{n - 1}$$

The estimators σ^* given above will be used later (Chapter 6) in constructing goodness-of-fit statistics for testing the exponential $E(\theta, \sigma)$ when θ and σ are unknown. It is of interest to note that the estimators θ^* and σ^* for complete samples ($r_1 = r_2 = 0$) are both MVUEs and BLUEs.

Uniform Distribution

The density function of a uniform distribution $U(\theta_1, \theta_2)$ is given by

$$f(x) = \frac{1}{\theta_2} \qquad \theta_1 \le x \le \theta_1 + \theta_2$$

If $z_{(i)} = (x_{(i)} - \theta)/\theta_2$, $\theta = (\theta_1 + \frac{1}{2}\theta_2)$, then

$$\mu_{i:n} = \frac{i}{n+1} \quad \text{and} \quad \sigma_{i,j:n} = \frac{i(n - j + 1)}{(n + 1)^2(n + 2)} \qquad i \le j \qquad (2.5.10)$$

[(B.13) and (B.14)]. Substituting these values in (2.5.3), we obtain the following BLUE of θ_1 and θ_2:

$$\theta_1^* = x_{(r_1+1)} - \frac{r_1 + 1}{n - r_1 - r_2 - 1}(x_{(n-r_2)} - x_{(r_1+1)})$$

and

$$\theta_2^* = \frac{n + 1}{n - r_1 - r_2 - 1}(x_{(n-r_2)} - x_{(r_1+1)})$$

(Sarhan and Greenberg, 1955). The variances of θ_1^* and θ_2^* and their covariance are given by

$$V(\theta_1^*) = \frac{(r_1 + 1)(n - r_2)\theta_2^2}{(n + 1)(n + 2)(n - r_1 - r_2 - 1)}$$

$$V(\theta_2^*) = \frac{(r_1 + r_2 + 2)\theta_2^2}{(n + 2)(n - r_1 - r_2 - 1)} \tag{2.5.11}$$

and

$$\operatorname{cov}(\theta_1^*, \theta_2^*) = -\frac{(r_1 + 1)\theta_2^2}{(n + 2)(n - r_1 - r_2 - 1)}$$

For complete samples ($r_1 = r_2 = 0$),

$$\theta_1^* = x_{(1)} - \frac{1}{n - 1}(x_{(n)} - x_{(1)}) \quad \text{and} \quad \theta_2^* = \frac{n + 1}{n - 1}(x_{(n)} - x_{(1)})$$

The estimators θ_2^* given above will be used later (Chapter 6) in constructing a goodness-of-fit statistic for testing the uniform $U(\theta_1, \theta_2)$ when θ_1 and θ_2 are unknown.

Some other papers dealing with linear unbiased estimation of location and scale parameters are Abe (1971), Ali and Chan (1964), Balakrishnan and Kocherlakota (1985a), Barnett et al. (1967), Bartholomew (1963), Basu (1965), Bennett (1952), Beyer et al. (1976), Birnbaum and Laska (1967), Birnbaum et al. (1971), Birnbaum and Make (1970), Blom (1962), Chan (1967a,b, 1969, 1970a,b, 1970a,b, 1971, 1974), Chan et al. (1971), D'Agostino (1971a), D'Agostino and Cureton (1973), D'Agostino and Lee (1976), Downton (1954, 1966a,b), Dyer and Whisenand (1973a,b), Englehardt and Bain (1974), Epstein (1956), Epstein and Sobel (1953), Goldberger (1962), Govindarajulu (1968), Govindarajulu and Eisenstat (1965), Govindarajulu and Joshi (1968), Hamouda and Leone (1974), Jung (1955, 1962), Kabir and Ahsanullah (1974), Kulldorff and Vanman (1973), Lawless (1971, 1975, 1977), Lwin (1976), McCool (1965), Mann (1967, 1968, 1971), Mann and Fertig (1973), Menon (1963), Nair (1950), Plackett (1958), Saleh (1966), Sinha (1973), Watterson (1959), White (1964), and Zelen and Dannemiller (1961).

2.6 MAXIMUM LIKELIHOOD ESTIMATORS (MLEs)

Suppose that the underlying distribution (probability density function) is of the type $(1/\sigma)f((x - \mu)/\sigma)$, μ and σ being the location and scale parameters, respectively. Let $F(z) = \int_{-\infty}^{z} f(z)\,dz$, where $z = (x - \mu)/\sigma$. Then the likelihood function L based on the type II censored sample $x_{(i)}$ ($i = r_1 + 1, r_1 + 2, \ldots, n - r_2$) is given by

$$L = \frac{n!}{r_1!\,r_2!}\sigma^{-(n-r_1-r_2)}\left[\prod_{i=r_1+1}^{n-r_2} f(z_{(i)})\right][F(z_{(r_1+1)})]^{r_1}[1 - F(z_{(n-r_2)})]^{r_2} \tag{2.6.1}$$

where $z_{(i)} = (x_{(i)} - \mu)/\sigma$.

The MLEs are the solutions of the equations

$$\frac{\partial \log L}{\partial \mu} = 0 \quad \text{and} \quad \frac{\partial \log L}{\partial \sigma} = 0 \tag{2.6.2}$$

For most distributions, however, the equations (2.6.2) do not admit explicit solutions. Consider, for example, the normal $N(\mu, \sigma)$ distribution. Here $f(z) = (2\pi)^{-\frac{1}{2}} \exp(-z^2/2)$, and

$$\frac{\partial \log L}{\partial \mu} = \frac{n}{\sigma}\left[\frac{1}{n}\sum_{i=r_1+1}^{n-r_2} z_{(i)} - q_1 g_1(z_{(r_1+1)}) + q_2 g_2(z_{(n-r_2)})\right] = 0 \tag{2.6.3}$$

and

$$\frac{\partial \log L}{\partial \sigma} = \frac{n}{\sigma}\left[-(1 - q_1 - q_2) + \frac{1}{n}\sum_{i=r_1+1}^{n-r_2} z_{(i)}^2 - q_1 z_{(r_1+1)} g_1(z_{(r_1+1)}) + q_2 z_{(n-r_2)} g_2(z_{(n-r_2)})\right] = 0 \tag{2.6.4}$$

where $q_1 = r_1/n$, $q_2 = r_2/n$, $g_1(z) = f(z)/F(z)$, and $g_2(z) = f(z)/[1 - F(z)]$.

Equations (2.6.3) and (2.6.4) do not admit explicit solutions and have to be solved by iterative methods. Suitable iterative methods for solving these equations have been developed by Cohen (1957, 1961) and Harter and Moore (1966). These methods work well and the iterations converge to the true solutions fairly quickly. Due to the implicit nature of these iterations, however, it is difficult to make any analytical study of the resulting MLE, especially for small samples. We will, therefore, modify (2.6.2) and define modified maximum likelihood estimators as explicit functions of sample observations (Tiku, 1967; Tiku and Stewart, 1977).

Modified Maximum Likelihood Estimator (MMLE)

Let $f(x;\theta)$ be the probability density function of x, θ being an unknown parameter. Suppose that the ML equation (derivative of the log-likelihood) based on a sample of size n (Tiku, 1967; Tiku and Stewart, 1977),

$$\xi(\underset{\sim}{x}, n; \theta) = \frac{d \log L}{d\theta} = 0$$

has no explicit solution (MLE). To obtain the MMLE, we seek a function $\xi^*(\underset{\sim}{x}, n; \theta)$, and hence L^*, such that

$$\xi^*(\underset{\sim}{x}, n; \theta) = \frac{d \log L^*}{d\theta} = 0$$

has an explicit solution (MMLE) and is such that for all θ and $\underset{\sim}{x}' = (x_1, x_2, \ldots, x_n)$,

$$\lim_{n \to \infty} \{ \xi(\underset{\sim}{x}, n; \theta) = \xi^*(\underset{\sim}{x}, n; \theta) \} \tag{2.6.5}$$

Then it is clear that the MMLE is asymptotically identical with the MLE. Furthermore, if for a given $\epsilon > 0$ there exists an N such that

$$| \xi(\underset{\sim}{x}, n; \theta) - \xi^*(\underset{\sim}{x}, n; \theta) | < \epsilon \qquad \text{(for all } \underset{\sim}{x} \text{ and } \theta) \tag{2.6.6}$$

for $n \geq N$, then the MMLE will be almost identical with the MLE for all sample sizes $n \geq N$.

The modification above can in a straightforward fashion be extended in case of two or more unknown parameters (Tiku, 1967, 1973). Other types of modifications of the ML equations exist and for these the reader is referred to Mehrotra and Nanda (1974), Persson and Rootzen (1977), and Cohen and Whitten (1980, 1982).

2.7 MMLEs FOR THE NORMAL

It is clear that (2.6.3) and (2.6.4) do not admit explicit solutions because of the terms involving the functions $g_1(z_{(r_1+1)})$ and $g_2(z_{n-r_2})$. It is, therefore, pertinent to replace them by simpler functions in such a way that the resulting MML equations admit explicit solutions and satisfy the condition (2.6.5) and also, if possible, the condition (2.6.6).

It can be verified that the points $g(z) = f(z)/[1 - F(z)]$ over an interval $a \leq z \leq b$ of finite length lie very close to the line $g(z) = \alpha + \beta z$ (Tiku, 1967), where

$$\beta = \frac{g(b) - g(a)}{b - a} \quad \text{and} \quad \alpha = g(z) - a\beta \tag{2.7.1}$$

For example, the values of $\max |\epsilon|$ obtained from the linear equation $g(z) = \alpha + \beta z + \epsilon$, for

$(a, b) = (0.00, 0.49), (0.50, 0.99), (1.00, 1.49), (1.50, 1.99), (2.00, 2.49)$

are

$\max |\epsilon| = 0.003, \quad 0.003, \quad 0.002, \quad 0.003, \quad 0.007$

where

$g(z) = 0.804, \quad 1.141, \quad 1.534, \quad 2.108, \quad 2.541$

To modify (2.6.3) and (2.6.4), we therefore consider the linear approximations

$$g_1(z_{(r_1+1)}) \simeq \alpha_1 - \beta_1 z_{(r_1+1)} \quad \text{and} \quad g_2(z_{(n-r_2)}) \simeq \alpha_2 + \beta_2 z_{(n-r_2)} \tag{2.7.2}$$

Note that $g_1(z) = g_2(-z)$. In these equations,

$$\beta_1 = -\frac{g_1(k_1) - g_1(h_1)}{k_1 - h_1} \quad \text{and} \quad \alpha_1 = g_1(h_1) + h_1\beta_1 \tag{2.7.3}$$

and

$$\beta_2 = \frac{g_2(k_2) - g_2(h_2)}{k_2 - h_2} \quad \text{and} \quad \alpha_2 = g_2(h_2) - h_2\beta_2 \tag{2.7.4}$$

where $g_1(z) = f(z)/F(z)$ and $g_2(z) = f(z)/[1 - F(z)]$, and the intervals (h_1, k_1) and (h_2, k_2) are wide enough to cover both t_1 and $z_{(r_1+1)}$ and t_2 and $z_{(n-r_2)}$, respectively, and where t_1 and t_2 are determined by the equations

$$F(t_1) = q_1 \quad \text{and} \quad 1 - F(t_2) = q_2 \qquad \left(q_1 = \frac{r_1}{n},\ q_2 = \frac{r_2}{n}\right) \tag{2.7.5}$$

The difficulty is that we do not know the exact positions of $z_{(r_1+1)}$ and $z_{(n-r_2)}$, but $E(z_{(r_1+1)}) = t_1$ and $E(z_{(n-r_2)}) = t_2$ for large n (David, 1981, p. 80); hence, for reasonably large n, the interval (h_2, k_2) is likely to cover both t_2 and $z_{(n-r_2)}$ if h_2 and k_2 are determined by the equations

$$C_1 = 1 - F(h_2) = q_2 + \left[\frac{q_2(1 - q_2)}{n}\right]^{\frac{1}{2}} \quad \text{and} \quad C_2 = 1 - F(k_2) = q_2 - \left[\frac{q_2(1 - q_2)}{n}\right]^{\frac{1}{2}}$$

and similarly for (h_1, k_1). To simplify the computations, one could take C_1 equal to the tabulated value (Pearson and Hartley, 1972) just greater than $q_2 + [q_2(1 - q_2)/n]^{\frac{1}{2}}$ and C_2 equal to the tabulated value just smaller than $q_2 - [q_2(1 - q_2)/n]^{\frac{1}{2}}$.

The values of (α_1, β_1) and (α_2, β_2) can thus easily be calculated and are given in Table 2.7.1 for n = 10 and 20. Note that for $q_2 = q$, $\alpha_2 = \alpha$ and

Table 2.7.1 Values of α and β for Calculating the MMLE; q = r/n

	n = 10		n = 20		n = ∞	
q	α	β	α	β	α	β
0	0	1	0	1	0	1
0.1	0.6737	0.8611	0.6873	0.8395	0.6902	0.8309
0.2	0.7493	0.7914	0.7456	0.7859	0.7422	0.7814
0.3	0.7862	0.7402	0.7796	0.7379	0.7733	0.7355
0.4	0.8074	0.6891	0.7990	0.6898	0.7915	0.6882
0.5	0.8160	0.6335	0.8067	0.6351	0.7979	0.6366

$\beta_2 = \beta$, and for $q_1 = q$, $\alpha_1 = \alpha$ and $\beta_1 = \beta$. A comprehensive table, calculated by using the IMSL routine MDNRIS for evaluating the inverse of the normal probability integral $F(z) = p$, $0 < p < 1$, and hence C_1 and C_2, of the values of α and β is given in Table B.1.

As n tends to infinity, h_1 and k_1 both tend to t_1 and h_2 and k_2 both tend to t_2, in which case

$$\beta_1 = -\left[\frac{d}{dt} g_1(t)\right]_{t=t_1} = \frac{f(t_1)}{q_1}\left[t_1 + \frac{f(t_1)}{q_1}\right] \text{ and } \alpha_1 = \frac{f(t_1)}{q_1} + \beta_1 t_1 \quad (2.7.6)$$

and

$$\beta_2 = \left[\frac{d}{dt} g_2(t)\right]_{t=t_2} = -\frac{f(t_2)}{q_2}\left[t_2 - \frac{f(t_2)}{q_2}\right] \text{ and } \alpha_2 = \frac{f(t_2)}{q_2} - \beta_2 t_2 \quad (2.7.7)$$

The values of α and β calculated from these asymptotic equations by using the IMSL routine mentioned above are given in Table 2.7.1 under $n = \infty$. It is clear from this table that for a given $q = r/n$, the values of (α, β) for all $n \geq 10$ are not much different from each other. Therefore, the values of (α_1, β_1) and (α_2, β_2) may be obtained from the asymptotic equations (2.7.6) and (2.7.7), respectively, for all $n \geq 10$.

Using the linear approximations (2.7.2), we obtain the following MML (modified maximum likelihood) equations from the ML equations (2.6.3) and (2.6.4):

$$\frac{\partial \log L}{\partial \mu} \simeq \frac{\partial \log L^*}{\partial \mu} = \frac{1}{\sigma}\left[\sum_{i=r_1+1}^{n-r_2} z_{(i)} - r_1(\alpha_1 - \beta_1 z_{(r_1+1)}) + r_2(\alpha_2 + \beta_2 z_{(n-r_2)})\right]$$

$$= \frac{m}{\sigma^2}(K + D\sigma - \mu) = 0 \quad (2.7.8)$$

and

$$\frac{\partial \log L}{\partial \sigma} \simeq \frac{\partial \log L^*}{\partial \sigma} = \frac{1}{\sigma}\left[-(n - r_1 - r_2) + \sum_{i=r_1+1}^{n-r_2} z_{(i)}^2 - r_1 z_{(r_1+1)}(\alpha_1 - \beta_1 z_{(r_1+1)}) + r_2 z_{(n-r_2)}(\alpha_2 + \beta_2 z_{(n-r_2)})\right]$$

$$= -\frac{A}{\sigma^3}\left(\sigma - \frac{B(\mu) + \{[B(\mu)]^2 + 4AC(\mu)\}^{\frac{1}{2}}}{2A}\right)\left(\sigma - \frac{B(\mu) - \{[B(\mu)]^2 + 4AC(\mu)\}^{\frac{1}{2}}}{2A}\right)$$

$$= -\frac{1}{\sigma^3}[(A\sigma^2 - B\sigma - C) - m(K - \mu)(K + D\sigma - \mu)]$$

$$= 0 \quad (2.7.9)$$

where

$$A = n - r_1 - r_2 \qquad m = n - r_1 - r_2 + r_1\beta_1 + r_2\beta_2 \qquad D = \frac{r_2\alpha_2 - r_1\alpha_1}{m}$$

$$K = \frac{\Sigma_{i=r_1+1}^{n-r_2} x_{(i)} + r_1\beta_1 x_{(r_1+1)} + r_2\beta_2 x_{(n-r_2)}}{m}$$

$$B = r_2\alpha_2(x_{(n-r_2)} - K) - r_1\alpha_1(x_{(r_1+1)} - K)$$

$$C = \sum_{i=r_1+1}^{n-r_2} (x_{(i)} - K)^2 + r_1\beta_1(x_{(r_1+1)} - K)^2 + r_2\beta_2(x_{(n-r_2)} - K)^2$$

$$= \sum_{i=r_1+1}^{n-r_2} x_{(i)}^2 + r_1\beta_1 x_{(r_1+1)}^2 + r_2\beta_2 x_{(n-r_2)}^2 - mK^2$$

The expressions for $B(\mu)$ and $C(\mu)$ are exactly the same as B and C, respectively, but with K replaced by μ. It may be noted that both (2.7.8) and (2.7.9) satisfy the condition (2.6.5) since asymptotically $z_{(r_1+1)} = t_1$ and $z_{(n-r_2)} = t_2$, in which case

$$g_1(z_{(r_1+1)}) = \alpha_1 - \beta_1 z_{(r_1+1)} \quad \text{and} \quad g_2(z_{(n-r_2)}) = \alpha_2 + \beta_2 z_{(n-r_2)} \tag{2.7.10}$$

The MML equations (2.7.8) and (2.7.9) also satisfy the condition (2.6.6) since $|\epsilon|$ in $g(z) = \alpha + \beta z + \epsilon$ is quite small, as we have shown earlier.

Solutions of (2.7.8) and (2.7.9), with correction for bias in the estimator of σ, yield the MML estimators $\hat{\mu}_c$ and $\hat{\sigma}_c$ as

$$\hat{\mu}_c = K + D\hat{\sigma}_c$$

and (2.7.11)

$$\hat{\sigma}_c = \frac{B + (B^2 + 4AC)^{\frac{1}{2}}}{2[A(A-1)]^{\frac{1}{2}}}$$

For $r_1 = r_2 = 0$ (no censoring on either side of the sample), $\hat{\mu}_c$ and $\hat{\sigma}_c^2$ reduce to the sample mean $\bar{x} = \Sigma_{i=1}^{n} x_i/n$ and sample variance $s^2 = \Sigma_{i=1}^{n} (x_i - \bar{x})^2/(n-1)$, respectively. Now, the equation

$$\frac{\partial \log L^*}{\partial \mu} = \frac{m}{\sigma^2}(K + D\sigma - \mu)$$

can easily be solved to give

$$L^* = \exp\left[-\frac{m}{2\sigma^2}(K + D\sigma - \mu)^2 + g(\underset{\sim}{x}, \sigma)\right] \tag{2.7.12}$$

where $g(\underset{\sim}{x}, \sigma)$ is a function of $\underset{\sim}{x}' = (x_{(r_1+1)}, \ldots, x_{(n-r_2)})$ and σ but is free of μ. From (2.7.9) and (2.7.12), we obtain

$$\left(\frac{\partial \log L^*}{\partial \sigma}\right)_{\mu=K+D\sigma} = \frac{\partial}{\partial \sigma} g(\underset{\sim}{x}, \sigma) = \frac{1}{\sigma^3}(C + B\sigma - A\sigma^2)$$

which gives $g(\underset{\sim}{x}, \sigma) = (-B/\sigma - C/2\sigma^2 - A \log \sigma)H(\underset{\sim}{x})$, where $H(\underset{\sim}{x})$ is a function of $\underset{\sim}{x}$ and is free of μ and σ. It follows that

$$L^* = \sigma^{-A} \exp\left[-\frac{B}{\sigma} - \frac{C}{2\sigma^2} - \frac{m}{2\sigma^2}(K + D\sigma - \mu)^2\right] H(\underset{\sim}{x}) \tag{2.7.13}$$

Alternatively, by using the linear approximations in (2.7.2) and expanding $\Sigma_{i=r_1+1}^{n-r_2} z_{(i)}^2$, $\log F(z_{(r_1+1)})$, and $\log [1 - F(z_{(n-r_2)})]$ in Taylor series around μ and σ, it can be shown that L reduces to L* as given above. [A derivation along these lines was given by Tan (1984a) for the case $r_1 = r_2 = r$; a similar derivation can easily be carried out for $r_1 \neq r_2$.]

From (2.7.13), one sees clearly that (K, B, C) forms a minimum set of sufficient statistics for (μ, σ). It follows that if one uses L*, optimal estimation and testing procedures for μ and σ are functions of K, B, and C. Note that both (2.7.8) and (2.7.9) satisfy conditions (2.6.5) and (2.6.6) since the linear approximations in (2.7.2) are very close approximations, and therefore L* approximates L very closely so that inference procedures for (μ, σ) based on L* would be almost as efficient as those based on L. Note further that asymptotically L* is identical to L, which establishes the asymptotic optimality of the MML estimators. One also notes from (2.7.13) that the partial derivatives of log L*, with respect to μ and σ, of all orders exist and are continuous functions. Using standard results of maximum likelihood estimation, one then has the following lemma, giving the asymptotic distribution of the MML estimators $(\hat{\mu}_c, \hat{\sigma}_c)$ of (μ, σ):

LEMMA 2.7.1 Asymptotically, that is, for large $A = n - r_1 - r_2$ (hence large n), $(\hat{\mu}_c, \hat{\sigma}_c)$ is bivariate normal with expected values $[E(\hat{\mu}_c) = \mu$, $E(\hat{\sigma}_c) = \sigma]$ and variance-covariance matrix

$$V = \begin{pmatrix} J_{11} & J_{12} \\ J_{21} & J_{22} \end{pmatrix}^{-1} = \begin{pmatrix} V_{11} & V_{12} \\ V_{21} & V_{22} \end{pmatrix} \qquad (J_{12} = J_{21})$$

where

$$J_{11} = -E\left(\frac{\partial^2 \log L^*}{\partial \mu^2}\right) \quad J_{12} = -E\left(\frac{\partial^2 \log L^*}{\partial \mu\, \partial \sigma}\right) \quad J_{22} = -E\left(\frac{\partial^2 \log L^*}{\partial \sigma^2}\right)$$

For a proof, see Rao (1975) or Kendall and Stuart (1973).

Now, from (2.7.8) and (2.7.9),

$$\frac{\partial^2 \log L^*}{\partial \mu^2} = -\frac{m}{\sigma^2} \tag{2.7.14}$$

and

$$\frac{\partial^2 \log L^*}{\partial \mu\, \partial \sigma} = -\frac{m}{\sigma^3}[2(K + D\sigma - \mu) - D\sigma] \tag{2.7.15}$$

Since for large A, $AC(\mu)$ dominates over $[B(\mu)]$ (see Example 2.7.1, for instance), $[B(\mu)]^2/AC(\mu) \simeq 0$ and consequently,

$$\frac{\partial \log L^*}{\partial \sigma} \simeq \frac{A}{\sigma^3}\left[\frac{C(\mu)}{A} - \sigma^2\right] = 0 \quad \text{and} \quad \frac{\partial^2 \log L^*}{\partial \sigma^2} \simeq \frac{A}{\sigma^2}\left[-\frac{3}{\sigma^2}\frac{C(\mu)}{A} + 1\right] \tag{2.7.16}$$

It then follows from (2.7.8) and (2.7.14)-(2.7.16) that

$$E(K) = \mu - D\sigma \quad \text{and} \quad E\left[\frac{C(\mu)}{A}\right] \simeq \sigma^2$$

and (2.7.17)

$$J_{11} = \frac{m}{\sigma^2} \qquad J_{12} = J_{21} = -\frac{mD}{\sigma^2} = -\frac{r_2\alpha_2 - r_1\alpha_1}{\sigma^2} \qquad J_{22} = \frac{2A}{\sigma^2}$$

For large A, therefore, the variances and the covariance of the MML estimators $\hat{\mu}_c$ and $\hat{\sigma}_0 = [(A-1)/A]^{\frac{1}{2}}\hat{\sigma}_c$ are given by the elements of the matrix

$$\frac{\sigma^2}{m}\begin{pmatrix} 1 & -D \\ -D & 2A/m \end{pmatrix}^{-1} = \begin{pmatrix} \nu_{11} & \nu_{12} \\ \nu_{21} & \nu_{22} \end{pmatrix} \qquad (\nu_{12} = \nu_{21})$$

If $q_1 = r_1/n$ and $q_2 = r_2/n$ are fixed, then D tends to a constant as A tends infinity, in which case $\hat{\mu}_c$ and $\hat{\sigma}_c$ are not uncorrelated even asymptotically. However, if r_1 and r_2 are fixed, then D tends to zero as A tends to infinity, in which case $\hat{\mu}_c$ and $\hat{\sigma}_c$ are asymptotically independent. Note, however, that in either case $\hat{\mu}_c$ and $\hat{\sigma}_c$ are asymptotically unbiased and minimum variance estimators, with their variances attaining the Fisher-Cramér minimum variance bounds.

Symmetric Censoring

If $r_1 = r_2 = r$ (symmetric censoring), then $\alpha_1 = \alpha_2 = \alpha$, $\beta_1 = \beta_2 = \beta$ and $D = 0$, in which case m, K, A, B, and C reduce to

$$m = n - 2r + 2r\beta$$

$$\hat{\mu} = K = \frac{\Sigma_{i=r+1}^{n-r} x_{(i)} + r\beta(x_{(r+1)} + x_{(n-r)})}{m}$$

$$A = n - 2r$$

$$B = r\alpha(x_{(n-r)} - x_{(r+1)}) \tag{2.7.18}$$

and

$$C = \sum_{i=r+1}^{n-r} (x_{(i)} - \hat{\mu})^2 + r\beta[(x_{(n-r)} - \hat{\mu})^2 + (x_{(r+1)} - \hat{\mu})^2]$$

$$= \sum_{i=r+1}^{n-r} x_{(i)}^2 + r\beta(x_{(r+1)}^2 + x_{(n-r)}^2) - m\hat{\mu}^2$$

The MMLE for μ and σ (with correction for bias) are given by $\hat{\mu}$ and

$$\hat{\sigma} = \frac{B + (B^2 + 4AC)^{\frac{1}{2}}}{2\,[A(A-1)]^{\frac{1}{2}}} \qquad (2.7.19)$$

The estimators $\hat{\mu}$ and $\hat{\sigma}$ are a useful pair of robust estimators of location and scale parameters; this will be discussed in Chapter 4. Since $\mu_{i:n} = -\mu_{n-i+1:n}$ for the normal distribution (in fact, for any symmetric distribution), where $\mu_{i:n}$ is the expected value of $z_{(i)} = (x_{(i)} - \mu)/\sigma$, the estimator $\hat{\mu}$ is obviously unbiased. The variance of $\hat{\mu}$ is given by

$$V(\hat{\mu}) = \frac{(\underset{\sim}{u}'\underset{\sim}{V}\underset{\sim}{u})\sigma^2}{m^2} \qquad (2.7.20)$$

where $\underset{\sim}{u}' = (1 + r\beta, 1, \ldots, 1, 1 + r\beta)$ is a vector of $n - 2r$ elements and $\sigma^2 V$ is the variance-covariance matrix of $x_{(i)}$, $i = r + 1, \ldots, n - r$. Now from (2.7.8),

$$\frac{\partial \log L}{\partial \mu} \simeq \frac{\partial \log L^*}{\partial \mu} = \frac{m}{\sigma^2}(\hat{\mu} - \mu)$$

Hence $\hat{\mu}$ is not only unbiased but also is, for large n, a minimum variance estimator of μ with its variance attaining the Fisher-Cramér-Rao lower bound $\mathrm{var}(\hat{\mu}) \simeq \sigma^2/m = 1/[-E(\partial^2 \log L^*/\partial\mu^2)]$. Further, $E(\partial^2 \log L^*/\partial\mu\,\partial\sigma) = 0$ implies that $\hat{\mu}$ and $\hat{\sigma}$ are asymptotically independent. Thus asymptotically, $\hat{\mu}$ is the most efficient estimator of μ and is independent of $\hat{\sigma}$. Since L^* approximates L very closely, similar results also hold for small n. In this context it is very interesting to note that the asymptotic formula $\mathrm{var}(\hat{\mu}) \doteq \sigma^2/m$ gives remarkably close approximations to the variance of $\hat{\mu}$ for all $n \geq 5$ (see Table 2.7.2). For comparing $\hat{\mu}$ with the best linear unbiased estimator μ^*, we also give the exact values of the variance of μ^* (Sarhan and Greenberg, 1962, pp. 252-265) in Table 2.7.2. It is clear that $\hat{\mu}$ is as efficient as the best linear unbiased estimator μ^*. This is not surprising since L^* approximates L very closely and L is the likelihood function that is known to give unbiased and minimum variance estimator of μ for all sample sizes $n - r_1 - r_2$ (Harter and Moore, 1966).

From methods exactly similar to the ones used in case of unsymmetric censoring, the following additional results can easily be established:

Table 2.7.2 Comparison of $V(\mu^*)/\sigma^2$, $1/m$, and var $(\mu^*)/\sigma^2$

	n = 5			n = 10			n = 20		
q	$\frac{V(\mu^*)}{\sigma^2}$	$\frac{V(\hat{\mu})}{\sigma^2}$	$\frac{1}{m}$	$\frac{V(\mu^*)}{\sigma^2}$	$\frac{V(\hat{\mu})}{\sigma^2}$	$\frac{1}{m}$	$\frac{V(\mu^*)}{\sigma^2}$	$\frac{V(\hat{\mu})}{\sigma^2}$	$\frac{1}{m}$
0.0	0.2	0.2	0.2	0.1	0.1	0.1	0.05	0.05	0.05
0.1				0.1043	0.1043	0.1029	0.0520	0.0520	0.0517
0.2	0.2258	0.2258	0.2170	0.1113	0.1113	0.1091	0.0552	0.0552	0.0547
0.3				0.1218	0.1218	0.1185	0.0602	0.0602	0.0593
0.4				0.1383	0.1383	0.1331	0.0679	0.0679	0.0665

1. $\hat{\mu}$ and $\hat{\sigma}$ are uncorrelated for all sample sizes $A = n - 2r$; in fact, they are asymptotically independent.
2. For large A, $\hat{\sigma}$ is an unbiased estimator of σ and its variance is given by

$$\text{var}(\hat{\sigma}) \simeq \frac{\sigma^2}{2(A-1)} \tag{2.7.21}$$

3. For large A, $\hat{\sigma}$ is a minimum variance estimator with its variance attaining the Fisher-Cramér-Rao minimum variance bound.

To determine the bias in $\hat{\sigma}$ for moderate values of A, $E(\hat{\sigma})$ may be calculated from the equation

$$E(\hat{\sigma}) \simeq \frac{E(B) + [E(B^2) + 4AE(C)]^{\frac{1}{2}}}{2[A(A-1)]^{\frac{1}{2}}} \tag{2.7.22}$$

The exact values of the expression on the right-hand side of (2.7.22) are given in Table 2.7.3; it is clear from these values that $\hat{\sigma}$ is almost unbiased. Also given are the simulated values (based on 10,000 Monte Carlo runs) of

Table 2.7.3 Simulated Values of $V(\hat{\sigma})/\sigma^2$ and Exact Values of $V(\sigma^*)/\sigma^2$

	n = 10				n = 20			
q	$E(\hat{\sigma})/\sigma$	$V(\hat{\sigma})/\sigma^2$	$1/2(A-1)$	$V(\sigma^*)/\sigma^2$	$E(\hat{\sigma})/\sigma$	$V(\hat{\sigma})/\sigma^2$	$1/2(A-1)$	$V(\sigma^*)/\sigma^2$
0.1	0.96	0.075	0.071	0.082	0.99	0.034	0.033	0.038
0.2	0.93	0.104	0.100	0.129	0.99	0.049	0.045	0.058

var $(\hat{\sigma})$; it is clear that even for moderate values of A, (2.7.21) gives accurate approximations to var $(\hat{\sigma})$. For comparing $\hat{\sigma}$ with the BLUE σ^*, we give the exact values of var $(\hat{\sigma})$ in Table 2.7.3. As expected, $\hat{\sigma}$ is considerably more efficient than σ^*.

The following lemma establishes the distributions of $\hat{\mu}$ and $\hat{\sigma}$:

LEMMA 2.7.2 For large $A = n - 2r$ (hence large n), $\sqrt{m}(\hat{\mu} - \mu)/\sigma$ is distributed as normal $N(0, 1)$ and $(A - 1)\hat{\sigma}^2/\sigma^2$ is independently distributed as central chi-square distribution with $A - 1$ degrees of freedom.

Proof: It has already been shown that $\hat{\mu}$ is $N(\mu, \sigma^2/m)$ for large samples. Thus $\sqrt{m}(\hat{\mu} - \mu)/\sigma$ is $N(0, 1)$ for large A. Now for large A, $\hat{\sigma}^2 \simeq C/(A - 1)$ [since the term AC in (2.7.9) dominates over B^2 so that $B^2/AC \simeq 0$], in which case the modified likelihood function based on symmetric type II censored samples is given by

$$L^* \simeq (\sigma^2)^{-(A-1)/2} \exp\left[-\frac{(A - 1)\hat{\sigma}^2}{2\sigma^2} - \frac{m}{2\sigma^2}(\hat{\mu} - \mu)^2\right] H(\underset{\sim}{x}) \qquad (2.7.23)$$

where $H(\underset{\sim}{x})$ is free of μ and σ^2. In (2.7.23), making the transformation $(\hat{\mu} = \hat{\mu}(\underset{\sim}{x}),\ \hat{\sigma}^2 = \hat{\sigma}^2(\underset{\sim}{x}),\ w_3 = x_{(r+3)}, \ldots, w_{n-r} = x_{(n-r)})$ and integrating out $(w_3, \ldots, w_{n-r})$, one obtains the joint density of $(\hat{\mu},\ Z = (A - 1)\hat{\sigma}^2/\sigma^2)$ as

$$f(\hat{\mu}, Z) = f_1\left(\hat{\mu}; \mu, \frac{\sigma^2}{m}\right) f_2(Z; A - 1) R(\hat{\mu}, Z) \qquad (2.7.24)$$

where $f_1(\hat{\mu}; \mu, \sigma^2/m)$ is the density of $N(\mu, \sigma^2/m)$, $f_2(Z; A - 1)$ the density of χ^2_{A-1} and $R(\hat{\mu}, Z)$, an analytic function of $\hat{\mu}$ and Z, which is free of μ and σ^2. This follows readily from the fact that $-\infty < \hat{\mu} < \infty$ and $\hat{\sigma} > 0$ and that $H(\underset{\sim}{x})$ is free of the parameters. Now $(\hat{\mu}, Z)$ are complete sufficient statistics from (2.7.24) (see Hogg and Craig, 1978). Put $U(\hat{\mu}, Z) = R(\hat{\mu}, Z) - 1$; then

$$\int_{-\infty}^{\infty} \int_0^{\infty} f_1\left(\hat{\mu}; \mu, \frac{\sigma^2}{m}\right) f_2(Z; A - 1) U(\hat{\mu}, Z)\, d\hat{\mu}\, dZ = 0$$

for all $-\infty < \mu < \infty$ and $\sigma^2 > 0$. Hence $U(\hat{\mu}, Z) = 0$ almost surely, or $R(\hat{\mu}, Z) = 1$ almost surely. Thus

$$f(\hat{\mu}, Z) = f_1\left(\hat{\mu}; \mu, \frac{\sigma^2}{m}\right) f_2(Z; A - 1)$$

and that proves the lemma.

Alternatively, writing

$$C(\mu) = \sum_{i=r+1}^{n-r} (x_{(i)} - \mu)^2 + r\beta[(x_{(r+1)} - \mu)^2 + (x_{(n-r)} - \mu)^2]$$

then like

$$\sum_{i=1}^{n} (x_i - \mu)^2 = \sum_{i=1}^{n} (x_i - \bar{x})^2 + n\bar{x}^2$$

one also has $C(\mu) = C(\hat{\mu}) + m(\hat{\mu} - \mu)^2$, where $m\hat{\mu} = \Sigma_{i=r+1}^{n-r} x_{(i)} + r\beta(x_{(r+1)} + x_{(n-r)})$. Further, all partial derivatives of L* with respect to μ and σ, as obtained from (2.7.8) and (2.7.9), are exactly of the same form as the respective partial derivatives of the likelihood function L based on a random sample of size n from a normal population. Thus L* has exactly the same behavior as the likelihood function based on a random sample from a normal population (Tiku, 1982a, pp. 626-627). Consequently, $\hat{\mu}$ and $\hat{\sigma}^2$ have, for large A (hence large n), distributions exactly similar to those of the mean $\bar{x}$ and variance s^2 of a normal random sample.

EXAMPLE 2.7.1 Consider the following Darwin's well-known data (Fisher, 1966), which represent the differences (in heights) between cross- and self-fertilized plants of the same pair grown together in one pot:

49 -67 8 16 6 23 28 41 14 29 56 24 75 60 -48

The sample mean $\bar{x} = 20.9333$ and the sample standard deviation $s = 37.744$.

For purposes of achieving "robustness" (this will be explained in Chapter 4), however, we censor (give zero weights to) the $r = [0.5 + 0.1n] = [0.5 + 0.1(15)] = 2$ smallest and largest observations; the remaining observations, arranged in ascending order of magnitude,

6 8 14 16 23 24 28 29 41 49 56

constitute a symmetric type II censored sample of size $n - 2r = 11$. Pretend that the underlying distribution is normal $N(\mu, \sigma)$. To calculate the MML estimates, we have (Table B.1)

$$n = 15 \qquad q = \frac{r}{n} = \frac{2}{15} \qquad \alpha = 0.710 \qquad \beta = 0.823$$

$$\hat{\mu} = \frac{294 + 102.3}{14.3} = 27.713$$

and

$$\hat{\sigma} = \frac{72 + (5184 + 209044.616)^{\frac{1}{2}}}{20.976} = 25.498$$

Note that to compute $\hat{\mu}$ and $\hat{\sigma}$ all the observations in the sample need not be ordered; one simply has to locate the $r + 1$ smallest and largest observations.

The BLUE estimates of μ and σ are (Sarhan and Greenberg, 1962, pp. 233-256)

$$\mu^* = 6(0.1835) + 8(0.0701) + 14(0.0703) + 16(0.0704) + 23(0.0705) + 24(0.0705)$$
$$+ 28(0.0705) + 29(0.0704) + 41(0.0703) + 49(0.0701) + 56\ 0.1835)$$
$$= 27.695$$

and

$$\sigma^* = 6(-0.4139) + 8(-0.0838) + 14(-0.0607) + 16(-0.0395) + 23(-0.0195)$$
$$+ 24(0.0) + 28(0.0195) + 29(0.0395) + 41(0.0607) + 49(0.0838) + 56(0.4139)$$
$$= 26.381$$

The standard errors of $\hat{\mu}$ and $\hat{\sigma}$ are, respectively,

$$\frac{25.498}{\sqrt{14.3}} = 6.743 \quad \text{and} \quad \frac{25.498}{\sqrt{20}} = 5.702$$

The standard errors of μ^* and σ^* are, respectively,

$$26.381\sqrt{0.0706} = 7.009 \quad \text{and} \quad 26.381\sqrt{0.0594} = 6.430$$

The standard errors of $\bar{x}$ and s are, respectively,

$$\frac{37.744}{\sqrt{15}} = 9.745 \quad \text{and} \quad \frac{37.744}{\sqrt{28}} = 7.133$$

which are substantially larger than the corresponding standard errors of the MMLE and BLUE estimates based on the censored sample above. Perhaps, some of the extreme observations in the complete sample are outliers, thus rendering $\bar{x}$ (and s) inefficient.

Asymmetric Censoring

For small samples, the estimators $\hat{\mu}_c$ and $\hat{\sigma}_c$ based on asymmetric censored samples (i.e., $r_1 \neq r_2$) are biased. To study this bias, Tiku (1970) conducted a Monte Carlo investigation and concluded that unless r_1 and r_2 (or q_1 and q_2) are very different from each other, the bias in $\hat{\mu}_c$ and $\hat{\sigma}_c$ are negligible. We give in Table 2.7.4 the results of his investigation (based on 1000 Monte Carlo runs) of (1) the mean of $\hat{\mu}_c$, $M(\hat{\mu}_c)$; (2) the mean of $\hat{\sigma}_c$, $M(\hat{\sigma}_c)$; (3) the variance of $\hat{\mu}_c$, $V(\hat{\mu}_c)$; (4) the covariance of $\hat{\mu}_c$ and $\hat{\sigma}_c$, cov $(\hat{\mu}_c, \hat{\sigma}_c)$; (5) the variance of $\hat{\sigma}_c$, var $(\hat{\sigma}_c)$; (6) the conditional (σ known) mean of $\hat{\mu}_c$, $M(\hat{\mu}_c | \sigma)$; (7) the conditional (σ known) mean of $\hat{\sigma}_c$, $M(\hat{\sigma}_c | \mu)$; (8) the conditional variance of $\hat{\mu}_c$, $V(\hat{\mu}_c | \sigma)$; and (9) the conditional variance of $\hat{\sigma}_c$, $V(\hat{\sigma}_c | \mu)$. The standard errors of $M(\hat{\mu}_c)$ and $M(\hat{\sigma}_c)$ are well within ± 0.004.

It may be noted that Tiku (1970, Tables 1 and 2) gives values for $\hat{\sigma}_c = [B + (B^2 + 4AC)^{\frac{1}{2}}]/2A$; we have multiplied his results for $M(\hat{\sigma}_c)$ and cov $(\hat{\mu}_c, \hat{\sigma}_c)$ by $[A/(A - 1)]^{\frac{1}{2}}$ and the results for $V(\hat{\sigma}_c)$ by $A/(A - 1)$, to obtain the results given in Table 2.7.4.

Table 2.7.4 Simulated Values of the Means and Variances and Covariances of the Estimators $\hat{\mu}_c$ and $\hat{\sigma}_c$; $\mu = 0$ and $\sigma = 1$

n	q_1	q_2	(1)	(2)	(3)	(4)	(5)	(6)	(7)	(8)	(9)
10	0.0	0.1	-0.009	0.971	0.104	0.006	0.061	-0.003	0.978	0.103	0.052
		0.2	-0.017	0.965	0.107	0.016	0.070	-0.002	0.975	0.104	0.059
		0.4	-0.056	0.942	0.135	0.047	0.106	-0.004	0.969	0.116	0.077
	0.1	0.2	-0.012	0.956	0.110	0.010	0.083	-0.002	0.974	0.109	0.073
		0.3	-0.028	0.943	0.119	0.024	0.106	-0.002	0.972	0.114	0.087
	0.2	0.3	-0.019	0.927	0.121	0.017	0.137	-0.002	0.971	0.119	0.112
20	0.0	0.1	-0.007	0.988	0.052	0.003	0.031	-0.005	0.990	0.052	0.029
		0.2	-0.010	0.985	0.053	0.005	0.035	-0.004	0.990	0.053	0.033
		0.4	-0.028	0.977	0.063	0.018	0.052	-0.004	0.986	0.057	0.044
	0.1	0.2	-0.007	0.969	0.054	0.003	0.044	-0.002	0.978	0.054	0.041
		0.4	-0.030	0.952	0.063	0.017	0.069	-0.002	0.969	0.059	0.059
	0.2	0.3	-0.010	0.951	0.057	0.005	0.071	-0.000	0.969	0.057	0.064
		0.4	-0.026	0.933	0.063	0.017	0.095	-0.000	0.962	0.060	0.078
	0.3	0.4	-0.020	0.910	0.064	0.012	0.132	-0.002	0.959	0.063	0.107

For comparing $(\hat{\mu}_c, \hat{\sigma}_c)$ with the least-squares estimator (μ^*, σ^*), we define the deficiency (efficiency = 1/deficiency) of $(\hat{\mu}, \hat{\sigma})$ as

$$\operatorname{def}(\hat{\mu}_c, \hat{\sigma}_c) = \operatorname{MSE}(\hat{\mu}_c) + \operatorname{MSE}(\hat{\sigma}_c) \tag{2.7.25}$$

and, similarly, $\operatorname{def}(\mu^*, \sigma^*) = V(\mu^*) + V(\sigma^*)$ for (μ^*, σ^*), since μ^* and σ^* are unbiased. It is clearly seen from Table 2.7.5 that $\operatorname{def}(\hat{\mu}_c, \hat{\sigma}_c)$ is smaller than $\operatorname{def}(\mu^*, \sigma^*)$; that is, the estimators $\hat{\mu}_c$ and $\hat{\sigma}_c$ are jointly more efficient than the BLUEs μ^* and σ^*. Further, one may note that $\hat{\mu}_c$ and $\hat{\sigma}_c$ have simple algebraic forms and, unlike μ^* and σ^*, one does not need the expected values and variances and covariances of order statistics to compute them. Comparing these with the ML estimators, Tiku (1968a, b, c) and Smith et al. (1973) concluded that the MML estimates are almost identical to the ML estimates, as expected, even for small sample sizes.

Table 2.7.5 Deficiencies of the MML and BLU Estimators

n	q_1	q_2	$V(\hat{\mu}_c)$	$V(\hat{\sigma}_c)$	Def.	$V(\mu^*)$	$V(\sigma^*)$	Def.
10	0.0	0.1	0.104	0.061	0.165	0.103	0.068	0.171
		0.2	0.107	0.070	0.178	0.107	0.081	0.188
		0.4	0.135	0.106	0.247	0.134	0.168	0.302
	0.2	0.3	0.121	0.137	0.263	0.118	0.171	0.289
20	0.0	0.1	0.052	0.031	0.083	0.051	0.032	0.083
		0.2	0.053	0.035	0.088	0.053	0.037	0.090
		0.4	0.063	0.052	0.116	0.065	0.055	0.120
	0.2	0.3	0.057	0.071	0.130	0.058	0.075	0.133
		0.4	0.063	0.095	0.163	0.065	0.103	0.168
	0.3	0.4	0.064	0.132	0.204	0.066	0.156	0.222

EXAMPLE 2.7.2 Consider the data of Example 2.2.1. Here $n = 10$; $q_1 = 0$, $\alpha_1 = 0$, $\beta_1 = 1$; $q_2 = 0.3$, $\alpha_2 = 0.7855$, $\beta_2 = 0.7405$;

$$\hat{\mu}_c = \frac{11.933 + 3.9482}{9.2206} + \frac{2.3585}{9.2206}\sigma$$

$$= 1.7224 + 0.2558\sigma$$

$A = 10 - 3 = 7$, $B = 2.3586(1.778 - 1.7224) = 0.1311$, $C = 20.3669 + 7.0199 - 27.3544 = 0.0324$, and

$$\hat{\sigma}_c = \frac{(0.1311 + (0.0172 + 0.9072)^{\frac{1}{2}}}{2\sqrt{42}} = 0.0843$$

Therefore, the MML estimates of μ and σ are

$$\hat{\mu}_c = 1.744 \quad \text{and} \quad \hat{\sigma}_c = 0.0843$$

with the following standard errors obtained from (2.7.17)

$$SE(\hat{\mu}_c) = 0.0843\sqrt{0.114} = 0.028 \quad \text{and} \quad SE(\hat{\sigma}_c) = 0.0843\sqrt{0.0877} = 0.025$$

The ML estimates obtained by iterative methods are (Gupta, 1952, p. 271)

$$\hat{\hat{\mu}} = 1.743 \quad \text{and} \quad \hat{\hat{\sigma}} = 0.078$$

with

$$SE(\hat{\hat{\mu}}) = 0.028 \quad \text{and} \quad SE(\hat{\hat{\sigma}}) = 0.025$$

To reduce the bias in the ML estimates, we multiplied Gupta's estimate 0.072 of σ by $[A(A - 1)]^{\frac{1}{2}} = \sqrt{7/6}$ to obtain the ML estimate $\hat{\hat{\sigma}} = 0.078$, and similarly for the standard error. The BLU estimates are (David, 1981) p. 271)

$$\mu^* = 1.746 \quad \text{and} \quad \sigma^* = 0.091$$

with

$$SE(\mu^*) = 0.031 \quad \text{and} \quad SE(\sigma^*) = 0.029$$

As expected, the MML estimates are hardly any different from the ML estimates and also have the same efficiencies; the BLU estimates have somewhat larger standard errors, and furthermore, one needs special tables to compute them (Sarhan and Greenberg, 1962).

EXAMPLE 2.7.3 In a time-mortality experiment (Cohen, 1957), the first observation is delayed until the collapse of a fixed time interval, with the result that r_1 sample specimens die before observation begins. The experiment is subsequently terminated at a predetermined time with r_2 specimens remaining alive. Actual survival times are recorded for the $n - r_1 - r_2$ specimens that die during the period of full observation. For a specific sample of this type in which x, the log-survival time in days, is assumed to be normally distributed, $n=47$, $r_1 = 2$, $r_2 = 5$, $\Sigma_{i=3}^{42} x_{(i)} = 64.80444$, $x_{(3)} = 1.301030$, $x_{(42)} = 1.903090$, and $\Sigma_{i=3}^{42} x_{(i)}^2 = 105.85996$. Here $q_1 = 2/47$, $\alpha_1 = 0.62689$, $\beta_1 = 0.87523$; $q_2 = 5/47$, $\alpha_2 = 0.69364$, and $\beta_2 = 0.83065$. Substituting these values in (2.7.11), we obtain

$$\hat{\mu}_c = 1.64336 \quad \text{and} \quad \hat{\sigma}_c = 0.20545$$

with estimated variances and covariance

$$V(\hat{\mu}_c) = 0.0009208 \qquad \text{cov}(\hat{\mu}_c, \hat{\sigma}_c) = 0.000261 \qquad V(\hat{\sigma}_c) = 0.0005419$$

These estimates may be compared with the corresponding ML estimates, obtained by iterative methods (Cohen, 1957),

$$\hat{\hat{\mu}} = 1.64331 \quad \text{and} \quad \hat{\hat{\sigma}} = 0.20536$$

with estimated variances and covariance

$$V(\hat{\hat{\mu}}) = 0.0009192 \qquad \text{cov}(\hat{\hat{\mu}}, \hat{\hat{\sigma}}) = 0.000262 \qquad V(\hat{\hat{\sigma}}) = 0.0005451$$

The agreement between the two is amazingly close, as expected.

Nearly Unbiased Estimators

We have seen above that if r_1 and r_2 are very different from each other, the bias in $\hat{\mu}_c$ and $\hat{\sigma}_c$ for small samples could be considerable. To reduce this bias, we notice that conditional on σ, a Taylor series expansion gives

$$0 = \left(\frac{\partial \log L^*}{\partial \mu}\right)_{\mu=\hat{\mu}_c} \simeq \frac{\partial \log L^*}{\partial \mu} + (\hat{\mu}_\sigma - \mu)\frac{\partial^2 \log L^*}{\partial \mu^2} \tag{2.7.26}$$

so that

$$E(\hat{\mu}_\sigma | \sigma) = \mu + \frac{E(\partial \log L^*)/\partial \mu}{R^2(\mu)} \tag{2.7.27}$$

where $\hat{\mu}_\sigma = K + \sigma D$ and $R^2(\mu) = -E(\partial^2 \log L^*/\partial \mu^2) = m/\sigma^2$. Similarly, let $\hat{\sigma}_\mu$ satisfy $(\partial \log L^*/\partial \sigma)_{\sigma=\hat{\sigma}_\mu} = 0$. Then from a Taylor series expansion,

$$0 = \left(\frac{\partial \log L^*}{\partial \sigma}\right)_{\sigma=\hat{\sigma}_\mu} \simeq \frac{\partial \log L^*}{\partial \sigma} + (\hat{\sigma}_\mu - \sigma)\frac{\partial^2 \log L^*}{\partial \sigma^2} \tag{2.7.28}$$

which gives

$$E(\hat{\sigma}_\mu | \mu) \simeq \sigma + \frac{E(\partial \log L^*/\partial \sigma)}{R^2(\sigma)} \tag{2.7.29}$$

where $R^2(\sigma) = -E(\partial^2 \log L^*/\partial \sigma^2)$.

From these results one may reduce the bias in $(\hat{\mu}_c, \hat{\sigma}_c)$ if one chooses (α_1, β_1) and (α_2, β_2), say (α_1', β_1') and (α_2', β_2'), to satisfy $E(\partial \log L^*/\partial \mu) = E(\partial \log L^*/\partial \sigma) = 0$. Obviously, the values of (α_1', β_1') and (α_2', β_2') can be obtained by solving the following sets of equations:

$$\begin{aligned} E_{11} &= \sum_{i=1}^{n-r_2} E(z_{(i)}) + r_2[\alpha_2' + \beta_2' E(z_{(n-r_2)})] = 0 \\ E_{21} &= \sum_{i=1}^{n-r_2} E(z_{(i)}^2) + r_2[\alpha_2' E(z_{(n-r_2)}) + \beta_2' E(z_{(n-r_2)}^2)] - (n - r_2) = 0 \end{aligned} \tag{2.7.30}$$

and

$$\begin{aligned} E_{12} &= \sum_{i=1}^{r_1} E(z_{(i)}) + r_1[\alpha_1' - \beta_1' E(z_{(r_1+1)})] = 0 \\ E_{22} &= \sum_{i=1}^{r_1} E(z_{(i)}^2) + r_1[\alpha_1' E(z_{(r_1+1)}) - \beta_1' E(z_{(r_1+1)}^2)] - r_1 = 0 \end{aligned} \tag{2.7.31}$$

where $z_{(i)} = (x_{(i)} - \mu)/\sigma$ and $x_{(i)}$ is the ith-order statistic in a random sample of size n from $N(0,1)$. One may note that $E_{11} = E_{21} = E_{12} = E_{22} = 0$ imply that $E(\partial \log L^*/\partial \mu) = (E_{11} - E_{12})/\sigma = 0$ and $E(\partial \log L^*/\partial \sigma) = (E_{21} - E_{22})/\sigma = 0$. To evaluate E_{ij}'s, however, one needs tables of expected values and variances of normal order statistics. These are readily available for samples of size up to 50 (see Tietjen et al., 1977).

Table 2.7.6 Variances and Covariances of MMLEs $\hat{\mu}_c'$ and $\hat{\sigma}_c'$ and BLUEs μ^* and σ^*

n	q_1	q_2	(1)[a]	(2)	(3)	(4)	(5)	(6)
10	0.0	0.2	0.107	0.012	0.073	0.108	0.013	0.081
		0.4	0.132	0.044	0.114	0.134	0.047	0.124
	0.1	0.3	0.116	0.019	0.103	0.117	0.024	0.128
		0.4	0.131	0.041	0.138	0.134	0.050	0.168
20	0.0	0.2	0.0534	0.0055	0.0339	0.0534	0.0060	0.0374
		0.4	0.0645	0.0189	0.0509	0.0651	0.0204	0.0553
	0.1	0.4	0.0640	0.0177	0.0605	0.0652	0.0217	0.0740
	0.2	0.5	0.0785	0.0394	0.1198	0.0822	0.0505	0.1537

[a]Columns: (1) $V(\hat{\mu}_c')/\sigma^2$; (2) $\text{cov}(\hat{\mu}_c', \hat{\sigma}_c')/\sigma^2$; (3) $V(\hat{\sigma}_c')/\sigma^2$; (4) $V(\mu^*)/\sigma^2$; (5) $\text{cov}(\mu^*, \sigma^*)/\sigma^2$; (6) $V(\sigma^*)/\sigma^2$.

The MML estimators $\hat{\mu}_c'$ and $\hat{\sigma}_c'$ obtained by using (α_1', β_1') and (α_2', β_2') in place of (α_1, β_1) and (α_2, β_2), respectively, are nearly unbiased as estimators of μ and σ. Note that asymptotically, $\alpha_i' = \alpha_i$ and $\beta_i' = \beta_i$ $(i = 1, 2)$, in which case $\hat{\mu}_c' = \hat{\mu}_c$ and $\hat{\sigma}_c' = \hat{\sigma}_c$.

To have an idea about the efficiencies of the estimators $\hat{\mu}_c'$ and $\hat{\sigma}_c'$ relative to the BLUEs μ^* and σ^*, we give in Table 2.7.6 the variances and covariances of these estimators. It may be noted that $\hat{\mu}_c'$ and $\hat{\sigma}_c'$ are asymptotically identical with $\hat{\mu}_c$ and $\hat{\sigma}_c$, respectively, and hence with the MLEs. It is clear that $\hat{\mu}_c'$ and $\hat{\sigma}_c'$ are more efficient than μ^* and σ^*, besides being much easier to compute.

By approximating the likelihood function, Persson and Rootzen (1977) obtained different simplified estimators of μ and σ and showed that their estimators have properties similar to those of the MLE and are also numerically close to the MLE (see also Mehrotra and Nanda, 1974).

It may be noted that all the methods above are similar to a more general method of estimation (based on approximating the likelihood function) devised by Plackett (1958) [see also Cohen and Whitten (1980, 1982), who discuss a different modification of the likelihood function].

EXAMPLE 2.7.4 To compare the estimators above, consider the following data from an accelerated life-test experiment involving specimens of electrical insulation (Lawless, 1982, p. 226): 10 specimens were put on test and the test was terminated at the time of the eighth failure. The eight observed log-failure times

6.00 6.43 6.77 7.07 7.40 7.66 8.10 8.40 ... ...

constitute a type II censored sample of size 8. The methods above yield the following estimates:

		μ	σ	
Cohen	MLE	7.59	1.069	(Corrected for bias)
Tiku	MMLE	7.5895	1.0696	[equations (2.7.11)]
Lloyd	BLUE	7.6009	1.1416	[equations (2.5.3)]
Persson and Rootzen		7.57	0.99	(See Lawless, 1982, p. 226)
Lawless		7.60	1.07	(Based on a normal probability plot)

The agreement between these estimates is amazingly close, although Persson and Rootzen's estimates are a little too low.

The MMLEs, however, have a distinct advantage over other estimators—they are simple and explicit functions of sample observations and no special tables are needed to compute them and are as efficient as the MLEs and, consequently, more efficient than other estimators. Besides, they can easily be generalized to more complex situations and also to various other distributions (Chapters 3 and 5). We now proceed to show that the MMLEs $\hat{\mu}$ and $\hat{\sigma}$ have remarkably high efficiencies compared to some well-known robust estimators.

2.8 ROBUST ESTIMATORS AND COMPARISONS WITH THE MMLEs

Let $x_1, x_2, \ldots, x_n$ be a random sample from a distribution with location parameter μ and scale parameter σ. A pair of estimators, say $(\tilde{\mu}, \tilde{\sigma})$, is called a matching pair if $\lim_{n\to\infty} [\sqrt{n}\, \text{var}(\tilde{\mu})/E(\tilde{\sigma})] = \text{constant}$. As an example, we note that the MMLE $(\hat{\mu}, \hat{\sigma})$ calculated from a symmetric type II censored normal sample is a matching pair since asymptotically $\text{var}(\hat{\mu}) = \sigma^2/m$ $(m = n - 2r + 2r\beta)$ and $E(\hat{\sigma}) = \sigma$ so that $\lim_{n\to\infty} [\sqrt{n}\, \text{var}(\hat{\mu})/E(\hat{\sigma})] =$ $(1 - 2q + 2q\beta)^{-\frac{1}{2}}$ $(q = r/n \text{ fixed})$ [e.g., $\lim_{n\to\infty} \sqrt{m}\, \text{var}(\hat{\mu})/E(\hat{\sigma}) = 1$]. Other well-known matching pairs are the sample mean and sample standard deviation $(\bar{x}, s)$ and the following trimmed sample estimators $(\hat{\mu}_T, \hat{\sigma}_T)$ and the Huber estimators $(\hat{\mu}_H, \hat{\sigma}_H)$, among others.

Trimmed Sample Mean

Let $x_{(j)}$ be the jth-order statistic in a random sample of size n. The r-trimmed sample mean ($\hat{\mu}_T$) is given by (Dixon and Tukey, 1968)

$$\hat{\mu}_T = \sum_{i=r+1}^{n-r} \frac{x_{(i)}}{n - 2r} \tag{2.8.1}$$

The estimator $\hat{\sigma}_T$ which matches with $\hat{\mu}_T$ is

$$\hat{\sigma}_T^2 = \frac{\Sigma_{i=r+1}^{n-r}(x_{(i)} - \hat{\mu}_T)^2 + r(x_{(r+1)} - \hat{\mu}_T)^2 + r(x_{(n-r)} - \hat{\mu}_T)^2}{n - 2r - 1} \tag{2.8.2}$$

It was shown by Huber (1970) that under certain regularity conditions, $\lim_{n\to\infty}[\sqrt{(n-2r)\ \mathrm{var}(\hat{\mu}_T)}/E(\hat{\sigma}_T)] = 1$ for fixed $q = r/n$.

To evaluate the performance of $\hat{\mu}_T$ against $\bar{x}$, we consider the family of long-tailed symmetric distributions (2.3.1). This family constitutes very realistic alternatives to a normal distribution (Sprott, 1982; Tan, 1982; Tiku, 1980a) and has kurtosis $\beta_2^* = \mu_4/\mu_2^2 = 3(p - 3/2)/(p - 5/2)$, $p > 5/2$. Note that $t = \sqrt{\nu}\,(x - \mu)/\sigma\sqrt{k}$ has a central t-distribution with $\nu = 2p - 1$ degrees of freedom.

For the family above, the usual sample mean $\bar{x} = \Sigma_{i=1}^{n} x_i/n$ is an unbiased estimator of μ and has variance $V(\bar{x}) = \sigma^2/n$ always. If the underlying distribution is normal (i.e., $p = \infty$), $\bar{x}$ is the optimal estimator. In practice, however, one does not know the exact value of p.

For comparing $\hat{\mu}_T$ with $\bar{x}$, note first that $\hat{\mu}_T$ is also unbiased for symmetric distributions; this follows from the fact that for symmetric distributions, $E(z_i) = -E(z_{n-i+1})$, $z_i = (x_{(i)} - \mu)/\sigma$, $i = 1, 2, \ldots, n$. The variance of $\hat{\mu}_T$ is $V(\hat{\mu}_T) = (\underset{\sim}{u}'\underset{\sim}{V}\underset{\sim}{u})\sigma^2$, where $(n - 2r)\underset{\sim}{u} = \underset{\sim}{1}_{n-2r}$, a $(n - 2r) \times 1$ column of 1's, and $\underset{\sim}{V}$ is the variance-covariance matrix of z_i, $i = r + 1, r + 2, \ldots, n - r$, which are available from the tables of Tiku and Kumra (1981). Given in Table 2.8.1 are the exact values of $V(\bar{x})$, $V(\hat{\mu}_T)$, and the Fisher-Cramér-Rao lower bound MVB(μ) for n = 10 and 20, where

$$\mathrm{MVB}(\mu) = \frac{1}{-E(\partial^2 \log L/\partial\mu^2)} = \frac{\sigma^2(p + 1)(p - 3/2)}{np(p - 1/2)}$$

It is clear from Table 2.8.1 that $\hat{\mu}_T$ is remarkably efficient and is considerably more efficient overall than $\bar{x}$. Besides, $\hat{\mu}_T$ is almost optimal for the normal $N(\mu, \sigma)$ distribution.

To quote another example, we consider Dixon's (1953) outlier model wherein a random sample of size n has n - 1 observations coming from a normal $N(\mu, \sigma)$ distribution and one observation from a normal $N(\mu, \lambda\sigma)$, $\lambda > 0$. The sample mean $\bar{x}$ and the trimmed sample mean $\hat{\mu}_T$ are both unbiased estimators of μ. For n = 5, we give in Table 2.8.2 the exact values of $V(\bar{x})$

Table 2.8.1 Values of $\mathrm{MVB}(\mu)/\sigma^2$, $V(\bar{x})/\sigma^2$, and $V(\hat{\mu}_T)/\sigma^2$ for the Family (2.3.1); q = r/n = 0.1

	p: 2	2.5	3.5	4.5	5	10	∞
				n = 10			
$\mathrm{MVB}(\mu)$	0.0500	0.0700	0.0857	0.0917	0.0933	0.0984	0.1000
$\bar{x}$	0.1000	0.1000	0.1000	0.1000	0.1000	0.1000	0.1000
$\hat{\mu}_T$	0.0599	0.0771	0.0894	0.0943	0.0958	0.1013	0.1053
				n = 20			
$\mathrm{MVB}(\mu)$	0.0250	0.0350	0.0429	0.0458	0.0467	0.0492	0.0500
$\bar{x}$	0.0500	0.0500	0.0500	0.0500	0.0500	0.0500	0.0500
$\hat{\mu}_T$	0.0283	0.0372	0.0439	0.0466	0.0474	0.0505	0.0530

and $V(\hat{\mu}_T)$, calculated from the variances and covariances of normal order statistics in the presence of an outlier given by David et al. (1977); $\bar{x} = \Sigma_{i=1}^{5} x_i/5$ and $\hat{\mu}_T = \Sigma_{i=2}^{4} x_{(i)}/3$. Again, it is clear from Table 2.8.2 that $\hat{\mu}_T$ is considerably more efficient overall than $\bar{x}$.

Note that there is only one outlier in the foregoing random sample of size n = 5, the other four being "good" observations. The estimator $\hat{\mu}_T$ is, however, based only on three observations (the middle ones) and, thereby sacrifices one "good" observation and hopefully the outlier, which perhaps needs to be sacrificed; $\hat{\mu}_T$ being considerably more efficient overall than $\bar{x}$ indicates, therefore, that sacrificing a small number of "good" observations might not do too much harm as long as all the outliers or "bad" observations are left out (or their influence deflated in some other way).

To compare the MML estimator $\hat{\mu}$, given by (2.7.18), with the estimator $\hat{\mu}_T$, we consider the family of distributions (2.3.1) and also the following seven symmetric distributions (L stands for the logistic distribution) which

Table 2.8.2 Values of $V(\bar{x})/\sigma^2$ and $V(\hat{\mu}_T)/\sigma^2$ for n = 5

	λ: 0.5	2	3	4
$V(\bar{x})/\sigma^2$	0.170	0.320	0.520	0.800
$V(\hat{\mu}_T)/\sigma^2$	0.177	0.291	0.322	0.339

Table 2.8.3 Values of $V(\hat{\mu})/\sigma^2$ and $V(\hat{\mu}_T)/\sigma^2$ for the Family (2.3.1)

q	p:	2	2.5	3.5	4.5	5[a]	10	∞	
	β_2^*:	∞	∞	6	4.5	4.2	3.4	3	Sum
					n = 10				
0.1	$\hat{\mu}$	0.0653	0.0814	0.0921	0.0960	0.0973	0.1013	0.1043	0.6377
	$\hat{\mu}_T$	0.0599	0.0771	0.0894	0.0943	0.0958	0.1013	0.1053	0.6231
0.2	$\hat{\mu}$	0.0578	0.0767	0.0912	0.0972	0.0990	0.1059	0.1113	0.6391
	$\hat{\mu}_T$	0.0553	0.0747	0.0904	0.0970	0.0991	0.1071	0.1133	0.6369
0.3	$\hat{\mu}$	0.0563	0.0774	0.0950	0.1026	0.1050	0.1143	0.1218	0.6724
	$\hat{\mu}_T$	0.0559	0.0774	0.0955	0.1035	0.1060	0.1158	0.1238	0.6779
					n = 20				
0.1	$\hat{\mu}$	0.0314	0.0398	0.0455	0.0476	0.0482	0.0504	0.0520	0.3149
	$\hat{\mu}_T$	0.0283	0.0372	0.0439	0.0466	0.0474	0.0505	0.0530	0.3069
0.2	$\hat{\mu}$	0.0279	0.0374	0.0448	0.0479	0.0488	0.0524	0.0552	0.3144
	$\hat{\mu}_T$	0.0264	0.0363	0.0444	0.0480	0.0491	0.0534	0.0568	0.3144
0.3	$\hat{\mu}$	0.0270	0.0375	0.0464	0.0502	0.0515	0.0563	0.0602	0.3291
	$\hat{\mu}_T$	0.0268	0.0376	0.0469	0.0511	0.0525	0.0577	0.0621	0.3347

[a]This distribution has its first five moments common with the Logistic (Tiku and Jones, 1971) and the two distributions are, therefore, almost identical.

are frequently encountered in practice (Huber, 1972; Subrahmaniam et al., 1975; Tan, 1982; Wegman and Carroll, 1977):

(1) $0.95N(0,\sigma) + 0.05N(0,3\sigma)$, (2) $0.90N(0,\sigma) + 0.10N(0,3\sigma)$,

(3) $0.95L(0,\sigma) + 0.05L(0,3\sigma)$, (4) $0.90L(0,\sigma) + 0.10L(0,3\sigma)$

Outlier models: (5) $(n - 1)N(0,\sigma)$ and $1N(0,3\sigma)$, (6) $(n - 2)N(0,1)$ and $2N(0,3\sigma)$ (7) uniform

For all these distributions, $\hat{\mu}$ is also unbiased. For the family (2.3.1), the variance $V(\hat{\mu})$ can readily be computed by making use of the tables given by Tiku and Kumra (1981). For the distribution (7), $V(\hat{\mu})$ can be calculated from

Table 2.8.4 Simulated Values of $V(\hat{\mu})/\sigma^2$ and $V(\hat{\mu}_T)/\sigma^2$ for $n = 20$ and $q = r/n = 0.1$; $V(\bar{x})/\sigma^2 = 0.05$

	Distribution						
	(1)	(2)	(3)	(4)	(5)	(6)	(7)
$\hat{\mu}$	0.042	0.038	0.037	0.036	0.042	0.037	0.058
$\hat{\mu}_T$	0.041	0.037	0.037	0.034	0.042	0.036	0.067

the variances and covariances given by (2.5.10). However, for distributions (1) to (6) variances and covariances of order statistics are not available, and therefore we simulated (from 10,000 Monte Carlo runs) the values of the variances of $\hat{\mu}$ and $\hat{\mu}_T$ using the Cyber 170 computer at McMaster University. Given in Tables 2.8.3 and 2.8.4 are the values of $V(\hat{\mu})/\sigma^2$ and $V(\hat{\mu}_T)/\sigma^2$ for the seven distributions above, respectively.

For the family (2.3.1), it is clear from Table 2.8.3 that $\hat{\mu}$ is slightly more efficient than $\hat{\mu}_T$ for normal and near-normal distributions but is slightly less efficient for very long-tailed distributions. Both $\hat{\mu}$ and $\hat{\mu}_T$ have almost the highest efficiency overall for $q = r/n = 0.1$, that is, 10% censoring on either side of the ordered sample.

From Tables 2.8.3 and 2.8.4, and from other numerical results (Tiku, 1980a), it is clear that as an estimator of μ, $\hat{\mu}$ is almost as efficient overall as $\hat{\mu}_T$; furthermore, for symmetric distributions ranging from long-tailed (such as t_3) to short-tailed (such as uniform) distributions, $\hat{\mu}$ and $\hat{\mu}_T$ (with $r = [0.5 + 0.1n]$) are considerably more efficient overall than $\bar{x}$.

For comparing the deficiencies, $\text{def}(\tilde{\mu}, \tilde{\sigma}) = \text{MSE}(\tilde{\mu}) + \text{MSE}(\tilde{\sigma})$, of the matching pairs $(\bar{x}, s)$, $(\hat{\mu}_T, \hat{\sigma}_T)$, and $(\hat{\mu}, \hat{\sigma})$, Tiku (1980a) simulated the mean-square errors of s, $\hat{\sigma}_T$, and $\hat{\sigma}$ (from 10,000 Monte Carlo runs) for the following seven representative symmetric distributions:

The family (2.3.1) with: (1) $p = \infty$, (2) $p = 5$, (3) $p = 2.5$,

(4) $p = 2$; mixture model: (5) $0.90N(0,\sigma) + 0.10N(0,3\sigma)$;

outlier model: (6) $(n - 1)N(0,\sigma)$ and $1N(0,3\sigma)$, (7) $(n-2)N(0,\sigma)$ (2.8.3)

and $2N(0,3\sigma)$

Note that for symmetric distributions, $\bar{x}$, $\hat{\mu}_T$, and $\hat{\mu}$ are uncorrelated with s, $\hat{\sigma}_T$, and $\hat{\sigma}$, respectively.

Given in Table 2.8.5 are the simulated values of $\text{MSE}(\tilde{\sigma})/\sigma^2$ and $\text{def}(\tilde{\mu}, \tilde{\sigma})/\sigma^2$ for $n = 20$ and $q = r/n = 0.1$ based on 10,000 Monte Carlo runs. From Table 2.8.5 it is clear that the MML pair $(\hat{\mu}, \hat{\sigma})$ is jointly more efficient overall than the pairs $(\hat{\mu}_T, \hat{\sigma}_T)$ and $(\bar{x}, s)$. Similar results hold for other sample sizes (Tiku, 1980a).

Huber Estimators

Let $x_1, x_2, \ldots, x_n$ be a random sample from a distribution of the type $(1/\sigma)f[(x - \mu)/\sigma]$; μ and σ are the location and scale parameters, respectively. Huber (1964) proposed a new approach to estimate μ, assuming in particular that the function f represents a symmetric and long-tailed distribution (i.e., its Pearson coefficient $\beta_2^* = \mu_4/\mu_2^2$ is greater than 3). The log-likelihood function based on the random sample above is

$$\log L = \sum_{i=1}^{n} \log f\left(\frac{x_i - \mu}{\sigma}\right) = \sum_{i=1}^{n} \rho\left(\frac{x_i - \mu}{\sigma}\right)$$

If the functional form of f(x) is known, the MLE of μ for given σ is obtained from the equation

$$\frac{\partial \log L}{\partial \mu} = -\frac{1}{\sigma}\sum_{i=1}^{n} \rho'(z_i) = \frac{1}{\sigma}\sum_{i=1}^{n} \psi(z_i) = 0 \qquad (2.8.4)$$

where $z_i = (x_i - \mu)/\sigma$.

For the normal and double-exponential distributions, for example, the $\rho(z)$ and $\psi(z)$ functions are given by $[\rho(z) = \frac{1}{2}\log(2\pi) + \frac{1}{2}z^2,\ \psi(z) = z]$ and $[\rho(z) = \log 2 + |z|$ and $\psi(z) = \mathrm{sgn}(z),\ z \neq 0]$, respectively. Putting $w_i = w_i(z) = \psi(z_i)/z_i$, one may note that (2.8.4) reduces to $\Sigma_{i=1}^{n} w_i(x_i - \mu) = 0$, so that

Table 2.8.5 Simulated Values of $\mathrm{MSE}(\tilde{\sigma})/\sigma^2$ and $\mathrm{def}(\tilde{\mu}, \tilde{\sigma})/\sigma^2$ for $n = 20$ and $q = r/n = 0.1$

	Distribution						
	(1)	(2)	(3)	(4)	(5)	(6)	(7)
$\mathrm{MSE}(\tilde{\sigma})/\sigma^2$							
s	0.026	0.037	0.087	0.194	0.075	0.053	0.059
$\tilde{\sigma}$	0.034	0.040	0.062	0.115	0.068	0.043	0.062
$\hat{\sigma}_T$	0.040	0.055	0.090	0.144	0.089	0.065	0.084
$\mathrm{def}(\tilde{\mu}, \tilde{\sigma})/\sigma^2$							
$(\bar{x}, s)$	0.076	0.087	0.137	0.244	0.125	0.103	0.109
$(\hat{\mu}, \hat{\sigma})$	0.086	0.088	0.102	0.146	0.106	0.085	0.099
$(\hat{\mu}_T, \hat{\sigma}_T)$	0.093	0.102	0.127	0.172	0.126	0.107	0.120

$$\mu = \frac{\Sigma_{i=1}^{n} w_i x_i}{\Sigma_{i=1}^{n} w_i} \tag{2.8.5}$$

Given σ and $\psi(z)$, one may solve for μ either by applying (2.8.5) iteratively or by applying Newton-Raphson's procedure (see Low, 1959) to (2.8.4); Gross (1976) has in fact used a one-step Newton-Raphson procedure to solve for μ for various $\psi(z)$ functions.

Now, to solve for μ from either (2.8.4) or (2.8.5), one would need to know $\psi(z)$ and σ but, in practice, both the functional form of f(x) [and hence $\psi(x)$] and σ are not known. By combining the normal distribution in the middle with the double-exponential distribution in the tails, Huber (1964) proposed a function $\psi(z)$ as

$$\psi(z) = \begin{cases} z & \text{if } |z| \le c \\ c \operatorname{sgn}(z) & \text{if } |z| > c \end{cases} \tag{2.8.6}$$

[or equivalently, $\psi(z) = \max(-c, \min(c, z))$; sgn (z) is the sign function defined by $\operatorname{sgn}(z) = -1$ if $z < 0$ and $= 1$ if $z > 0$]. In (2.8.6), the popular choice of the c values have been 1.345 (Birch and Myers, 1982), 1.5, and 2.0 because these values correspond roughly to 10%, 5%, and 2.5% censoring on either side of a standard normal distribution. One would therefore expect that most observations in a random sample from a symmetric long-tailed distribution will satisfy the condition that $|(x - \mu)/\sigma| \le c$.

For the unknown σ, Huber (1964, 1977), Gross (1976, 1977) and many others have used $\tilde{\sigma}_0$ = mad to estimate σ, where mad = median$|x_i - \text{median}(x_j)|$; but Huber (1981) and Birch and Myers (1982) have suggested changing $\tilde{\sigma}_0$ to $\tilde{\sigma}_0$ = mad/0.6745 to make it an asymptotically unbiased estimator of σ in the case of a normal distribution.

The solutions of (2.8.4), which of course give different estimators of μ for different choices of c, have been called "hubers" in the extensive Princeton study (Andrews et al., 1972); the solutions corresponding to c = 1.5 and c = 2.0 have been denoted by H15 and H20, respectively. The solutions of (2.8.4) have been referred to as Huber's M estimators (Huber, 1977, 1981), to be denoted by $\hat{\mu}_H$.

To find the matching estimator $\hat{\sigma}_H$, Huber (1977, 1981) showed that the asymptotic variance of the M estimator $\hat{\mu}_H$ is $(1/n)\sigma^2 E(\psi^2(z))/[E(\psi'(z))]^2$, where $z = (x - \hat{\mu}_H)/\sigma$. Hence the matching estimator of scale is given by

$$\hat{\sigma}_H = \left\{ n\tilde{\sigma}_0^2 \left[\sum_{i=1}^{n} \psi^2\left(\frac{x_i - \hat{\mu}_H}{\tilde{\sigma}_0}\right)\right] \Big/ \left[\sum_{i=1}^{n} \psi'\left(\frac{x_i - \hat{\mu}_H}{\tilde{\sigma}_0}\right)\right]^2 \right\}^{\frac{1}{2}} \tag{2.8.7}$$

Descending ψ Functions

Consider the Cauchy distribution with density

$$f(x) = \frac{1}{\pi\sigma}\left[1 + \left(\frac{x-\mu}{\sigma}\right)^2\right]^{-1} \qquad -\infty < x < \infty$$

Then $\rho(z) = \log(1 + z^2)$ and $\psi(z) = 2z/(1 + z^2)$. This $\psi(z)$ function decreases as $|z|$ increases, so that it is called a descending ψ function. When the functional form of f(x) is not known and one wishes to downgrade the effects of tails, one may then attempt to approximate $\psi(z)$ by descending functions. Several important descending functions which have been considered in the Princeton study and elsewhere, are:

1. The wave function (Andrews et al., 1972; Andrews, 1974),

$$\psi(z) = \begin{cases} \sin(z) & \text{if } |z| \le \pi \\ 0 & \text{if } |z| > \pi \end{cases} \tag{2.8.8}$$

2. The bisquare function (Beaton and Tukey, 1974),

$$\psi(z) = \begin{cases} z(1 - z^2)^2 & \text{if } |z| \le 1 \\ 0 & \text{if } |z| > 1 \end{cases} \tag{2.8.9}$$

3. The Hampel piecewise linear function (Hampel, 1974),

$$\psi(z) = \text{sgn}(z) \times \begin{cases} |z| & 0 \le |z| < a \\ a & a \le |z| < b \\ \dfrac{c - |z|}{c - b} & b \le |z| < c \\ 0 & c \le |z| \end{cases} \tag{2.8.10}$$

This results in different estimators for different values of a, b, and c. For $a = 1.7$, $b = 3.4$, and $c = 8.5$, for example, the solution of (2.8.10) is denoted by A17 in the Princeton study.

In an extensive numerical study, Gross (1976) examined 25 representative estimators (out of 65 discussed by Andrews et al., 1972) of location μ and scale σ and concluded that the three descending $\psi(z)$ above with certain specified adjusting constants h (named w24, BS82, and H22) were generally the most efficient. Given below are these matching pairs as given by Gross (1976). Note that except for the $\tan^{-1}$ in the wave estimator w24, the estimators $\hat{\mu}_W$, $\hat{\mu}_B$, and $\hat{\mu}_H$ are in fact one-step Newton-Raphson solutions; the $\tan^{-1}$ in w24 was introduced to speed up convergence so that stability of solution is achieved in one step (see Gross, 1976):

$$T_0 = \text{median}\{x_i\} \text{ and } S_0 = \text{median}\{|x_i - T_0|\} \qquad i = 1, 2, \ldots, n$$

w24:

$$\hat{\mu}_W = T_0 + (hS_0)\tan^{-1}\left[\frac{\Sigma \sin(z_i)}{\Sigma \cos(z_i)}\right]$$

$$\hat{\sigma}_W = (hS_0)\left[n \frac{\Sigma \sin^2(z_i)}{[\Sigma \cos(z_i)]^2}\right]^{\frac{1}{2}} \qquad (2.8.11)$$

h = 2.4, and the summations include only those i such that

$$|z_i| < \pi, \quad z_i = \frac{x_i - T_0}{hS_0}$$

BS82:

$$\hat{\mu}_B = T_0 + (hS_0)\frac{\Sigma \psi(z_i)}{\Sigma \psi'(z_i)}$$

$$\hat{\sigma}_B = (hS_0)\left[n \frac{\Sigma \psi^2(z_i)}{[\Sigma \psi'(z_i)]^2}\right]^{\frac{1}{2}}$$

$h = 8.2$, $z_i = \dfrac{x_i - T_0}{hS_0}$, and $\psi(z)$ is the Beaton and Tukey's (2.8.12)
(1974) bisquare function defined in (2.8.9) and $\psi'(z)$ is the derivative of $\psi(z)$

H22:

$$\hat{\mu}_H = T_0 + \frac{S_0 \, \Sigma \psi(z_i)}{\Sigma \psi'(z_i)}$$

$$\hat{\sigma}_H = \left[\frac{n(S_0)^2 \, \Sigma \psi^2(z_i)}{[\Sigma \psi'(z_i)]^2}\right]^{\frac{1}{2}}$$

$z_i = \dfrac{x_i - T_0}{S_0}$, and $\psi(z)$ is the Hampel piecewise linear function (2.8.13)
defined in (2.8.10) and $\psi'(z)$ is the derivative of $\psi(z)$; $a = 2.25$, $b = 3.75$, and $c = 15.0$

From (2.8.11) to (2.8.13) it is clear that these estimators also censor extreme (smallest or largest) observations implicitly, although it is difficult to tell how many. One may note that for symmetric distributions, the estimators $\hat{\mu}_W$, $\hat{\mu}_B$, and $\hat{\mu}_H$ are unbiased and uncorrelated with the respective matching estimators $\hat{\sigma}_W$, $\hat{\sigma}_B$, and $\hat{\sigma}_H$; this clearly follows from symmetry. The M estimators above have, of course, many other interesting features, which have been discussed in detail in the authoritative works of Huber (1981), Hoaglin et al. (1983), and Rey (1983).

Comparisons of the MMLE $(\hat{\mu}, \hat{\sigma})$ with $(\hat{\mu}_W, \hat{\sigma}_W)$, $(\hat{\mu}_B, \hat{\sigma}_B)$, and $(\hat{\mu}_H, \hat{\sigma}_H)$

Since (2.8.11) to (2.8.13) are rather complex functions of sample observations, it is almost impossible to evaluate the variances of these estimators analytically. To compare the joint efficiencies of these estimators with the MML10 (MML with 10% censoring on either side) estimators $\hat{\mu}$ and $\hat{\sigma}$, therefore, Tiku (1980a) simulated the means and mean-square errors of w24, BS82, and H22 for the seven symmetric distributions considered in (2.8.3). The simulated values of the means, MSEs, and deficiencies (based on 10,000 Monte Carlo runs) are given in Table 2.8.6; since the estimators H22 were found to be slightly less efficient than w24 and BS82 (see also Gross, 1976, p. 414), only the values for w24 and BS82 are reproduced. Note that the mean-square errors of $\hat{\mu}_W$ and $\hat{\mu}_B$ are equal to their respective variances since they are both unbiased.

From Table 2.8.6, the estimators w24 and BS82 [and others, w18, BS74, BS90, H15, H22, and H25; Tiku (1980a, p. 131)] seem to be hardly any more efficient than the MML10 estimators.

For further comparisons between the MMLE $(\hat{\mu}, \hat{\sigma})$, $(\hat{\mu}_W, \hat{\sigma}_W)$, and $(\hat{\mu}_B, \hat{\sigma}_B)$, we consider the following long-tailed symmetric disaster distributions, considered by Andrews et al. (1972), which have nonexistent means or variances, or the samples have a large (20% or more) proportion of "bad" observations:

Table 2.8.6 Simulated Means, Mean-Square Errors, and Deficiencies of w24 and BS82 for n = 20

Distribution[a]	Mean $\hat{\sigma}_W$	Mean $\hat{\sigma}_B$	MSE $\hat{\mu}_W$	MSE $\hat{\mu}_B$	MSE $\hat{\sigma}_W$	MSE $\hat{\sigma}_B$	Deficiency w24	Deficiency BS82	Deficiency MML10[b]
(1)	0.98	0.98	0.052	0.052	0.031	0.031	0.083	0.083	0.086
(2)	0.93	0.93	0.048	0.048	0.040	0.041	0.088	0.089	0.088
(3)	0.83	0.82	0.037	0.037	0.065	0.066	0.102	0.103	0.102
(4)	0.71	0.71	0.028	0.027	0.113	0.114	0.141	0.141	0.146
(5)	0.82	0.82	0.038	0.038	0.065	0.065	0.103	0.103	0.106
(6)	0.87	0.87	0.042	0.042	0.043	0.044	0.085	0.086	0.085
(7)	0.81	0.81	0.037	0.037	0.062	0.062	0.099	0.099	0.099

[a]These distributions are given in (2.8.3).
[b]Borrowed from Table 2.8.5.

Table 2.8.7 Empirical Means and Variances of w24, BS82, and MML30 Estimators; Sample Size n = 20

		Distribution[a]				
	Estimator	(1)	(2)	(3)	(4)	(5)
Mean	$\hat{\sigma}_W$	1.33	1.39	1.37	1.80	2.53
	$\hat{\sigma}_B$	1.32	1.38	1.36	1.79	2.51
	$\hat{\sigma}$	1.15	1.26	1.15	1.41	2.09
Variance	$\hat{\mu}_W$	0.098	0.105	0.106	0.185	0.362
	$\hat{\mu}_B$	0.097	0.104	0.105	0.183	0.359
	$\hat{\mu}$	0.092	0.113	0.095	0.167	0.329

[a]These distributions are given in (2.8.14).

Outlier models: (1) $(n - k)N(0,1)$ and $kN(0,3)$, (2) $(n - k)N(0,1)$ and $kN(0,10)$ ($k \geq$ the integer value of $0.5 + 0.2n$), (3) Student's t_2, (4) Cauchy, (5) normal/uniform (2.8.14)

These disaster distributions do arise in practice, not as plausible alternatives to normality perhaps, but on their own right. Such distributions can easily be detected by formal goodness-of-fit tests (Chapter 6) or plotting techniques (Lawless, 1982, Sec. 2.4; Cleveland, 1984). For such a family of distributions, the MML estimators based on samples with a proportion substantially higher than 10% of smallest and largest observations censored usually provide more efficient estimators than the w24 and BS82 estimators (Tiku, 1980a). The MML30 estimators (i.e., the estimators $\hat{\mu}$ and $\hat{\sigma}$ based on samples with 30% smallest and largest observations censored) have high efficiencies. To illustrate this, we reproduce the simulated values (based on 10,000 Monte Carlo runs), given by Tiku (1980a, Table 6), of the variances of w24, BS82, and MML30 estimators of μ in Table 2.8.7. Also given are the simulated values of the expected values of the corresponding matching estimators of σ (these will be used later in evaluating the power functions of certain tests based on these estimators). Since the distributions are symmetric, $\hat{\mu}_W$, $\hat{\mu}_B$, and $\hat{\mu}$ are unbiased estimators of μ and are uncorrelated with the corresponding matching estimators $\hat{\sigma}_W$, $\hat{\sigma}_B$, and $\hat{\sigma}$, respectively. The population values of μ and σ are taken to be 0 and 1, respectively, without any loss of generality.

It is clear from Table 2.8.7 that for disaster distributions represented by (2.8.14), the MML30 estimator $\hat{\mu}$ is more efficient overall than the corresponding w24 and BS82 estimators.

Incidentally, the trimmed estimators TM30 (i.e., the estimators $\hat{\mu}_T$ and $\hat{\sigma}_T$ based on samples with 30% smallest and largest observations censored) are considerably less efficient jointly than the MML30 estimators (Tiku, 1980a).

It may be noted that the empirical means and mean-square errors are meaningful measures of the bias and efficiencies because all these estimators are, under some general regularity conditions, normally distributed for large n (Huber, 1981, Chap. 6; David, 1981, Chap. 9).

From the discussions above and from other available results (Tiku, 1980a), we draw the following conclusions:

1. For distributions that are deceptively close to a normal and are difficult to detect by goodness-of-fit tests [all tests of normality have low power against such distributions (Chapter 6)] or by plotting techniques, the MML10 estimator $\hat{\mu}$ has high efficiency overall and its use together with the matching MML10 estimator $\hat{\sigma}$ is recommended in preference to $\bar{x}$ and s.
2. Disaster distributions (e.g., symmetric distributions with infinite mean or variance) do occur in practice (Spjøtvoll and Aastveit, 1980; Rocke et al., 1982) and these can easily be detected by goodness-of-fit tests or plotting techniques. For estimating the location parameter (mean might not exist) of this family, the MML30 estimator $\hat{\mu}$ has high efficiency overall and its use together with the matching MML30 estimator $\hat{\sigma}$ is recommended, in preference to the MML10 estimators; $\bar{x}$ and s should never be used.
3. Extreme short-tailed symmetric distributions (the uniform distribution, for example) can also be easily detected (see Chapter 6). For such distributions, the MML0 estimators (i.e., $\bar{x}$ and s) should be used in preference to the MML10 estimators.
4. Very skewed distributions (chi-squares with low degrees of freedom, for example) are again easy to detect (see Chapter 6). For estimating the mean of this family, the MML10 estimator $\hat{\mu}$ has considerable bias and should not be used in preference to $\bar{x}$, which is unbiased. For estimating the difference $\mu_1 - \mu_2$ of two means, however, the MML10 estimator $\hat{\mu}_1 - \hat{\mu}_2$ is unbiased if sample sizes n_1 and n_2 are equal and almost unbiased if $n_1 \neq n_2$ and is somewhat more efficient than $\bar{x}_1 - \bar{x}_2$; an estimator based on samples censored only on one side (in the direction of the long tail), which is considerably more efficient than $\bar{x}_1 - \bar{x}_2$, will be developed in Chapter 4.

It is quite clear that the MML estimators have in a way the very desirable feature of adaptability in the Sprott (1982) and Hogg (1982) sense; that is, a knowledge about the type of underlying distribution or the proportion of outliers to be expected, or information provided by the sample to that end, can be utilized in choosing the appropriate level of censoring for constructing the "optimal" MML estimators.

2.9 RANDOM CHOICE OF THE PROPORTION OF CENSORING

It can be argued that in routine data analysis one has no opportunity to explore what type of underlying distribution one is dealing with, although one might be justified in assuming that the underlying distribution is symmetric. For example, Andrews et al. (1972) assume that the underlying distribution is symmetric and long-tailed (kurtosis $\mu_4/\mu_2^2 > 3$). In such situations, the "random" MML estimators $\hat{\mu}^*$ and $\hat{\sigma}^*$ are recommended; these are exactly the same as the MML estimators $\hat{\mu}$ and $\hat{\sigma}$, respectively, but with the constant proportion $q = r/n$ replaced by the random proportion $q^* = r^*/n$ of censoring, where

$$r^* = \begin{cases} 0 & \text{if } \dfrac{k^*}{n} = 0 \\ [0.5 + 0.1n] & \text{if } 0 < \dfrac{k^*}{n} \leq 0.1 \\ [0.5 + 0.3n] & \text{if } \dfrac{k^*}{n} > 0.1 \end{cases} \tag{2.9.1}$$

k^* is the number of values of

$$|z_i| = \frac{|x_i - \text{med}(x_i)|}{\tilde{\sigma}_0} \qquad (i = 1, 2, \ldots, n) \tag{2.9.2}$$

which exceed 3.0; $\tilde{\sigma}_0$ is given by $\tilde{\sigma}_0 = \text{med}\,|x_i - \text{med}(x_j)|/0.6745$.

Tiku (1980a) simulated the means and mean-square errors of $\hat{\mu}^*$ and $\hat{\sigma}^*$ and concluded that they are on the whole at least as efficient as w24 and BS82. Like the estimators w24 and BS82, however, the small-sample distributions of $\hat{\mu}^*$ and $\hat{\sigma}^*$ are difficult to determine, even under the assumption of normality, and that is seen as a drawback in such random estimators. Besides, certain important test procedures based on these estimators (including w24, BS82, etc.) might not possess robustness of validity (stability of type I error), at any rate for small samples; this will be illustrated in Chapter 4.

EXAMPLE 2.9.1 Consider the data of Example 2.7.1. Here

$$\text{med}(x_i) = 24 \qquad \text{med}\,|x_i - \text{med}(x_i)| = 17 \qquad \tilde{\sigma}_0 = 25.2038$$

$$\frac{k^*}{n} = \frac{1}{15} \text{ and, therefore, } r^* = [0.5 + 0.1(15)] = 2$$

The random MML estimates $\hat{\mu}^*$ and $\hat{\sigma}^*$ are, therefore, exactly the same as the MML10 estimates $\hat{\mu} = 27.713$ and $\hat{\sigma} = 25.498$, respectively, of Example 2.7.1.

EXAMPLE 2.9.2 (Lehmann, 1975, p. 45) The effectiveness of vitamin C in orange juice and in synthetic ascorbic acid was compared in 20 guinea pigs (divided at random into two groups of 10) in terms of the length of the odontoblasts after 6 weeks, with the following results:

Orange juice:	8.2	9.4	9.6	9.7	10.0	14.5	15.2	16.1	17.6	21.5
Ascorbic acid:	4.2	5.2	5.8	6.4	7.0	7.3	10.1	11.2	11.3	11.5

Here, for the first sample:

$$\text{med}(x_i) = 12.25 \qquad \tilde{\sigma}_0 = 4.2995 \qquad \frac{k^*}{n} = 0 \qquad r^* = 0$$

for the second sample:

$$\text{med}(x_i) = 7.15 \qquad \tilde{\sigma}_0 = 3.6323 \qquad \frac{k^*}{n} = 0 \qquad r^* = 0$$

The random MML estimates are, therefore, the sample means $\bar{x}_1 = 13.18$ and $\bar{x}_2 = 8.0$ and the sample standard deviations $s_1 = 4.438$ and $s_2 = 2.768$, and these are known to be good estimates in this particular situation (Lehmann, 1975).

Although this method seems to give satisfactory estimates, it is difficult to determine the standard errors of these estimates analytically even under the assumption of normality, due to the randomness of r^*.

It may be noted that the estimator $\hat{\mu}^*$ is adaptive in the Hogg sense and resembles the adaptive estimators developed by Hogg (1967, 1972, 1974, 1982) and Hogg et al. (1972, 1975) essentially along the following lines (see also Hogg et al., 1984).

2.10 ADAPTIVE ESTIMATORS OF LOCATION

Consider a family (not necessarily finite) of distributions of the type $f(x - \theta)$, θ being a location parameter. Let $f_i(x - \theta)$, $i = 1, 2, \ldots, k$, be a representative set of distributions from this family. Suppose that for each of these distributions f_i, one can find a good estimator $\hat{\theta}_i$ of θ. From a given random sample $x_1, x_2, \ldots, x_n$, calculate an appropriate selector statistic, say Q, to determine which one of the distributions f_i is closest to the distribution from which the sample arose. (The goodness-of-fit statistics given in Chapter 6 may in particular be used as selector statistics.) If Q picks f_i, then $\hat{\theta}_i$ is the adaptive estimator of θ. Or possibly Q might indicate some distribution between f_i and f_j, in which case the adaptive estimator of θ is a combination of $\hat{\theta}_i$ and $\hat{\theta}_j$. More generally, Q could indicate an estimator $\hat{\theta}$ which is a combination of $\hat{\theta}_1, \hat{\theta}_2, \ldots, \hat{\theta}_k$, say

$$\hat{\theta} = \sum_{i=1}^{k} w_i \hat{\theta}_i \qquad \sum_{i=1}^{k} w_i = 1 \tag{2.10.1}$$

where the weights are functions of the selector statistic Q. A number of location- and scale-invariant selector statistics were proposed by Hogg, who also studied the efficiency properties of the resulting estimators primarily through Monte Carlo methods. He showed that for symmetric distributions, some of these adaptive estimators can be at least as efficient as Huber's M estimators. However, it is difficult to determine the standard errors, let alone the sampling distributions, of these adaptive estimators analytically, especially for finite sample sizes, and that is seen as a drawback. Besides, large sample sizes are required to give a selector statistic Q enough sensitivity (power) to be able to discriminate between various distributions, unless of course these distributions are very different from each other, normal versus Cauchy, for example; see Hogg (1982, pp. 2538-2539) and also Chapter 6, where the problem of model discrimination is discussed (see also D'Agostino and Austin, 1977).

2.11 CONCLUSIONS AND SUMMARY

In this chapter we have introduced the concept of censoring and have established a link between censoring and robustness. We have illustrated the effects of censoring on population standard cumulants, particularly λ_3 and λ_4, of a t-distribution family and a gamma-distribution family. The BLUEs and the MMLEs of the location and scale parameters (μ, σ) are then introduced. It is shown that for type II censored normal samples, the MMLEs $(\hat{\mu}, \hat{\sigma})$ are almost as efficient as the MLEs $(\hat{\hat{\mu}}, \hat{\hat{\sigma}})$ and are jointly more efficient than the BLUEs (μ^*, σ^*). We have shown that the MMLEs are jointly more efficient overall than the trimmed estimators and at least as efficient as some of the celebrated robust estimators (Huber's M estimators), including the wave estimators, the bisquare estimators, and the Hampel estimators, over a wide range of symmetric distributions. We have emphasized that it is both the estimators $(\hat{\mu}, \hat{\sigma})$ in the MMLE matching pair (not just the estimator of μ) that are accountable for the robustness and high efficiency of the MMLE procedures. Specifically, $\hat{\sigma}$ in $(\hat{\mu}, \hat{\sigma})$ provides the "tuning effect" to yield type I robustness and high power of the t and F tests which are developed in Chapter 4. It is our opinion that the apparent lack of "tuning effect" in Huber's M estimators, particularly for short-tailed and skew distributions, make them less attractive than the MMLEs for developing robust statistical procedures for testing hypotheses on location and scale parameters; this is, however, discussed in detail in Chapter 4.

EXERCISES

2.1 Consider the Cauchy distribution

$$f(z) = \frac{1}{\pi(1 + z^2)} \qquad -\infty < z < \infty$$

(a) Show that all its moments are nonexistent.

(b) Find the value of z_0 for which $F(z_0) = \int_{-\infty}^{z_0} f(z)\, dz = 0.70$, 0.80, and 0.90.

(c) Show that the truncated distribution

$$\text{constant} \times \frac{1}{(1 + z^2)} \qquad -z_0 < z < z_0$$

is symmetric and all its moments exist and are finite. Calculate its fourth standard cumulant $\lambda_4 = (\mu_4/\mu_2^2) - 3$ for the values of z_0 above. How close are these values to the λ_4 value of a normal distribution?

2.2 Let $x_1, x_2, \ldots, x_n$ be a random sample of n independent observations, $(n - 1)$ observations coming from a normal $N(\mu, \sigma)$ and one (we do not know which one) from a normal $N(\mu, \lambda\sigma)$, for example, the Dixon outlier model

$$(n - 1)N(\mu, \sigma) \quad \text{and} \quad 1N(\mu, \lambda\sigma) \qquad \lambda > 0$$

(a) Show that, like the sample mean $\bar{x}$, the MML10 estimator is an unbiased estimator of μ.

(b) Using David et al. (1977) tables of the variances and covariances of order statistics, calculate the values of the relative efficiency $\sqrt{V(\hat{\mu})/V(\bar{x})}$ of $\hat{\mu}$ for $\lambda = 0.5$, 1, 2, 3, and 5, and $n = 5$ and 10. Show that $\hat{\mu}$ is only slightly less efficient than $\bar{x}$ for $\lambda = 0.5$ and 1, but considerably more efficient for other values of λ.

2.3 Consider the following well-known data by Cushney and Peebles (1905) on the prolongation of sleep by two soporific drugs as ordered differences:

0.0 0.8 1.0 1.2 1.3 1.3 1.4 1.8 2.4 4.6

(a) Calculate the sample mean $\bar{x}$ and the robust estimates MML10, TM10, w24, and BS82 of the location parameter that represents the difference of the average effects of the two drugs.

(b) Calculate the approximate standard errors of the robust estimates and show that all of them are considerably smaller than the standard error of the sample mean.

2.4 (a) If $\hat{\theta}_i$, $i = 1, 2, \ldots, k$, are k unbiased estimators of a parameter θ, show that the linear unbiased estimator with minimum variance is given by

$$\sum_{i=1}^{k} w_i \hat{\theta}_i \qquad \text{where } w_i = \frac{1}{V(\hat{\theta}_i)} \Bigg/ \sum_{i=1}^{k} \frac{1}{V(\hat{\theta}_i)}$$

(b) Following are four sets of eight measurements each of the smoothness of a certain type of paper, obtained in four different laboratories A, B, C, and D:

A:	30.7	41.5	43.8	44.5	45.5	46.0	47.7	66.2
B:	39.2	39.3	39.7	41.4	41.8	42.9	43.3	45.8
C:	34.0	35.0	39.0	40.0	43.0	43.0	44.0	45.0
D:	34.0	34.8	34.8	35.4	37.2	37.8	41.2	42.8

Calculate the classical and the MML10 estimates $(\bar{x}_i, s_i)$ and $(\hat{\mu}_i, \hat{\sigma}_i)$, $i = 1, 2, 3, 4$. Assuming that the four laboratories have a common mean μ and a common variance σ^2, calculate the classical (based on $\bar{x}_i$) and MML10 estimates of the common mean. Show that the MML10 estimate has considerably smaller standard error than the classical estimate.

2.5 Given k independent samples of size n_i from k symmetric populations with means μ_i, $i = 1, 2, \ldots, k$, and a common variance σ^2, obtain the classical and the MML10 estimators of a linear contrast $\Sigma_{i=1}^{k} \ell_i \mu_i$, $\Sigma_{i=1}^{k} \ell_i = 0$. How will you obtain the standard errors of the classical and the MML estimates in this case?

APPENDIX B: Order Statistics

Distribution of Order Statistics

Let $x_{(1)}, x_{(2)}, \ldots, x_{(n)}$ be the order statistics of a random sample of size n from a distribution of the type $(1/\sigma) f((x - \mu)/\sigma$. If we write

$$Z_r = \frac{x_{(r)} - \mu}{\sigma} \qquad (r = 1, 2, \ldots, n) \tag{B.1}$$

then the probability density function (PDF) of $u = Z_r$ is given by (David, 1981, p. 9)

$$p(u) = \frac{n!}{(r-1)!(n-r)!}[F(u)]^{r-1}[1 - F(u)]^{n-r} f(u) \tag{B.2}$$

where

$$F(y) = \int_c^y f(X)\, dX \qquad X = \frac{x - \mu}{\sigma} \tag{B.3}$$

c is the lower limit on X.

The joint PDF of $u = Z_r$ and $v = Z_s$ $(r < s)$ is given by (David, 1981, p. 10)

$$p(u, v) = \frac{n!}{(r-1)!(s-r-1)!(n-s)!}[F(u)]^{r-1}[F(v) - F(u)]^{s-r-1}[1 - F(v)]^{n-s} f(u) f(v) \tag{B.4}$$

For example, for the uniform distribution $U(\theta_1, \theta_2)$, we have

$$f(x) = 1 \text{ and } F(y) = \int_0^y dX = y$$

and therefore,

$$p(u) = \frac{n!}{(r-1)!(n-r)!} u^{r-1}(1-u)^{n-r} \qquad 0 < u < 1 \tag{B.5}$$

which is the beta distribution $B(r, n - r + 1)$, and

$$p(u, v) = \frac{n!}{(r-1)!(s-r-1)!(n-s)!} u^{r-1}(v-u)^{s-r-1}(1-v)^{n-s} \qquad 0 < u < v < 1 \tag{B.6}$$

LEMMA B.1 The PDF of $w = F(u)$ $(u = Z_r)$ is the beta distribution $B(r, n - r + 1)$ always.

Proof: Clearly,

$$\frac{du}{dw} = \frac{1}{F'(u)} = \frac{1}{f(u)}$$

Substituting this in (B.2), we obtain the PDF of w as

$$p(w) = \frac{n!}{(r-1)!(n-r)!} w^{r-1}(1-w)^{n-r} \qquad 0 < w < 1 \tag{B.7}$$

and hence the result.

Moments of Order Statistics

The jth raw moment of Z_r $(1 \le r \le n)$ and the first-order product moment of Z_r and Z_s $(1 \le r < s \le n)$ are given by

$$E(u^j) = \int_c^d u^j p(u)\, du \tag{B.8}$$

and

$$E(uv) = \iint_{c<u<v<d} uv\, p(u, v)\, du\, dv \tag{B.9}$$

respectively, where p(u) and p(u, v) are given by (B.5) and (B.6); c and d are the lower and upper limits on $X = (x - \mu)/\sigma$, respectively. Denoting the expected value of Z_r by $\mu_{r:n}$, variance of Z_r by $\sigma_{r,r:n}$, and the covariance of Z_r and Z_s by $\sigma_{r,s:n}$, it easily follows that

$$E(x_{(r)}) = \mu + \sigma\mu_{r:n} \qquad (1 \le r \le n) \tag{B.10}$$

$$V(x_{(r)}) = \sigma^2\sigma_{r,r:n} \tag{B.11}$$

and

$$\operatorname{cov}(x_{(r)}, x_{(s)}) = \sigma^2 \sigma_{r,s:n} \qquad (1 \le r < s \le n) \tag{B.12}$$

LEMMA B.2 For the uniform distribution,

$$\mu_{r:n} = \frac{r}{n+1} \qquad \sigma_{r,r:n} = \frac{r(n-r+1)}{(n+1)^2(n+2)} \tag{B.13}$$

and

$$\sigma_{r,s:n} = \frac{r(n-s+1)}{(n+1)^2(n+2)} \tag{B.14}$$

Proof: These results are easily derived by using the marginal and joint densities of order statistics given by (B.5) and (B.6). In fact, for the uniform distribution, similar algebraic expressions can also be derived for the higher-order moments (see David, 1981, pp. 35-36).

LEMMA B.3 For the exponential distribution $E(0, \sigma)$,

$$\mu_{r:n} = \sum_{i=1}^{r} \frac{1}{n-i+1} \tag{B.15}$$

and

$$\sigma_{r,r:n} = \sigma_{r,s:n} = \sum_{i=1}^{r} \frac{1}{(n-i+1)^2} \qquad (r < s) \tag{B.16}$$

Proof: Realizing that the PDF of Z_r is given by (B.2) and the joint PDF of Z_r and Z_s is given by (B.4), the results can be obtained exactly on the same lines as Lemma B.2; the algebra is a little involved, though (Sarhan, 1955). However, a very simple derivation is possible in view of the fact that if we write

$$y_{i+1} = \frac{1}{\sigma} D_i \qquad (i = 0, 1, 2, \ldots, n-1)$$

where

$$\frac{1}{\sigma} D_i = \frac{1}{\sigma}(n-i)[x_{(i+1)} - x_{(i)}] \qquad (x_{(0)} = 0)$$

$$= (n-i)(Z_{i+1} - Z_i) \qquad (Z_0 = 0)$$

then $y_1, y_2, \ldots, y_n$ are independent, identically distributed (IID) exponential $E(0, 1)$; this result follows immediately from the fact that [D_i are usually called exponential sample spacings] the joint probability density function of $y_1, y_2, \ldots, y_n$ is given by $\exp(-\Sigma_{i=1}^{n} y_i)$, $0 < y_i < \infty$, $i = 1, 2, \ldots, n$

(see David, 1981, Ex. 6.4.1). In particular, $E(y_i) = 1$, $V(y_i = 1)$, and $\text{cov}(y_i, y_j) = 0$, $i \neq j$. Since

$$Z_r = \frac{\Sigma_{i=1}^{r} y_i}{n - i + 1} \quad \text{and} \quad Z_s = \frac{\Sigma_{i=1}^{s} y_i}{n - i + 1} \tag{B.17}$$

the result follows.

For the exponential distribution, similar explicit and simple algebraic expressions are not available for higher-order moments of order statistics. However, these could be obtained systematically from a set of recurrence relations given by Balakrishnan and Joshi (1984) and Joshi (1978, 1982).

Explicit algebraic expressions for $\mu_{r:n}$, $\sigma_{r,r:n}$, and $\sigma_{r,s:n}$ can be obtained for a few other distributions; Malik (1967) and Huang (1975), for example, have obtained these expressions for the power-function and Pareto distributions, respectively (see also David, 1981, pp. 49-50, and Balakrishnan and Joshi, 1981, 1982).

In general, however, it is not possible to obtain explicit algebraic expressions for $\mu_{r:n}$, $\sigma_{r,r:n}$, and $\sigma_{r,s:n}$. Consider, for example, the family of distributions (2.3.1) with PDF

$$f(p;z) = C\left(1 + \frac{z^2}{k}\right)^{-p} \qquad -\infty < x < \infty \tag{B.18}$$

$z = (x - \mu)/\sigma$. Here (Johnson and Kotz, 1970; Tiku and Kumra, 1981)

$$F(y) = \frac{1}{2} + \frac{1}{\pi}\left\{\theta + \left[\cos(\theta) + \frac{2}{3}\cos^3(\theta) + \cdots + \frac{(2)(4)\cdots(\nu-3)}{(3)(5)\cdots(\nu-2)}\cos^{\nu-2}(\theta)\right] \sin(\theta)\right\}$$

$$\text{for } \nu = 2p - 1 \text{ odd and } > 1 \tag{B.19}$$

$$= \frac{1}{2} + \frac{1}{\pi}\left\{\left[1 + \frac{1}{2}\cos^2(\theta) + \frac{(1)(3)}{(2)(4)}\cos^4(\theta) + \cdots + \frac{(1)(3)\cdots(\nu-3)}{(2)(4)\cdots(\nu-2)}\cos^{\nu-2}(\theta)\right] \sin(\theta)\right\}$$

$$\text{for } \nu \text{ even} \tag{B.20}$$

$\theta = \tan^{-1}(y/\sqrt{\nu})$

These expressions are substituted in (B.2) and (B.4) and the values of $\mu_{r:n}$, $\sigma_{r,r:n}$, and $\sigma_{r,s:n}$ obtained from the integrals (B.8) and (B.9). These integrals, however, have to be evaluated numerically [see Tiku and Kumra (1981), who have discussed several such numerical procedures].

Of course, a number of authors have developed and discussed numerical procedures to evaluate the integrals of the type (B.8) and (B.9) and have given tables of the expected values and variances and covariances of order statistics for various distributions (see, e.g., Balakrishnan, 1984; Balakrishnan and Kocherlakota, 1985a,b; Balakrishnan and Malik, 1985a;

Barnett, 1966; David et al., 1977; Govindarajulu, 1966; Gupta et al., 1967; Harter, 1964; Mann et al., 1973; Tietjen et al., 1977; Tiku and Kumra, 1981; Shah, 1966, 1970).

A very important feature of order statistics is that their moments satisfy certain interesting recurrence relations and identities and that renders the computation of these moments manageable. These recurrence relations can also be used to verify the accuracy of the values obtained numerically [see Balakrishnan and Malik (1984b), Govindarajulu (1963), Joshi (1971), Joshi and Balakrishnan (1982), and Prescott (1974), who have developed certain interesting recurrence relations and identities]. Two very useful recurrence relations that hold for any arbitrary distribution are (David, 1981, pp. 46-49)

$$r\mu_{r+1:n} + (n-r)\mu_{r:n} = n\mu_{r:n-1} \qquad (1 \le r \le n-1) \tag{B.21}$$

and

$$(r-1)\mu_{r,s:n} + (s-r)\mu_{r-1,s:n} + (n-s+1)\mu_{r-1,s-1:n} = n\mu_{r-1,s-1:n-1} \tag{B.22}$$

for $1 \le r < s \le n$; $\mu_{r,s:n} = E(Z_r Z_s) = \sigma_{r,s:n} + \mu_{r:n}\mu_{s:n}$.

Several other recurrence relations and identities for the product moments $\mu_{r,s:n}$ are derived by Govindarajulu (1963); see also Joshi and Balakrishnan (1982), who arrive at the following interesting result. To find the first and second moments and the first order product moments of order statistics in a sample of size n from an arbitrary distribution, given these moments in samples of sizes $n - 1$ and less, one has to evaluate at most two single integrals and $(n-2)/2$ double integrals if n is even; and two single integrals and $(n-1)/2$ double integrals if n is odd.

For distributions symmetric about zero, we further have

$$\mu_{n-r+1:n} = -\mu_{r:n} \qquad \sigma_{n-r+1,n-r+1:n} = \sigma_{r,r:n} \tag{B.23}$$

and $(r < s)$

$$\sigma_{n-s+1,n-r+1:n} = \sigma_{r,s:n} \tag{B.24}$$

It may also be noted that for any arbitrary distribution,

$$\sum_{r=1}^{n}\sum_{s=1}^{n} \sigma_{r,s:n} = n \tag{B.25}$$

That is, the elements of the variance-covariance matrix of the order statistics of a random sample of size n from any arbitrary distribution add to $n\sigma^2$. This result follows from the fact that

$$V\left(\frac{1}{n}\sum_{i=1}^{n} x_{(i)}\right) = \left(\frac{\sigma}{n}\right)^2 \sum_{r=1}^{n}\sum_{s=1}^{n} \sigma_{r,s:n}$$

$$= \frac{\sigma^2}{n}$$

since $\Sigma_{i=1}^{n} x_{(i)} = \Sigma_{i=1}^{n} x_i = n\bar{x}$ and the variance of $\bar{x}$ is always equal to σ^2/n.

For the normal distribution, the following relation also holds

$$\sum_{s=1}^{n} \sigma_{r,s:n} = 1 \qquad (r = 1, 2, \ldots, n) \tag{B.26}$$

That is, the elements of any row (column) of the variance-covariance matrix of the order statistics $x_{(i)}$ $(i = 1, 2, \ldots, n)$ of a random sample of size n add up to σ^2. This follows from the well-known independence of $\bar{x}$ and $x_{(r)} - \bar{x}$, $(r = 1, 2, \ldots, n)$, in which case

$$\operatorname{cov}(\bar{x}, x_{(r)} - \bar{x}) = \operatorname{cov}\left(x_{(r)}, \frac{1}{n}\sum_{i=1}^{n} x_{(i)}\right) - V(\bar{x})$$

$$= \frac{\sigma^2}{n}\sum_{s=1}^{n} \sigma_{r,s:n} - \frac{\sigma^2}{n} = 0$$

The independence of $\bar{x}$ and $x_{(r)} - \bar{x}$ follows from the fact that for normal samples, the sample mean is independent of any location-free statistic (David, 1981, Sec. 2.7). Some additional relations and identities for moments of order statistics from the normal distribution are available (see Govindarajulu, 1963; Joshi and Balakrishnan, 1981).

Approximations to Moments

A tabulation of the expected values and variances and covariances of order statistics becomes voluminous very quickly. Therefore, most of the tables cover only the sample sizes $n \leq 20$. For large n (say $n > 20$), the following equations (David and Johnson, 1954) usually provide accurate approximations for $\mu_{r:n} = E(Z_r)$, $\sigma_{r,r:n} = V(Z_r)$, and $\sigma_{r,s:n} = \operatorname{cov}(Z_r, Z_s)$, $r < s$; $Z_i = (x_{(i)} - \mu)/\sigma$ $(i = 1, 2, \ldots, n)$:

$$E(Z_r) \simeq Q_r + \frac{p_r q_r}{2(n+2)} Q_r'' + \frac{p_r q_r}{(n+2)^2}\left[\frac{1}{3}(q_r - p_r)Q_r''' + \frac{1}{8}p_r q_r Q_r''''\right] \tag{B.27}$$

$$V(Z_r) \simeq \frac{p_r q_r}{n+2}(Q_r')^2 + \frac{p_r q_r}{(n+2)^2}\left[2(q_r - p_r)Q_r' Q_r'' + p_r q_r (Q_r' Q_r''' + \frac{1}{2}(Q_r'')^2\right] \tag{B.28}$$

and

$$\operatorname{cov}(Z_r, Z_s) \simeq \frac{p_r q_s}{n+2} Q_r' Q_s' + \frac{p_r q_s}{(n+2)^2} \Big[(q_r - p_r) Q_r'' Q_s' + (q_s - p_s) Q_r' Q_s'' + \frac{1}{2}(p_r q_r Q_r''' Q_s' + p_s q_s Q_r' Q_s''' + p_r q_s Q_r'' Q_s'') \Big] \qquad \text{(B.29)}$$

where $p_r = r/(n+1)$, $q_r = 1 - p_r$, and $Q_r = Q(p_r)$ is given by

$$F(Q_r) = \int_c^{Q_r} f(y)\, dy = p_r \qquad y = \frac{x - \mu}{\sigma} \qquad \text{(B.30)}$$

c being the lower limit on y; Q_r', Q_r'', and so on, are the successive derivatives of Q_r with respect to p_r. Conditions for the validity of the expressions (B.27) to (B.29) are discussed by Blom (1958, Chap. 5).

The expressions for Q_r', Q_r'', and so on, are easy to work out for most distributions and are given below for a few distributions.

Normal Distribution. Write

$$u = F(y) = (2\pi)^{-\frac{1}{2}} \int_{-\infty}^{y} e^{-t^2/2}\, dt$$

Then

$$y = Q(u) = Q \qquad Q' = \frac{1}{du/dy} = (2\pi)^{\frac{1}{2}} e^{Q^2/2}$$

$$Q'' = Q(Q')^2 \qquad Q''' = Q'[(Q')^2 + 2QQ''] = (Q')^3(1 + 2Q^2) \qquad \text{(B.31)}$$

$$Q'''' = 5(Q')^2 Q'' + 2Q[(Q'')^2 + Q'Q'''] = Q(Q')^4(7 + 6Q^2)$$

Gamma Distribution. Write

$$u = F(y) = \frac{1}{\Gamma(k)} \int_0^y e^{-t} t^{k-1}\, dt$$

Then

$$y = Q(u) = Q \qquad Q' = \Gamma(k)\, e^{Q} Q^{-k+1}$$

$$Q'' = (Q')^2\left(1 - \frac{k-1}{Q}\right) \qquad Q''' = 2\frac{(Q'')^2}{Q'} + (k-1)\frac{(Q')^3}{Q^2} \qquad \text{(B.32)}$$

$$Q'''' = \frac{7Q'''Q''}{Q'} - \frac{8(Q'')^3}{(Q')^2} - \frac{2(k-1)(Q')^4}{Q^3}$$

Logistic Distribution. Write

$$u = F(y) = \int_{-\infty}^{y} \frac{e^{-t}}{(1 + e^{-t})^2}\, dt = \frac{1}{1 + e^{-y}}$$

Then

$$y = Q(u) = Q = \log u - \log(1 - u) \qquad Q' = \frac{1}{u(1 - u)}$$

$$Q'' = (2u - 1)(Q')^2 \qquad Q''' = 2\left[\frac{(Q'')^2}{Q'} + (Q')^2\right] \tag{B.33}$$

$$Q'''' = 6\left[2Q''Q' + \frac{(Q'')^3}{(Q')^2}\right]$$

<u>Extreme-Value Distribution.</u> Write

$$u = F(y) = \int_{-\infty}^{y} e^{-e^{-y}} e^{-y}\, dy$$

Then

$$y = Q(u) = Q = -\log(-\log u) \qquad Q' = \frac{e^{Q}}{u}$$

$$Q'' = Q'\left(Q' - \frac{1}{u}\right) \qquad Q''' = 2\left[\frac{(Q'')^2}{Q'}\right] + \frac{(Q')^2}{u} \tag{B.34}$$

$$Q'''' = 3Q''\left(\frac{Q'''}{Q'} + \frac{Q'}{u}\right) - \frac{(Q')^2}{u^2}$$

<u>The Family (B.18).</u> Write $(k = 2p - 3,\ p \geq 2)$

$$u = F(y) = C\int_{-\infty}^{y} \left(1 + \frac{t^2}{k}\right)^{-p} dt$$

Then

$$y = Q(u) = Q \qquad Q' = \frac{1}{du/dy} = \frac{1}{C}\left(1 + \frac{y^2}{k}\right)^{p}$$

$$Q'' = \frac{(2p/k)Q(Q')^2}{1 + (Q^2/k)}$$

$$Q''' = \frac{2p}{k}(Q')^3\, \frac{1 + [2(2p - 1)/k]Q^2[1 + (Q^2/k)]^{-1}}{1 + (Q^2/k)}$$

$$Q'''' = \frac{4p}{k}Q(Q')^4\, \frac{(7p - 3) + [2(2p - 1)(3p - 2)/k]Q^2[1 + (Q^2/k)]^{-1}}{[1 + (Q^2/k)]^2}$$

Table B.1 Values of α and β for Calculating the MMLE

n		r: 1	2	3	4	5	6	7	8	9
5	α	0.7540								
	β	0.8108								
6	α	0.7303								
	β	0.8273								
7	α	0.7118	0.7867							
	β	0.8391	0.7508							
8	α	0.6968	0.7719							
	β	0.8481	0.7680							
9	α	0.6844	0.7594	0.7959						
	β	0.8552	0.7811	0.7242						
10	α	0.6737	0.7485	0.7855						
	β	0.8610	0.7915	0.7405						
11	α	0.6645	0.7391	0.7762						
	β	0.8658	0.8001	0.7535						
12	α	0.6564	0.7307	0.7677	0.7919					
	β	0.8700	0.8072	0.7643	0.7230					
13	α	0.6492	0.7232	0.7599	0.7847					
	β	0.8736	0.8133	0.7733	0.7354					
14	α	0.6428	0.7165	0.7529	0.7779					
	β	0.8767	0.8186	0.7809	0.7459					
15	α	0.6369	0.7103	0.7464	0.7715	0.7896				
	β	0.8795	0.8233	0.7876	0.7518	0.7223				
16	α	0.6316	0.7048	0.7405	0.7656	0.7840				
	β	0.8820	0.8274	0.7935	0.7626	0.7324				
17	α	0.6267	0.6996	0.7350	0.7600	0.7788				
	β	0.8843	0.8311	0.7987	0.7695	0.7411				
18	α	0.6221	0.6949	0.7299	0.7548	0.7737				
	β	0.8863	0.8344	0.8033	0.7755	0.7488				
19	α	0.6180	0.6905	0.7252	0.7500	0.7689	0.7836			
	β	0.8882	0.8374	0.8075	0.7810	0.7556	0.7302			
20	α	0.6141	0.6864	0.7208	0.7454	0.7644	0.7793			
	β	0.8899	0.8401	0.8113	0.7859	0.7617	0.7378			
21	α	0.6104	0.6826	0.7166	0.7411	0.7600	0.7751			
	β	0.8915	0.8427	0.8148	0.7903	0.7672	0.7445			

Table B.1 (continued)

n		r: 1	2	3	4	5	6	7	8	9
22	α	0.6070	0.6790	0.7127	0.7370	0.7559	0.7711	0.7832		
	β	0.8930	0.8450	0.8180	0.7944	0.7722	0.7506	0.7289		
23	α	0.6038	0.6756	0.7090	0.7331	0.7520	0.7672	0.7796		
	β	0.8944	0.8472	0.8209	0.7981	0.7768	0.7561	0.7354		
24	α	0.6008	0.6724	0.7055	0.7295	0.7483	0.7636	0.7760		
	β	0.8957	0.8492	0.8236	0.8016	0.7810	0.7611	0.7413		
25	α	0.5979	0.6694	0.7023	0.7260	0.7448	0.7600	0.7726	0.7829	
	β	0.8968	0.8510	0.8261	0.8048	0.7849	0.7657	0.7468	0.7278	
26	α	0.5952	0.6666	0.6991	0.7227	0.7414	0.7566	0.7693	0.7798	
	β	0.8980	0.8528	0.8285	0.8077	0.7885	0.7700	0.7518	0.7336	
27	α	0.5926	0.6638	0.6962	0.7196	0.7381	0.7534	0.7661	0.7767	
	β	0.8990	0.8544	0.8307	0.8105	0.7918	0.7740	0.7564	0.7389	
28	α	0.5902	0.6613	0.6933	0.7166	0.7350	0.7502	0.7630	0.7737	
	β	0.9000	0.8560	0.8328	0.8131	0.7950	0.7776	0.7607	0.7438	
29	α	0.5878	0.6588	0.6906	0.7137	0.7321	0.7472	0.7600	0.7708	0.7800
	β	0.9010	0.8574	0.8347	0.8155	0.7979	0.7811	0.7647	0.7484	0.7321
30	α	0.5856	0.6564	0.6880	0.7110	0.7292	0.7443	0.7571	0.7680	0.7772
	β	0.9019	0.8588	0.8365	0.8178	0.8006	0.7843	0.7684	0.7527	0.7370

3
MMLEs for Other Distributions

3.1 INTRODUCTION

In Chapter 2 we introduced the MMLEs of the location and scale parameters of a normal distribution based on type II censored samples. We have shown that these MMLEs are almost as efficient as the MLEs; in fact, the MMLEs are asymptotically identical to the MLEs. In this chapter we proceed to develop the MMLEs, based on type II censored samples, of the location and scale parameters of some other important distributions, and show that these MMLEs are also almost as efficient as the MLEs for small samples but asymptotically identical to the MLEs. The specific distributions we consider are the log-normal distribution, the logistic distribution, the exponential distribution, and the Rayleigh distribution; these distributions arise in numerous practical situations. We also extend these methods to progressively censored samples and truncated distributions. We emphasize that like the MMLEs given in Chapter 2, the MMLEs given here constitute robust estimators, robust to reasonable departures from the respective distribution.

3.2 MMLEs FOR LOG-NORMAL DISTRIBUTION

Data sets resulting from numerous biological and industrial experiments are assumed to follow the log-normal distribution $LN(\tau;\mu,\sigma^2)$ (see, e.g., Gaddum, 1933; Bliss, 1934; Cochran, 1938; Williams, 1937; Finney, 1941; Grundy, 1951); Aitchison and Brown, 1957). The probability density function h(x) of the log-normal distribution $LN(\tau;\mu,\sigma^2)$ is given by

$$h(x) = [\sqrt{2\pi}(x - \tau)\sigma]^{-1} \exp\left[- \frac{[\log(x - \tau) - \mu]^2}{2\sigma^2}\right] \quad \tau \leq x \leq \infty \qquad (3.2.1)$$

so that the distribution of $z = [\log(x - \tau) - \mu]/\sigma$ is normal N(0, 1); τ is called the threshold parameter and μ and σ are the location and scale parameters, respectively.

The estimation of τ, μ, and σ has been a troublesome affair. As has been shown by Hill (1963), the situation is complicated by the fact that the global maximization of the likelihood function L based on a complete random sample $x_1, x_2, \ldots, x_n$ can lead to estimates $\hat{\mu} = -\infty$, $\hat{\sigma} = \infty$, and $\hat{\tau} = x_{(1)}$, $x_{(1)}$ being the smallest observation in the sample. In spite of these difficulties, we attempt to derive the MML estimators, in a more complex setting when the available sample is type II censored [i.e., $x_{(a)} < \cdots < x_{(b)}$ ($a = r_1 + 1$, $b = n - r_2$)]; here $z_{(i)} = \{[\log(x_{(i)} - \tau) - \mu]\}/\sigma$.

In case of no censoring ($r_1 = r_2 = 0$), Harter and Moore (1966) obtained the MLE of (τ, μ, σ) by a scheme that included a modification involving the censoring of the smallest observation not previously censored in those instances when no value of the range parameter in the permissible interval $[-\infty, x_{(a)}]$ satisfied the equation $\partial \log L/\partial\tau = 0$. This was done to avoid the previously mentioned pathways where $\tau \to x_{(a)}$, $\mu \to -\infty$, and $\sigma \to \infty$. Evaluation of the MLE of (τ, μ, σ), however, is a formidable task, primarily due to the presence of the terms $g_1(z)$ and $g_2(z)$ [see (2.6.4)] in the ML equations. By using the linear approximation for $g_1(z_{(a)})$ and $g_2(z_{(b)})$ as given in (2.7.2), one may obtain the MMLE for (τ, μ, σ) by procedures similar to those given in Chapter 2. Now, conditional on τ, the MMLE $(\hat{\mu}(\tau), \hat{\sigma}(\tau))$ of (μ, σ), based on a type II censored sample, have already been obtained in Chapter 2 as given in (2.7.11). The MML equation for estimating τ is given by

$$\begin{aligned}\frac{\partial \log L}{\partial\tau} \simeq \frac{\partial \log L^*}{\partial\tau} \simeq \frac{1}{\sigma}\Bigg[&\sum_{i=a}^{b}(x_{(i)} - \tau)^{-1} + \sum_{i=a}^{b} z_{(i)}(x_{(i)} - \tau)^{-1} \\ &- r_1(x_{(a)} - \tau)^{-1}(\alpha_1 - \beta_1 z_{(a)}) + r_2(x_{(b)} - \tau)^{-1}(\alpha_2 + \beta_2 z_{(b)})\Bigg] \\ = 0 \qquad (3.2.2)\end{aligned}$$

Thus the following iterative procedure can be adopted to obtain the MMLE of (τ, μ, σ):

1. Choose an initial value δ_0 for $\delta = x_{(a)} - \tau_0 = \delta_0$ and obtain the MMLE of (μ, σ) corresponding to this δ_0, say (μ_0, σ_0).
2. The MMLE of (τ, μ, σ) is $(\tau_0, \hat{\mu}_0, \hat{\sigma}_0)$ if $S(\delta_0) = (\partial \log L^*/\partial\tau)_{\hat{\mu}_0, \hat{\sigma}_0, \hat{\tau}_0} = 0$.

   Otherwise, compute δ_1 by Newton-Raphson's method, where $\delta_1 = \delta_0 - S(\delta_0)/I(\delta_0)$ and $I(\delta_0) = (\partial^2 \log L^*/\partial\tau^2)_{\hat{\mu}_0, \hat{\sigma}_0, \hat{\tau}_0}$.
3. Use δ_1 as initial value and repeat step 2. Continue the process until $S(\delta) \simeq 0$.

EXAMPLE 3.2.1 Consider the following 20 order statistics of a random sample of size 20 (see Cohen, 1951, p. 210) from a log-normal distribution (with $\mu = 3.912$, $\sigma = 0.4$, and $\tau = 100$):

127.211	128.709	131.375	132.971	133.143	135.880	137.338
144.328	145.788	148.290	153.070	155.369	155.680	157.238
164.307	166.475	168.554	174.800	184.101	201.415	

Here $r_1 = r_2 = 0$, and the values of $S(\delta)$ are given by

δ:	1	3	5	7	9	11	13	15
$S(\delta)$:	-0.896	-0.180	-0.063	-0.022	-0.003	0.007	0.012	0.015

By graphical interpolation $S(\delta = 9.5) = 0$. For $\delta = \delta_0 = 9.5$, we obtain, from (2.7.11),

$$\hat{\mu} = 3.374 \quad \hat{\sigma} = 0.604 \quad \hat{\tau} = x_{(1)} - \delta_0 = 127.211 - 9.5 = 117.71$$

Substituting these values in (3.2.2), we obtain

$$S(\delta_0) = [S(\delta)]_{\delta=9.5} = 0.00005$$

Since $S(\delta_0)$ is very close to zero, there is perhaps no need to improve on the value of δ_0 above.

The estimates above may be compared with the ML estimates given by Cohen (1951):

$$\hat{\mu} = 3.374 \quad \hat{\sigma} = 0.60 \quad \hat{\tau} = 117.68$$

For various censoring of the data above, the following MML estimates were similarly obtained:

r_1	r_2	$\hat{\mu}$	$\hat{\sigma}$	$\hat{\tau}$
2	0	3.711	0.724	113.31
2	4	3.648	0.824	114.61
0	10	3.135	1.800	126.02

The bias in these estimates is apparent in case of heavy censoring.

The Asymptotic Variances and Covariances

Let $J_{ij} = -E(\partial^2 \log L^*/\partial\theta_i\, \partial\theta_j)$, where $\theta_1 = \mu$, $\theta_2 = \sigma$, and $\theta_3 = \tau$. Then the asymptotic variance-covariance matrix of the MMLE $(\hat{\mu}, \hat{\sigma}, \hat{\tau})$ is given by

$$V(\hat{\underline{\theta}}) = (\nu_{ij}) = (J_{ij})^{-1}$$

Since conditional on τ the estimation of (μ, σ) is exactly the same as in the case of the normal distribution, obviously the asymptotic values of J_{11}, $J_{12} = J_{21}$, and J_{22} are the same as those given in Chapter 2. For finding the variance $V(\hat{\theta}_3)$ and the covariances $\mathrm{cov}(\hat{\theta}_1, \hat{\theta}_3)$, and so on, we have

$$J_{13} = -E\left(\frac{\partial^2 \log L^*}{\partial\mu\,\partial\tau}\right)$$

$$= \frac{1}{\sigma^2} E\left[\sum_{i=a}^{b} (x_{(i)} - \tau)^{-1} - r_1\beta_1(x_{(a)} - \tau)^{-1} + r_2\beta_2(x_{(b)} - \tau)^{-1}\right] \quad (3.2.3)$$

$$J_{23} = -E\left(\frac{\partial^2 \log L^*}{\partial\sigma\,\partial\tau}\right)$$

$$= \frac{1}{\sigma^2} E\Big[2\sum_{i=a}^{b} z_{(i)}(x_{(i)} - \tau)^{-1} - r_1(x_{(a)} - \tau)^{-1}(\alpha_1 - 2\beta_1(z_{(a)})$$

$$+ r_2(x_{(b)} - \tau)^{-1}(\alpha_2 + 2\beta_2(z_{(b)})\Big] \quad (3.2.4)$$

and

$$J_{33} = -E\left(\frac{\partial^2 \log L^*}{\partial\tau^2}\right)$$

$$= \frac{1}{\sigma^2} E\Big\{\sum_{i=a}^{b} (x_{(i)} - \tau)^{-2} - r_1\beta_1(x_{(a)} - \tau)^{-2} + r_2\beta_2(x_{(b)} - \tau)^{-2}$$

$$- \sigma\left[\sigma \sum_{i=a}^{b} (x_{(i)} - \tau)^{-2} + \sum_{i=a}^{b} z_{(i)}(x_{(i)} - \tau)^{-2}\right]$$

$$- r_1(x_{(a)} - \tau)^{-2}(\alpha_1 - \beta_1 z_{(a)}) + r_2(x_{(b)} - \tau)^{-2}(\alpha_2 + \beta_2 z_{(b)})\Big\} \quad (3.2.5)$$

To simplify J_{13}, J_{23}, and J_{33}, note that asymptotically (Harter and Moore, 1966)

$$\frac{1}{n}\sum_{i=a}^{b} (x_{(i)} - \tau)^{-1} = \int_{t_1}^{t_2} e^{-(\mu+\sigma t)} f(t)\,dt$$

$$= e^{-\mu+\sigma^2/2}[F(t_2 + \sigma) - F(t_1 + \sigma)] \quad (3.2.6)$$

$$\frac{1}{n}\sum_{i=a}^{b} (x_{(i)} - \tau)^{-2} = \int_{t_1}^{t_2} e^{-2(\mu+\sigma t)} f(t)\,dt$$

$$= e^{2(\sigma^2-\mu)}[F(t_2 + 2\sigma) - F(t_1 + 2\sigma)] \quad (3.2.7)$$

$$\frac{1}{n}\sum_{i=a}^{b} z_{(i)}(x_{(i)} - \tau)^{-1} = \int_{t_1}^{t_2} e^{-(\mu+\sigma t)}\, t f(t)\,dt$$

$$= -e^{-\mu+(\sigma^2/2)}\{\sigma[F(t_2+\sigma) - F(t_1+\sigma)] + f(t_2+\sigma) - f(t_1+\sigma)\}$$

$$(3.2.8)$$

and

$$\frac{1}{n}\sum_{i=a}^{b}(\sigma z_{(i)} - 1)(x_{(i)} - \tau)^{-2} = \int_{t_1}^{t_2} e^{-2(\mu+\sigma t)}(\sigma t - 1)\, f(t)\, dt$$

$$= -e^{2(\sigma^2-\mu)}\{(2\sigma^2 + 1)[F(t_2 + 2\sigma) - F(t_1 + 2\sigma)] + \sigma[f(t_2 + 2\sigma) - f(t_1 + 2\sigma)]\} \quad (3.2.9)$$

t_1 and t_2 are given by (2.7.5).

On substituting (3.2.6)-(3.2.9) into (3.2.3)-(3.2.5), one can readily obtain the asymptotic variances and covariances of the MMLE $(\hat{\mu}, \hat{\sigma}, \hat{\tau})$, which work out exactly equal to the corresponding asymptotic variances and covariances of the ML estimators given by Harter and Moore (1966, p. 846), as expected. Incidentally, note the misprint in their expressions for $E\,\Sigma_{i=a}^{b}\, z_{(i)}(x_{(i)} - \tau)^{-1}$ and $E\,\Sigma_{i=a}^{b}\,(\sigma z_{(i)} - 1)(x_{(i)} - \tau)^{-2}$; these should be multiplied by a minus sign [equations (3.2.6)-(3.2.9) are the correct ones].

For finite n, J_{ij} and the corresponding equations for the ML estimators have to be evaluated by quadrature. This task has been admirably carried out by Smith et al. (1973) by high-powered Gauss-Legendre quadrature for numerous sample sizes. They also evaluated the corresponding equations for the Harter-Moore ML estimators and the Syler (1968) estimators, where Syler's estimators were obtained by approximating $g_i(z)$ by $g_i(z) \simeq \alpha + \beta z + \gamma z^2$ $(-3 \leq z \leq 3)$. For $n = 20$ and various censoring proportions $q_1 = r_1/n$ and $q_2 = r_2/n$, their results are given in Table 3.2.1. Also given in Table 3.2.1 are the values of the relative bias $b(\theta)$ for Tiku's and Syler's approximations, where $b(\theta)$ is defined as the nonzero value of $E(d \log L^*/d\theta)$ (see Smith et al., 1973).

From the evaluations given in Table 3.2.1, Smith et al. (1973, pp. 25-26) arrived at the following conclusions:

> $V(\tilde{\mu})$ values range from 0.2018 to 1.445 in the Harter-Moore equations. They are slightly more affected by censoring from the left than from the right. There are no appreciable differences between the methods with regard to $V(\tilde{\mu})$, except that it is smaller in the Tiku and Syler types for very heavy censoring from both sides.
>
> $V(\tilde{\sigma})$ values are very small throughout. They range from 0.035 to 0.216 in the Syler evaluations. They also appear to be only slightly affected by heavier censoring from the left than from the right. There does not seem to be any noticeable difference between the three techniques with respect to $V(\tilde{\sigma})$.
>
> $V(\tilde{\tau})$ values display the largest range and are most affected by censoring. They range from 412.915 to 3256.717 in Tiku's equations and are much larger with a greater increase in censoring from the left. The Tiku and Syler methods appear to have smaller $V(\tilde{\tau})$ values when censoring is very heavy.

Table 3.2.1 Information Matrices and Bias Vectors, Sample Size n = 20; μ = 3.912, σ = 0.4, τ = 100

	Harter-Moore	Tiku	Syler	Harter-Moore	Tiku	Syler
	$q_1 = 0,\ q_2 = 0$			$q_1 = 0,\ q_2 = 0.1$		
ν_{11}	0.202	0.202	0.202	0.246	0.246	0.246
ν_{22}	0.035	0.035	0.035	0.049	0.048	0.049
ν_{33}	412.920	412.920	412.920	519.830	518.630	520.740
ν_{12}	-0.077	-0.077	-0.077	-0.102	-0.102	-0.102
ν_{13}	-8.946	-8.946	-8.946	-11.120	-11.090	-11.140
ν_{23}	3.578	3.578	3.578	4.797	4.785	4.806
$b(\mu)$	0.000	0.000	0.000	0.000	0.014	-0.014
$b(\sigma)$	0.000	0.000	0.000	0.000	0.011	0.012
$b(\tau)$	0.000	0.000	0.000	0.000	-0.000	0.000
	$q_1 = 0.1,\ q_2 = 0$			$q_1 = 0.2,\ q_2 = 0.2$		
ν_{11}	0.425	0.424	0.421	1.445	1.400	1.314
ν_{22}	0.061	0.061	0.060	0.233	0.226	0.216
ν_{33}	920.310	918.220	911.840	3358.250	3256.720	3114.100
ν_{12}	-0.154	-0.153	-0.152	-0.570	-0.550	-0.526
ν_{13}	-19.580	-19.530	-19.400	-69.450	-67.320	-64.390
ν_{23}	7.195	7.177	7.127	27.500	26.650	25.470
$b(\mu)$	0.000	0.004	-0.014	0.000	0.000	0.000
$b(\sigma)$	0.000	0.011	0.012	0.000	0.031	0.068
$b(\tau)$	0.000	-0.000	-0.000	0.000	-0.000	-0.001

Sharp changes are noticed in the expected covariances between $\tilde{\mu}$ and $\tilde{\tau}$, and between $\tilde{\sigma}$ and $\tilde{\tau}$, with changes in the censored proportions. This is especially true when censoring is from the left side of the data. Cov $(\tilde{\mu}, \tilde{\tau})$ ranges from -8.946 at the non-increasing level to -69.45 in the $q_1 = q_2 = 0.2$ case of the Harter-Moore equations. Cov $(\tilde{\sigma}, \tilde{\tau})$ goes from 3.578 to 27.505. The terminal figures are smaller in the Syler evaluations than in the other two. Cov $(\tilde{\sigma}, \tilde{\tau})$ values are relatively small throughout. They do not seem to be affected greatly by moderate censoring.

The expected likelihood vectors appear to be small relative to the lower bounds on variances. It is observed that these biases are smaller using the Tiku approximations than those of Syler. This could mean that the Tiku approximations are slightly closer than those of Syler's remembering that Tiku's method computes a different approximation for the ratio of interest for each different censoring level, whereas Syler's method does not. The loss in bias appears to be slight for both methods.

Since the MML estimators $\hat{\mu}$, $\hat{\sigma}$, and $\hat{\tau}$ are also much easier to compute than the Harter-Moore and Syler estimators, they are the preferred estimators.

It is difficult to obtain the distributions of the estimators above. However, the conditional (τ known) distributions of the MML estimators $\hat{\mu}$ and $\hat{\sigma}$ are exactly the same as the distributions of the estimators given by (2.7.11). This is due to the fact that $\log(x - \tau)$ is normally distributed with mean μ and variance σ^2.

Harter-Moore Determinations of the Bias

Since all the partial derivatives of log L* with respect to (μ, σ, τ) exist and are continuous functions in its arguments almost surely, the MMLEs are asymptotically unbiased. For finite n, the values of $b(\theta)$ given in Table 3.2.1 clearly indicate that the bias in the MML estimators are not much different from those of the ML estimators, at any rate for large samples. Harter and Moore (1966) determined empirically the bias in the latter. Their results based on 500 independent random samples from the log-normal distribution with $\mu = 4$, $\sigma = 2$, and $\tau = 10$ are given in Table 3.2.2. Harter and Moore (1966, p. 850) also determined the conditional bias in the ML estimators and thereby arrived at the following conclusions:

> The ML estimator of μ from complete samples has a negative bias which is approximately proportional to n^{-1}, the reciprocal of the sample size. The bias appears to be unaffected by knowledge of σ, but it is small or non-existent if τ is known. The bias remains negative and increases in absolute value for moderate censoring, but may be positive in cases of severe censoring.

Table 3.2.2 Means of the ML Estimates Calculated from 500 Samples

n	q_1	q_2	(μ)	(σ)	(τ)
50	0.00	0.0	3.917	2.076	10.588
100	0.00	0.0	3.962	2.045	10.321
200	0.00	0.0	3.990	2.013	10.149
100	0.00	0.5	3.984	2.094	10.311
100	0.01	0.0	3.893	2.180	10.804
100	0.01	0.5	4.056	2.474	10.861
100	0.01[a]	0.5	4.073	2.520	10.864

[a]Results given on this line are for all 500 samples, including those omitted on the line above which required use of the modified procedure (censoring the second-order statistic), so that $q_1 = 0.02$.

The ML estimator of σ has a positive bias when τ is unknown and a negative bias when τ is known. The magnitude of bias is roughly proportional to n^{-1}, and appears to be unaffected by knowledge of μ. It is increased by censoring, especially from below.

The ML estimator of τ has a positive bias which is closely proportional to n^{-1}. The magnitude of the bias appears to be unaffected by knowledge of μ and/or σ and censoring from above, but it is markedly increased by censoring even a single observation from below.

As far as the variances of the ML estimators of μ, σ, and τ for complete samples are concerned, Harter and Moore's (1966, Table 2) empirical results suggest that for samples of size n in the range 50 to 200, these variances are approximately equal to $4.2/n$, $3.2/n$, and $1800/n^2$, respectively (the population values being $\mu = 4$, $\sigma = 2$, and $\tau = 10$). However, the conditional (τ known) variances of the ML estimators of μ and σ are approximately $4.1/n$ and $2.1/n$, respectively.

We might remark that the log-normal distribution also provides a useful theoretical model for certain economic populations involving income distributions, and others in which the standard deviation of individual observations is approximately proportional to the magnitude of the observations (see, e.g., Gibrat, 1930; Hatch, 1933; Hatch and Choute, 1929; Krumbein, 1936; Kolmogorov, 1941; Kottler, 1950; Wise, 1952).

3.3 MMLEs FOR THE LOGISTIC DISTRIBUTION

The logistic distribution $L(\mu,\sigma)$ has probability density function

$$f(x) = \frac{\pi}{\sigma\sqrt{3}} \frac{\exp[-\pi(x-\mu)/\sigma\sqrt{3}]}{\{1+\exp[-\pi(x-\mu)/\sigma\sqrt{3}]\}^2} \qquad -\infty \le x \le \infty \tag{3.3.1}$$

and has been used extensively as a growth curve model for economic and demographic data (see, e.g., Berkson, 1951; Fisk, 1961; Pearl, 1940; Perks, 1932; Reed and Berkson, 1929). Hence, if X is a $L(\mu,\sigma)$ variate, then $E(X) = \mu$, $\text{var}(X) = \sigma^2$, and the probability density function of $Y = \pi(X-\mu)/\sigma\sqrt{3}$ is $f(y) = e^{-y}/(1+e^{-y})^2$, so that $F(y) = 1/(1+e^{-y})$.

Consider now a type II censored sample

$$x_{(a)} < x_{(a+1)} < \cdots < x_{(b)} \qquad (a = r_1 + 1,\ b = n - r_2)$$

and let $y_i = \pi(x_{(i)} - \mu)/\sigma\sqrt{3}$. Then it is easy to see that the ML equations $\partial \log L/\partial\mu = 0$ and $\partial \log L/\partial\sigma = 0$, based on this type II censored sample, contain terms $g_1(z) = f(z)/F(z)$ and $g_2(z) = f(z)/[1-F(z)]$ and therefore do not admit explicit solutions. Thus only iterative procedures are available for calculating the MLE (see Harter and Moore, 1967). Consequently, it is difficult to study the analytic properties of the MLE. To obtain explicit approximations to the MLE of (μ,σ), Tiku (1968b) introduced the MMLE method by approximating $g_2(y_b)$, $g_1(y_a)$, and $g_1(y_i)$, $i = a+1, \ldots, b-1$, respectively, by

$$g_2(y_b) \simeq \alpha_2 + \beta_2 y_b \tag{3.3.2}$$

where $\beta_2 = [g_2(k_2) - g_2(h_2)]/(k_2 - h_2)$ and $\alpha_2 = g_2(h_2) - \beta_2 h_2$ with h_2 and k_2 being determined by $F(k_1) = q_1 + [q_1(1-q_1)/n]^{\frac{1}{2}}$ and $F(h_1) = q_2 - [q_2(1-q_2)/n]^{\frac{1}{2}}$ $(q_2 = r_2/n)$;

$$g_1(y_a) \simeq \alpha_1 - \beta_1 y_a \tag{3.3.3}$$

where $\beta_1 = -[g_1(k_1) - g_1(h_1)]/(k_1 - h_1)$ and $\alpha_1 = g_1(h_1) + \beta_1 h_1$, with h_1 and k_1 being determined by $F(k_1) = q_1 + [q_1(1-q_1)/n]^{\frac{1}{2}}$ and $F(h_1) = q_1 - [q_1(1-q_1)/n]^{\frac{1}{2}}$ $(q_1 = r_1/n)$;

$$g_1(y_i) \simeq \alpha_0 - \beta_0 y_i \qquad i = a+1, \ldots, b-1 \tag{3.3.4}$$

where $\beta_0 = -[g_1(k) - g_1(h)]/(k-h)$ and $\alpha_0 = g_1(h) + \beta_0 h$, with h and k being determined by $F(h) = q_1 = r_1/n$ and $1 - F(k) = q_2 = r_2/n$.

It is shown by Tiku (1968b) that (3.3.2)-(3.3.4) provide very close approximations so that the MMLEs, based on type II censored samples, are almost as efficient as the corresponding MLEs. Note that as $n \to \infty$, both h_1 and k_1 converge to h_0, satisfying $F(h_0) = q_1$ so that

$$\beta_1 \to f(h_0) \qquad \alpha_1 \to g_1(h_0) + \beta_1 h_0$$

and similarly

$$\beta_2 \to f(k_0) \qquad \alpha_2 \to g_2(k_0) - \beta_2 k_0$$

where k_0 is determined by $1 - F(k_0) = q_2$. Using the approximations (3.3.2)-(3.3.4), we have

$$\frac{\partial \log L}{\partial \mu} \simeq \frac{\partial \log L^*}{\partial \mu} = \frac{m\pi^2}{3\sigma^2}(K + D\sigma - \mu) = 0 \tag{3.3.5}$$

$$\frac{\partial \log L}{\partial \sigma} \simeq \frac{\partial \log L^*}{\partial \sigma} = \frac{\pi^2}{3\sigma^3}[(C + B\sigma - A\sigma^2) + m(K - \mu)(K + D\sigma - \mu)] = 0 \tag{3.3.6}$$

where

$$K = \frac{2\beta_0 \Sigma_{i=a}^{b} x_{(i)} + r_1\beta_1 x_{(a)} + r_2\beta_2 x_{(b)}}{m} \qquad m = 2(n - r_1 - r_2)\beta_0 + r_1\beta_1 + r_2\beta_2$$

$$B = \frac{\sqrt{3}}{\pi}\left\{ r_2\alpha_2(x_{(b)} - K) - r_1\alpha_1(x_{(a)} - K) + (1 - 2\alpha_0)\left[\sum_{i=a}^{b} x_{(i)} - (n - r_1 - r_2)K\right]\right\}$$

$$D = \frac{\sqrt{3}}{\pi}\,\frac{(1 - 2\alpha_0)(n - r_1 - r_2) + (r_2\alpha_2 - r_1\alpha_1)}{m} \qquad A = \frac{3}{\pi^2}(n - r_1 - r_2)$$

$$C = 2\beta_0 \sum_{i=a}^{b} (x_{(i)} - K)^2 + r_1\beta_1(x_{(a)} - K)^2 + r_2\beta_2(x_{(b)} - K)^2$$

$$= 2\beta_0 \sum_{i=a}^{b} x_{(i)}^2 + r_1\beta_1 x_{(a)}^2 + r_2\beta_2 x_{(b)}^2 - mK^2$$

Hence the MMLE of (μ, σ) is

$$\hat{\mu} = K + D\hat{\sigma}$$

and $\hat{\sigma}$ corrected for bias is given by

$$\hat{\sigma} = \frac{B + (B^2 + 4AC)^{\frac{1}{2}}}{2\{A[A - (3/\pi^2)]\}^{\frac{1}{2}}}$$

Further, solving the differential equations (3.3.5) and (3.3.6), we obtain

$$L^* = \sigma^{-(n-r_1-r_2)} \exp\left[-\frac{\pi^2 B}{3\sigma} - \frac{\pi^2 C}{6\sigma^2} - \frac{m\pi^2}{6\sigma^2}(K + D\sigma - \mu)^2\right] H(\underset{\sim}{x}) \tag{3.3.7}$$

where $H(\underset{\sim}{x})$ is a function of $\underset{\sim}{x}' = (x_{(a)}, \ldots, x_{(b)})$ but is free of μ and σ.

Comparing (3.3.7) with (2.7.13), it is pleasing to note that with (B, C, m) in (2.7.13) being replaced by $(\pi^2 B/3,\ \pi^2 C/3,\ m\pi^2/3)$, (3.3.7) reduces to

(2.7.13). Hence all results for the normal distribution as given in Chapter 2 also apply without change to the MMLE for the logistic distribution. Specifically, it follows that the MMLE are asymptotically unbiased and minimum variance estimators of (μ, σ) and also have very good efficiency properties for small n (Tiku, 1968b). Moreover, procedures based on these MMLE are expected to be robust to departures from a logistic distribution (see Chapter 4 for details). Note that for $q_1 = q_2$, $\alpha_0 = 0.5$, $\alpha_1 = \alpha_2$ and $\beta_1 = \beta_2$.

EXAMPLE 3.3.1 Sarhan and Greenberg (1962, p. 212) have given the measurements resulting from an experiment to measure strontium-90 concentrated in samples of milk. The test substance was supposed to contain 9.22 μCi/liter. Ten measurements were considered, but owing to the relatively larger measurement error known to exist at the extremes, the two smallest and the three largest observations were censored. The remaining five observations are, in order of magnitude,

8.2 8.4 9.1 9.8 9.9

Here $n = 10$, $q_1 = 0.2$, $q_2 = 0.3$; $\alpha_0 = 0.490$, $\beta_0 = 0.224$; $\alpha_1 = 0.572$, $\beta_1 = 0.140$; $\alpha_2 = 0.511$, $\beta_2 = 0.197$. Assuming that the underlying distribution is logistic, we substitute these values in the equations above and obtain the following MMLE:

$$\hat{\mu} = 9.29 \quad \text{and} \quad \hat{\sigma} = 1.71$$

These estimates may be compared with the following MLEs (Harter and Moore, 1967, p. 683) and BLUEs (Sarhan and Greenberg, 1962, p. 212):

$$\text{MLE:} \quad \hat{\hat{\mu}} = 9.27 \quad \text{and} \quad \hat{\hat{\sigma}} = 1.57$$

$$\text{BLUE:} \quad \mu^* = 9.29 \quad \text{and} \quad \sigma^* = 1.69$$

Note that the MMLE are the easiest to compute.

3.4 ESTIMATION OF PARAMETERS OF EXPONENTIAL DISTRIBUTION

The two-parameter exponential distribution $E(\theta, \sigma)$ has probability density function

$$f(x) = \frac{1}{\sigma} \exp\left(-\frac{x - \theta}{\sigma}\right) \qquad \theta \le x < \infty \tag{3.4.1}$$

This distribution has many applications, particularly in the area of life testing (see Epstein and Sobel, 1954, for example). Very often, it is reasonable to take $\theta = 0$, in which case (3.4.1) reduces to the one-parameter exponential distribution $E(0, \sigma)$.

Let $x_{(a)} < \cdots < x_{(b)}$ $(a = r_1 + 1,\ b = n - r_2)$ be a type II censored sample from $E(\theta, \sigma)$ which has mean $\theta + \sigma$ and variance σ^2. Our problem here is to estimate θ and σ from the censored sample above.

ML Estimators

Let $Z_i = (x_{(i)} - \theta)/\sigma$, $f(z) = e^{-z}$, and $F(z) = \int_0^z f(z)\,dz = 1 - e^{-z}$. The ML equations are given by (see also Kambo, 1978)

$$\frac{\partial \log L}{\partial \theta} = \frac{1}{\sigma}\left[(n - r_1) - r_1 \frac{f(Z_a)}{F(Z_a)}\right] = 0 \tag{3.4.2}$$

and

$$\frac{\partial \log L}{\partial \sigma} = \frac{1}{\sigma}\left[-(n - r_1 - r_2) + \sum_{i=a}^{b} Z_i - r\, Z_a \frac{f(Z_a)}{F(Z_a)} + r_2 Z_b\right] = 0 \tag{3.4.3}$$

The ML equations above do not admit explicit solutions individually (Kambo, 1978). However, they do admit explicit simultaneous solutions which can be obtained as follows:

From (3.4.2), $f(Z_a)/F(Z_a) = (n - r_1)/r_1$, $r_1 \neq 0$. Substituting this in (3.4.3), we get the equation

$$(n - r_1 - r_2) + \sum_{i=a}^{b} Z_i - (n - r_1)Z_a + r_2 Z_b = 0 \tag{3.4.4}$$

The solution of (3.4.4) gives the MLE $\hat{\hat{\sigma}}$ of σ,

$$\hat{\hat{\sigma}} = \frac{\Sigma_{i=r_1+1}^{n-r_2} x_{(i)} + r_2 x_{(n-r_2)} - (n - r_1)x_{(r_1+1)}}{n - r_1 - r_2 - 1} \tag{3.4.5}$$

This solution has in fact $n - r_1 - r_2$ as the divisor, but we replace it by $n - r_1 - r_2 - 1$ to correct for the bias.

From (3.4.2) we then obtain the MLE of θ as

$$\hat{\hat{\theta}} = x_{(r_1+1)} + \hat{\hat{\sigma}} \log(1 - q_1) \qquad q_1 = \frac{r_1}{n} \neq 0 \tag{3.4.6}$$

If $r_1 = 0$, then the ML estimators are given by

$$\hat{\hat{\theta}} = x_{(1)} - \frac{1}{n}\hat{\hat{\sigma}} \quad \text{[adjusted for the fact that } E(x_{(1)}) = \theta + \sigma/n] \tag{3.4.7}$$

and

$$\hat{\hat{\sigma}} = \frac{\Sigma_{i=1}^{n-r_2} x_{(i)} + r_2 x_{(n-r_2)} - n x_{(1)}}{n - r_2 - 1}$$

If θ is known (say, $\theta = \theta_0$) and $r_1 = 0$, then the MLE of σ is given by

$$\hat{\hat{\sigma}}_0 = \frac{\Sigma_{i=1}^{n-r_2} x_{(i)} + r_2 x_{(n-r_2)} - n\theta_0}{n - r_2} \tag{3.4.8}$$

If θ is known and $r_1 \neq 0$, then the MLE of σ has to be obtained from (3.4.3) iteratively. As with previous iterative procedures, one can face problems with these iterations; for example, if the data come from a distribution which is not exactly exponential, although we assumed it to be so (a real possibility from a practical point of view), these iterations might converge very slowly or not converge at all (see also Section 3.5).

MML Estimators

To obtain the MML estimators, we make use of the linear approximation

$$\frac{f(Z_a)}{F(Z_a)} \simeq \alpha - \beta Z_a \tag{3.4.9}$$

and thereby obtain the following MML equations:

$$\frac{\partial \log L}{\partial \theta} \simeq \frac{\partial \log L^*}{\partial \theta} = \frac{1}{\sigma}[(n - r_1) - r_1(\alpha - \beta Z_a)] = 0 \tag{3.4.10}$$

and

$$\frac{\partial \log L}{\partial \sigma} \simeq \frac{\partial \log L^*}{\partial \sigma} = \frac{1}{\sigma}\left[-(n - r_1 - r_2) + \sum_{i=a}^{b} Z_i - r_1 Z_a(\alpha - \beta Z_a) + r_2 Z_b\right] = 0 \tag{3.4.11}$$

Since expected values and variances of the order statistics of a random sample from the exponential E(0,1) assume neat algebraic expressions [equations (B.15) and (B.16)], one can easily determine α and β in such a way that

$$E\left(\frac{\partial \log L^*}{\partial \theta}\right) = 0 \quad \text{and} \quad E\left(\frac{\partial \log L^*}{\partial \sigma}\right) = 0$$

This will produce estimators which are conditionally unbiased (see Section 2.7). Thus the coefficients α and β are solutions of the equations

$$r_1[\alpha - \beta E(Z_a)] = n - r_1$$

and

$$r_1[\alpha E(Z_a) - \beta E(Z_a^2)] = -(n - r_1 - r_2) + \sum_{i=a}^{b} E(Z_i) + r_2 E(Z_b)$$

Realizing that $E(Z_i) = \Sigma_{j=1}^{i} [1/(n - j + 1)]$ and $V(Z_i) = \Sigma_{j=1}^{i} [1/(n - j + 1)^2]$, the values of α and β can easily be evaluated from the equations

$$\beta = \frac{(n - r_1 - r_2) - \Sigma_{i=a}^{b} E(Z_i) + (n - r_1)E(Z_a) - r_2 E(Z_b)}{r_1 V(Z_a)}$$

and

$$\alpha = \frac{n - r_1}{r_1} + \beta \sum_{j=1}^{r_1+1} \frac{1}{n - j + 1} \tag{3.4.12}$$

Note that α and β are both positive.

As $A = n - r_1 - r_2$ becomes large, Z_a tends to its expected value $E(Z_a) = \Sigma_{j=1}^{r_1+1} 1/(n - j + 1)$, in which case the expressions on both sides of (3.4.9) are exactly the same; this follows from the fact that $\alpha - \beta Z_a = (n - r_1)/r_1$ [from (3.4.12)] and $F(Z_a) = 1 - e^{-Z_a} = r_1/n$; $f(Z_a) = e^{-Z_a} = (n - r_1)/n$ and $[f(Z_a)/F(Z_a)] = (n - r_1)/r_1$. For large n, therefore, the MML equations above are identical with the corresponding ML equations. Note, however, that (3.4.10) and (3.4.11) admit explicit solutions individually.

The solutions of (3.4.10) and (3.4.11) are the MML estimators $\hat{\theta}$ and $\hat{\sigma}$:

$$\hat{\theta} = \begin{cases} x_{(r_1+1)} - \frac{1}{n}\hat{\sigma} & \text{if } r_1 = 0 \\ x_{(r_1+1)} - \sum_{j=1}^{r_1+1} \frac{1}{n - j + 1}\hat{\sigma} & \text{if } r_1 \neq 0 \end{cases} \tag{3.4.13}$$

and replacing the divisor $n - r_1 - r_2$ by $n - r_1 - r_2 - 1$ (to correct for bias)

$$\hat{\sigma} = \frac{\Sigma_{i=r_1+1}^{n-r_2} x_{(i)} + r_2 x_{(n-r_2)} - (n - r_1)x_{(r_1+1)}}{n - r_1 - r_2 - 1} \tag{3.4.14}$$

If θ is known ($\theta = \theta_0$), then the MMLE $\hat{\sigma}_0$ of σ is given by

$$\hat{\sigma}_0 = \frac{B + (B^2 + 4AC)^{\frac{1}{2}}}{2A} \tag{3.4.15}$$

where

$$A = n - r_1 - r_2 \qquad C = r_1\beta(x_{(r_1+1)} - \theta_0)^2$$

and

$$B = \sum_{i=r_1+1}^{n-r_2} (x_{(i)} - \theta_0) + r_2(x_{(n-r_2)} - \theta_0) - r_1\alpha(x_{(r_1+1)} - \theta_0)$$

For $r_1 = 0$, (3.4.15) reduces to (3.4.8).

Interestingly indeed, $\hat{\theta}$ and $\hat{\sigma}$ are exactly the same as the BLUEs θ^* and σ^* given by equations following (2.5.8). In particular, $\hat{\theta}$ and $\hat{\sigma}$ are unbiased and their variances and covariance are given by (2.5.9).

The MLE $\hat{\hat{\sigma}}$ is exactly the same as the MMLE $\hat{\sigma}$. The MLE $\hat{\hat{\theta}}$ is somewhat different from the MMLE $\hat{\theta}$ and is biased for small samples; for large samples, however, $\hat{\hat{\theta}}$ and $\hat{\theta}$ are identical, as expected, since

$$\lim_{n\to\infty}\left(\sum_{j=1}^{n}\frac{1}{j} - \log n\right) = \gamma \quad \text{(Euler's constant)}$$

and, consequently,

$$\lim_{n\to\infty} \sum_{j=1}^{r_1+1} \frac{1}{n-j+1} = -\log(1-q_1)$$

$q_1 = r_1/n$.

If θ is known ($\theta = \theta_0$, say) and $r_1 \neq 0$, then $\hat{\sigma}_0$ [given by (3.4.15)] is not linear and is different from the BLUE σ_0^*; σ_0^* is given by

$$\sigma_0^* = \frac{1}{K}\left\{\left[\frac{\Sigma_{j=1}^{r_1+1} 1/(n-j+1)}{\Sigma_{j=1}^{r_1+1} 1/(n-j+1)^2} - (n-r_1)\right] x_{(r_1+1)} + r_2 (x_{(n-r_2)} + \sum_{i=r_1+1}^{n-r_2} x_{(i)}\right\} \tag{3.4.16}$$

with variance $V(\sigma_0^*) = \sigma^2/K$;

$$K = \frac{[\Sigma_{j=1}^{r_1+1} 1/(n-j+1)]^2}{\Sigma_{j=1}^{r_1+1} 1/(n-j+1)^2} + (n - r_1 - r_2 - 1) \tag{3.4.17}$$

For large $A = n - r_1 - r_2$, $\hat{\sigma}_0$ is unbiased and its variance is given by $R^2(\sigma) = 1/[-E(d^2 \log L^*/d\sigma^2)]$;

$$-E\left(\frac{d^2 \log L^*}{d\sigma^2}\right) = \frac{1}{\sigma^2}\left\{-(n - r_1 - r_2) + 2\sum_{i=a}^{b} E(Z_i) + 2r_2 E(Z_b) - r_1[2\alpha E(Z_a) + 3\beta E(Z_a^2)]\right\} \tag{3.4.18}$$

Note that $E(d^2 \log L^*/d\sigma^2) = E(d^2 \log L/d\sigma^2)$ for large A.

We give below the values of the variances of σ^* and $\hat{\sigma}$, calculated from (3.4.17) and (3.4.18); $\hat{\sigma}_0$ is more efficient than σ_0^*, as expected:

$q_1 = r_1/n$	$q_2 = r_2/n$	$V(\hat{\sigma}_0)/\sigma^2$	$V(\sigma_0^*)/\sigma^2$
		n = 30	
0.1	0.0	0.03126	0.03334
	0.2	0.03847	0.04168
0.2	0.0	0.03129	0.03338
	0.2	0.03852	0.04173
		n = 100	
0.1	0.0	0.00980	0.01000
	0.2	0.01220	0.01250
0.2	0.0	0.00981	0.01001
	0.2	0.01221	0.01251

Realize, however, that $\hat{\sigma}_0$ will have a little bias for smaller samples; this bias can be worked out from the approximate equation

$$E(\hat{\sigma}_0) \simeq \frac{E(B) + [E(B^2) + 4AE(C)]^{\frac{1}{2}}}{2A}$$

and turns out to be negligible.

To examine the properties of the MMLE $(\hat{\theta}, \hat{\sigma})$ still further, we note that the MML equations (3.4.10) and (3.4.11) can be written as

$$\frac{\partial \log L^*}{\partial \theta} = \frac{r_1\beta}{\sigma^2}(x_{(a)} + D\sigma - \theta) \tag{3.4.19}$$

and

$$\frac{\partial \log L^*}{\partial \sigma} = \frac{1}{\sigma^3}[r_1\beta(x_{(a)} - \theta)(x_{(a)} + D\sigma - \theta) + \sigma(n - r_1 - r_2)(\hat{\sigma}^* - \sigma)] \tag{3.4.20}$$

where

$$D = -\sum_{j=1}^{a} \frac{1}{n - j + 1}$$

and

$$\hat{\sigma}^* = \frac{(n - r_1 - r_2 - 1)\hat{\sigma}}{n - r_1 - r_2} \simeq \hat{\sigma}$$

for large A. Solving (3.4.19) and (3.4.20), one obtains

$$L^* = \sigma^{-(n-r_1-r_2)} \exp\left[-\frac{\hat{\sigma}^*}{\sigma} - \frac{r_1\beta}{2\sigma^2}(x_{(a)} + D\sigma - \theta)^2\right] H(\underset{\sim}{x}) \tag{3.4.21}$$

where $H(\underset{\sim}{x})$ is a function of $\underset{\sim}{x}' = (x_{(a)}, \ldots, x_{(b)})$ but is free of (θ, σ). From (3.4.21), it is clear (see Lemma 2.7.2) that based on L^*, the conditional (given σ) distribution of $x_{(a)} + D\sigma$ is normal with mean θ and variance $\sigma^2/r_1\beta$ and that $2(n - r_1 - r_2)\hat{\sigma}^*/\sigma$ is a central chi-square variate with degrees of freedom $2(n - r_1 - r_2 - 1)$. Furthermore, since

$$(n - r_1 - r_2 - 1)\hat{\sigma} = \sum_{j=a}^{b-1} (n - j)(x_{(j+1)} - x_{(j)})$$

it follows that based on L^*, $D_j = (n - j)(x_{(j+1)} - x_{(j)})$, $j = a, \ldots, b - 1$, are independently and identically distributed as exponential $E(0, \sigma)$. Or the asymptotic distribution of $2(n - r_1 - r_2 - 1)\hat{\sigma}/\sigma$ is a central chi-square variate with degrees of freedom $2(n - r_1 - r_2 - 1)$, and also that asymptotically $2D_j/\sigma$, $j = a, a + 1, \ldots, b - 1$, are IID central chi-square variates with 2 degrees of freedom. Note, however, that the foregoing distributional results on D_j, $j = a, \ldots, b - 1$, and $\hat{\sigma}$ are in fact true for all values of $A = n - r_1 - r_2$ (see Epstein and Sobel, 1953; Pyke, 1965; Sarhan and Greenberg, 1962, pp. 361-371; Sukhatme, 1937).

EXAMPLE 3.4.1 Ten electronic components of each of the two types I and II were put to test. Due to some technical difficulty, recording of the failure times could not begin until two components of type I and one component of type II had already failed. The failure time (in hours) of the remaining components were recorded as follows:

I:	...	...	409	455	509	519	541	543	550	583
II:	...	400	429	463	526	527	542	560	567	598

The failure times are assumed to be exponential $E(\theta_1, \sigma)$ and $E(\theta_2, \sigma)$ for types I and II, respectively, and one wants to estimate θ_1, θ_2, and σ.

The MML (or BLU) estimates are given by

$$\hat{\theta}_1 = 409 - \sum_{j=1}^{3} \frac{1}{10 - j + 1}\hat{\sigma} = 409 - 0.33611\hat{\sigma}$$

and

$$\hat{\theta}_2 = 400 - \sum_{j=1}^{2} \frac{1}{10 - j + 1}\hat{\sigma} = 400 - 0.21111\hat{\sigma}$$

From the first set of data, the MMLE of σ is obtained as

$$\hat{\sigma}_1 = \frac{4109 - 8(409)}{7} = 119.57143$$

From the second set of data, the MMLE of σ is obtained as

$$\hat{\sigma}_2 = \frac{4612 - 9(400)}{8} = 126.5$$

Now $\hat{\sigma}_1$ and $\hat{\sigma}_2$ are independent unbiased estimators of σ with variances $V(\hat{\sigma}\ ) = \sigma^2/7$ and $V(\hat{\sigma}\ ) = \sigma^2/8$. Hence the combined estimate $\hat{\sigma}$ of σ is (Exercise 2.4)

$$\hat{\sigma} = \frac{[\hat{\sigma}_1/V(\hat{\sigma}_1) + \hat{\sigma}_2/V(\hat{\sigma}_2)]}{[1/V(\hat{\sigma}_1)] + [1/V(\hat{\sigma}_2)]}$$

$$= \frac{7(119.57143) + 8(126.5)}{15}$$

$$= 123.2667$$

The estimator $\hat{\sigma}$ is unbiased for σ and has smaller variance than both $\hat{\sigma}_1$ and $\hat{\sigma}_2$ (see Exercise 2.4). Given $\hat{\sigma} = 123.2667$, the estimates of θ_1 and θ_2 are then $\hat{\theta}_1 = 367.57$ and $\hat{\theta}_2 = 373.98$ hr.

Data similar to the above (e.g., censored on the left) often arise in the context of life insurance (Weissman, 1978). For a detailed discussion of numerous real-life examples dealing with data censored only on the right, see Epstein and Sobel (1953).

There is a huge literature on estimation and hypothesis testing problems related to the exponential $E(\theta, \sigma)$; some of the important references are the following: Bain and Weeks (1964), Bartholomew (1963), Barton et al. (1963), Basu (1965), Deemer and Votaw (1955), El-Sayyad (1967), Epstein (1958, 1960a,b,c), Epstein and Sobel (1954), Harter (1964), Kambo (1978), Lawless (1982), Mann et al. (1974), Patil (1963), Saleh (1966), Sarhan et al. (1963), Siddiqui (1963), Sukhatme (1937), Susara and Pathala (1965), and Varde (1969).

3.5 MML ESTIMATION OF PARAMETERS FOR RAYLEIGH DISTRIBUTION

We have already demonstrated that for the normal, log-normal, logistic, and exponential distributions, the MMLEs are asymptotically identical with the corresponding MLEs and almost identical for small samples. We now consider the Rayleigh distribution and show that the MMLEs can be used in an iterative procedure to obtain still better estimates. The Rayleigh distribution has density function

$$f(x) = \frac{2}{\sqrt{\sigma}} xe^{-x^2/\sigma} \qquad 0 < x < \infty$$

and arises in electromagnetic wave propagation through a scattering medium, communication engineering, and so on (Siddiqui, 1962); see also Ariyawansa and Templeton (1984).

Consider now a type II censored sample $x_{(a)} < \cdots < x_{(b)}$ $(a = r_1 + 1, b = n - r_2)$. The log-likelihood based on this type II censored sample is given by

$$\log L = \text{constant} - (n - r_1 - r_2) \log \sqrt{\sigma} + \sum_{i=a}^{b} \log z_i - \sum_{i=a}^{b} z_i^2$$
$$+ r_1 \log[F(z_a)] + r_2 \log[1 - F(z_b)]$$

where $z_i = x_{(i)}/\sqrt{\sigma}$, $F(z) = \int_0^z f(z)\,dz$, and $f(z) = 2ze^{-z^2}$, $0 < z < \infty$.

The ML equation for estimating σ is given by

$$\frac{d \log L}{d\sigma} = \frac{1}{\sigma}\left[-(n - r_1 - r_2) + \sum_{i=a}^{b} z_i^2 - \frac{1}{2} r_1 z_a g(z_a) + r_2 z_b^2\right] = 0 \qquad (3.5.1)$$

$g(z) = f(z)/F(z) = 2ze^{-z^2}/(1 - e^{-z^2})$.

Equation (3.5.1) does not admit an explicit solution and has to be solved iteratively. Lee et al. (1980) adopted an iterative procedure but, as expected, found it too time consuming and often slowly converging; with a little bit of heavy censoring, they experienced divergence even with the actual population parameter as a starting value (Lee et al., 1980, p. 16).

To avoid these pitfalls, one works out the MMLE by introducing the linear approximation

$$g(z_a) \simeq \alpha - \beta z_a \qquad (3.5.2)$$

The constants α and β are obtained from the equations

$$\beta = -\frac{g(h_2) - g(h_1)}{h_2 - h_1} \quad \text{and} \quad \alpha = g(h_1) + \beta h_1$$

h_1 and h_2 are determined by the equations $(q_1 = r_1/n)$

$$F(h_1) = q_1 - \left[\frac{q_1(1 - q_1)}{n}\right]^{\frac{1}{2}} \quad \text{and} \quad F(h_2) = q_1 + \left[\frac{q_1(1 - q_1)}{n}\right]^{\frac{1}{2}}$$

where $F(z) = 1 - e^{-z^2}$. Note that α and β are both positive.

As n tends to infinity, h_1 and h_2 both tend to h, determined by $F(h) = q_1$, in which case

$$\beta = -\left[\frac{d}{dz} g(z)\right]_{z=h} = -\frac{2e^{-h^2}}{1 - e^{-h^2}}\left(1 - \frac{2h^2}{1 - e^{-h^2}}\right) \quad \text{and} \quad \alpha = \frac{f(h)}{q_1} + \beta h$$

The MML equation is given by

$$\frac{d \log L}{d\sigma} \simeq \frac{d \log L^*}{d\sigma} = \frac{1}{\sigma}\left[-(n - r_1 - r_2) + \sum_{i=a}^{b} z_i^2 - \frac{1}{2} r_1 z_a(\alpha - \beta z_a) + r_2 z_b^2\right]$$

$$= -\frac{1}{2\sigma^{3/2}}(2A\sigma + B\sqrt{\sigma} - C) = 0 \qquad (3.5.3)$$

The solution of (3.5.3) is the MMLE of σ,

$$\sqrt{\hat{\sigma}} = \frac{-B + (B^2 + 8AC)^{\frac{1}{2}}}{4A}$$

that is,

$$\hat{\sigma} = \frac{(B^2 + 4AC) - B(B^2 + 8AC)^{\frac{1}{2}}}{8A^2}$$

where

$$A = n - r_1 - r_2 \qquad B = r_1 \alpha x_{(a)} \qquad C = 2\sum_{i=a}^{b} x_{(i)}^2 + r_1 \beta x_{(a)}^2 + 2r_2 x_{(b)}^2$$

As n tends to infinity, z_a tends to h ($q_1 = r_1/n$ fixed) and $g(z_a)$ tends to $f(h)/q_1$; consequently, $g(z_a) = \alpha - \beta z_a$, in which case the MMLE $\hat{\sigma}$ is identical to the MLE $\hat{\hat{\sigma}}$.

Approximate Bias and Variance

It is difficult to determine the exact bias and variance of $\hat{\sigma}$. The approximate bias and variance are given by (see Section 2.7)

$$\text{bias} = \frac{E(d \log L^*/d\sigma)}{R^2(\sigma)} \quad \text{and} \quad V(\hat{\sigma}) = \frac{1}{R^2(\sigma)}$$

$R^2(\sigma) = -E(d^2 \log L^*/d\sigma^2)$.

The expressions for $E(d \log L^*/d\sigma)$ and $-E(d^2 \log L^*/d\sigma^2)$ are given by ($a = r_1 + 1$, $b = n - r_2$)

$$E\left(\frac{d \log L^*}{d\sigma}\right) = \frac{1}{\sigma}\left[-(n - r_1 - r_2) + \sum_{i=a}^{b} E(z_i^2) - \frac{1}{2} r_1 \alpha E(z_a) + \frac{1}{2} r_1 \beta E(z_a^2) + r_2 E(z_b^2)\right]$$

and

$$-E\left(\frac{d^2 \log L^*}{d\sigma^2}\right) = \frac{1}{\sigma^2}\left[-(n - r_1 - r_2) + 2\sum_{i=a}^{b} E(z_i^2) - \frac{3}{4} r_1 \alpha E(z_a) + r_1 \beta E(z_a^2) + 2r_2 E(z_b^2)\right]$$

These equations can be evaluated from the values of $E(z_i)$ and $E(z_i^2)$ given by Lee et al. (1980, p. 19), namely,

Table 3.5.1 Values of the Linear Approximation in (3.5.2)

$q_1 = r_1/n$	$q_2 = r_2/n$	h_1	h_2	$g(x_{(r_1+1)}/\sqrt{\hat{\hat{\sigma}}})$	$\alpha + \beta(x_{(r_1+1)}/\sqrt{\hat{\hat{\sigma}}})$
			n = 10		
0.1	0.1	0.071	0.465	7.09	15.65
	0.2			6.33	13.82
0.2	0.1	0.276	0.628	4.16	4.98
	0.2			3.67	4.39
0.3	0.1	0.410	0.767	4.16	4.28
	0.2			3.67	3.94
	0.3			3.67	3.94
			n = 30		
0.1	0.1	0.215	0.410	8.65	8.84
	0.2			7.78	8.24
0.2	0.1	0.368	0.564	4.56	4.69
	0.2			4.04	4.24
0.3	0.1	0.494	0.695	4.06	3.89
	0.2			3.58	3.58
	0.3			3.27	3.34
			n = 100		
0.1	0.1	0.269	0.373	7.52	7.41
	0.2			7.10	7.11
0.2	0.1	0.417	0.523	4.13	4.17
	0.2			3.87	3.92
0.3	0.1	0.541	0.651	2.95	2.97
	0.2			2.74	2.76
	0.3			2.80	2.83

$$E(z_i) = \frac{n!}{(i-1)!(n-i)!} \sum_{j=0}^{i-1} \binom{i-1}{j} (-1)^j \frac{\sqrt{\pi}}{2(n-i+j+1)^{3/2}}$$

and

$$E(z_i^2) = \frac{n!}{(i-1)!(n-i)!} \sum_{j=0}^{i-1} \binom{i-1}{j} (-1)^j \frac{1}{(n-i+j+1)^2}$$

In particular, $\hat{\sigma}$ is asymptotically unbiased.

Accuracy of the Linear Approximation

To determine the accuracy of the linear approximation (3.5.2), Lee et al. (1980) worked with random samples of size n = 10, 30, and 100 from the Rayleigh distribution and evaluated, from type II censored samples, the values of $g(x_{(r_1+1)}/\sqrt{\hat{\hat{\sigma}}})$ and its linear approximation $\alpha + \beta(x_{(r_1+1)}/\sqrt{\hat{\hat{\sigma}}})$, $\hat{\hat{\sigma}}$ being the MLE, which they obtained iteratively. For $n \geq 30$, Lee et al. found the agreement between the two sides of the linear approximation (3.5.2) very close but not necessarily so for small samples ($n \leq 10$, say). We reproduce a part of their computations in Table 3.5.1.

To achieve greater accuracy in the linear approximation (3.5.2), Lee et al. (1980) suggested a second-linear-approximation procedure by using the MMLE. This procedure starts with using the approximation (3.5.2) and obtains the MMLE $\hat{\sigma}$ of σ. Now expand $g(z_a)$ in a Taylor series around $\hat{z}_a = x_{(a)}/\hat{\sigma}$ to obtain an improved approximation to $g(z_a)$, namely,

$$g(z_a) \simeq g(\hat{z}_a) - (\hat{z}_a - z_a)\left[\frac{dg(z)}{dz}\right]_{z=\hat{z}_a}$$

$$\simeq g(\hat{z}_a) + \beta'\hat{z}_a - \beta' z_a = \alpha' - \beta' z_a \qquad (3.5.4)$$

where $\beta' = -[dg(z)/dz]_{z=\hat{z}_a}$ and $\alpha' = g(\hat{z}_a) + \hat{z}_a\beta'$. These new α' and β' are then substituted in (3.5.3) to obtain the improved MMLE, say $\hat{\sigma}'$ of σ. One may note that the MMLE $\hat{\sigma}'$ is an explicit function of sample observations and is much easier to compute than the MLE of σ.

To illustrate the closeness of this improved MMLE $\hat{\sigma}'$ and the corresponding MLE $\hat{\hat{\sigma}}$, Lee et al. worked out their values from a type II censored sample and with various censoring proportions and found the agreement amazingly close. We reproduce in Table 3.5.2 a part of their computations (population value $\sigma = 3$).

Table 3.5.2 Comparison of $\hat{\sigma}$, $\hat{\sigma}'$, and $\hat{\hat{\sigma}}$; n = 10 and 30

q_1	q_2	$\hat{\sigma}'$	$\hat{\hat{\sigma}}$	$\hat{\sigma}'$	$\hat{\hat{\sigma}}$	$\hat{\sigma}$
		n = 10		n = 30		
0.1	0.1	2.5133	2.5123	2.94108	2.94108	2.934
	0.2	2.0379	2.0369	2.39997	2.39997	2.380
0.2	0.1	2.5337	2.5335	2.95687	2.95687	2.938
	0.2	2.0605	2.0603	2.41740	2.41738	2.390
0.3	0.1	2.5335	2.5335	2.95151	2.95148	2.985
	0.2	2.0603	2.0603	2.41096	2.41096	2.411
	0.3	2.0603	2.0603	2.09010	2.09010	2.076

3.6 MML ESTIMATION IN PROGRESSIVELY CENSORED SAMPLES

Progressively censored samples are particularly encountered in life and fatigue studies in the context of life testing and are defined by Cohen (1963, p. 328) as follows:

> Let N designate the total sample size and n the number of sample specimens which fail and which therefore result in completely determined life spans. Suppose that censoring occurs progressively in k stages at times T_i, i = 1, 2, ..., k, and, at the ith stage of censoring r_i sample specimens selected randomly from the survivors at time T_i are censored from further observations. It follows that
>
> $$N = n + \sum_{i=1}^{k} r_i$$
>
> In type I censoring, the T_i are fixed and the number of survivors at these times are random variables. In type II censoring, the T_i coincide with times of failure and are random variables, whereas the number of survivors at these times are fixed. For both types r_i are fixed.

Let $x_1, x_2, \ldots, x_n$ be a k-stage progressively censored sample of type I and assume that the underlying population is normal $N(\mu, \sigma)$. The log-likelihood function is given by

$$L = \text{constant} - n \log \sigma - \frac{1}{2\sigma^2} \sum_{i=1}^{n} (x_i - \mu)^2 + \sum_{i=1}^{k} r_i \log\{1 - F(y_i)\}$$

$y_i = (T_i - \mu)/\sigma$, $f(z) = \exp(-\frac{1}{2}z^2)/\sqrt{2\pi}\sigma$, and $F(z) = \int_{-\infty}^{z} f(t)\,dt$. The ML equations for estimating μ and σ are given by

$$\frac{\partial \log L}{\partial \mu} = \frac{1}{\sigma}\left[\frac{n}{\sigma}(\bar{x} - \mu) + \sum_{i=1}^{k} r_i g(y_i)\right] = 0 \qquad (3.6.1)$$

and

$$\frac{\partial \log L}{\partial \sigma} = \frac{1}{\sigma}\left\{\frac{n}{\sigma^2}[s^2 + (\bar{x} - \mu)^2] - 1 + \sum_{i=1}^{k} r_i y_i g(y_i)\right\} = 0 \qquad (3.6.2)$$

where $\bar{x} = \Sigma_{i=1}^{n} x_i/n$ and $s^2 = \Sigma_{i=1}^{n} (x_i - \bar{x})^2/n$, and $g(z) = f(z)/[1 - F(z)]$.

These equations do not admit explicit solutions and have to be solved iteratively, by a process that can become problematic (see Lawless, 1982, p. 222).

To adopt the MML method, we realize that for large n, $y_i = (T_i - \mu)/\sigma$ is likely to be covered by the interval (a_i, b_i), where

$$a_i = \frac{T_i - (\bar{x} + s/\sqrt{n})}{s} \quad \text{and} \quad b_i = \frac{T_i - (\bar{x} - s/\sqrt{n})}{s}$$

Of course, $\bar{x}$ and s^2 are biased estimators of μ and σ^2, but we are only interested in y_i being covered by (a_i, b_i), not that it should necessarily be the midpoint of this interval.

To work out the MML equations, we consider the linear approximations

$$g(y_i) \simeq \alpha_i + \beta_i y_i \qquad (i = 1, 2, \ldots, k) \qquad (3.6.3)$$

where

$$\beta_i = \frac{g(b_i) - g(a_i)}{b_i - a_i} \quad \text{and} \quad \alpha_i = g(a_i) - \beta_i a_i$$

Substituting (3.6.3) in (3.6.1) and (3.6.2), we obtain the MML equations

$$\frac{\partial \log L}{\partial \mu} \simeq \frac{\partial \log L^*}{\partial \mu} = \frac{1}{\sigma}\left[\frac{n}{\sigma}(\bar{x} - \mu) + \sum_{i=1}^{k} r_i(\alpha_i + \beta_i y_i)\right] = 0 \qquad (3.6.4)$$

and

$$\frac{\partial \log L}{\partial \sigma} \simeq \frac{\partial \log L^*}{\partial \sigma} = \frac{1}{\sigma}\left\{\frac{n}{\sigma^2}[s^2 + (\bar{x} - \mu)^2 - 1] + \sum_{i=1}^{k} r_i y_i(\alpha_i + \beta_i y_i)\right\} = 0 \qquad (3.6.5)$$

The solutions of (3.6.4) and (3.6.5) give the MML estimators as

$$\hat{\mu} = K + \hat{\sigma}D \tag{3.6.6}$$

and $\hat{\sigma}$ is the positive root of the equation ($q_i = r_i/n$)

$$\sigma^2 - \left[\sum_{i=1}^{k} q_i\alpha_i(T_i - \mu)\right]\sigma - \left[s^2 + (\bar{x} - \mu)^2 + \sum_{i=1}^{k} q_i\beta_i(T_i - \mu)^2\right] = 0 \tag{3.6.7}$$

or

$$\sigma^2 - \left[(1 - K)\sum_{i=1}^{k} q_i\alpha_i - D\bar{x}\right]\sigma - \left[s^2 + \bar{x}^2 + \sum_{i=1}^{k} q_i\beta_i T_i^2 - \left(1 + \sum_{i=1}^{k} q_i\beta_i\right)K^2\right] = 0 \tag{3.6.8}$$

according as μ is known or unknown. Here

$$K = \frac{\bar{x} + \Sigma_{i=1}^{k} q_i\beta_i T_i}{1 + \Sigma_{i=1}^{k} q_i\beta_i} \quad \text{and} \quad D = \frac{\Sigma_{i=1}^{k} q_i\alpha_i}{1 + \Sigma_{i=1}^{k} q_i\beta_i}$$

Like all the earlier MML equations for estimating σ, (3.6.7) and (3.6.8) are of the form $x^2 - bx - c = 0$, $c > 0$, and therefore have only one positive root.

A further simplification of the MMLE above is effected by replacing (α_i, β_i) ($i = 1, 2, \ldots, k$) by a common value (α, β). This can be done by noting that for large n, T_i's are likely to be covered, all of them, by the interval (a, b) where

$$a = \frac{T_1 - (\bar{x} + s)/\sqrt{n}}{s} \quad \text{and} \quad b = \frac{T_k - (\bar{x} - s)/\sqrt{n}}{s} \tag{3.6.9}$$

Thus

$$\beta = \frac{g(b) - g(a)}{b - a} \quad \text{and} \quad \alpha = g(a) - \beta a \tag{3.6.10}$$

Replacing (α_i, β_i) ($i = 1, 2, \ldots, k$) in (3.6.6)–(3.6.8) by (α, β), we obtain the simplified MML estimators.

The approximate bias, variances, and the covariance of these MML estimators can be obtained as usual from the expected values $E(\partial \log L^*/\partial\mu)$ and $E(\partial \log L^*/\partial\sigma)$ and the second-order derivatives of log L*, exactly on the same lines as in Section 2.7; $E(\partial \log L^*/\partial\mu)$, $E(\partial \log L^*/\partial\sigma)$, $-E(\partial^2 \log L^*/\partial\mu^2)$, and so on, will, however, work in terms of the unknown parameters μ and σ, but for large n they can be replaced by the MML estimators $\hat{\mu}$ and $\hat{\sigma}$, as suggested by Cohen (1963), Harter and Moore (1966), and Tiku (1968b, c), among others.

The MML equations for estimating μ and σ from progressively censored samples of type II are identical with (3.6.1) and (3.6.2) except that (α_i, β_i) ($i = 1, 2, \ldots, k$) are to be obtained as solutions of the equations

$$\beta_i = \frac{g(d_i) - g(c_i)}{d_i - c_i} \quad \text{and} \quad \alpha_i = g(c_i) - \beta_i c_i$$

where c_i and d_i are determined by the following equations:

$$F(c_i) = p_i - \left[\frac{p_i(1 - p_i)}{n}\right]^{\frac{1}{2}} \quad \text{and} \quad F(d_i) = p_i + \left[\frac{p_i(1 - p_i)}{n}\right]^{\frac{1}{2}}$$

($p_i = i/n$). The justification for this procedure follows from the fact that z_i is a random variable and as n tends to infinity, z_i tends to t_i determined by the equation $F(t_i) = i/n$. For large n, therefore, z_i is likely to be covered by the interval (c_i, d_i).

EXAMPLE 3.6.1 A total of 316 specimens were placed under observation and their life spans recorded in days (Cohen, 1963, p. 337). Ten specimens were censored after 36.5 days and 10 more were censored after 44.5 days. Data for this progressively censored sample of type I were summarized as follows: $N = 316$, $n = 296$, $q_1 = q_2 = 10/296 = 0.0338$, $T_1 = 36.5$, $T_2 = 44.5$, $\bar{x} = 39.2703$, and $s^2 = 20.1634$. Substituting these values in (3.6.9) and (3.6.10), we obtain

$$\beta = 0.6475 \quad \text{and} \quad \alpha = 0.8030$$

Using these values of α and β in place of α_i and β_i ($i = 1, 2$) in (3.6.6)-(3.6.8), we obtain the MML estimates

$$\hat{\mu} = 39.560 \quad \text{and} \quad \hat{\sigma} = 4.599$$

with the following variances and covariance evaluated at $\mu = 39.560$ and $\sigma = 4.599$ (Tiku, 1968c):

$$V(\hat{\mu}) = 0.0686 \qquad \text{cov}(\hat{\mu}, \hat{\sigma}) = 0.0018 \qquad V(\hat{\sigma}) = 0.0358$$

Cohen (1963) obtained the ML estimates iteratively,

$$\hat{\hat{\mu}} = 39.583 \quad \text{and} \quad \hat{\hat{\sigma}} = 4.611$$

with the following variances and covariance evaluated at $\mu = 39.583$ and $\sigma = 4.611$:

$$V(\hat{\hat{\mu}}) = 0.069 \qquad \text{cov}(\hat{\hat{\mu}}, \hat{\hat{\sigma}}) = 0.002 \qquad V(\hat{\hat{\sigma}}) = 0.036$$

The agreement between the two approaches is remarkably close, as expected.

Some of the interesting references dealing with estimation problems in life testing are the following: Bain (1972, 1978), Barlow and Proschan (1975), Bartholomew (1957), Basu and Ghosh (1980), Billman et al. (1972), Canfield and Borgman (1975), Cox (1959), Davis (1952), Dixon (1960), Engelhardt and Bain (1979), Epstein (1958), Epstein and Tsao (1953), Fleming and Harrington (1980), Gehan and Thomas (1969), Govindarajulu (1964), Harter (1969), Harter and Moore (1966, 1967, 1968, 1976), Irwin (1942), Johnson (1974), Lagakos (1979), Lawless (1971, 1979), Lawless and Singhal (1980), Mann (1969, 1972), Mantel and Myers (1971), Mendenhall and Hader (1958), Nelson (1972), Tallis and Light (1968), Wolynetz (1974), Wright et al. (1978), and Zacks (1971).

3.7 MML ESTIMATION FOR TRUNCATED DISTRIBUTIONS

As we mentioned in the beginning of Chapter 2, one way of reducing the effects of the standard cumulants of a continuous long-tailed distribution, particularly its third and fourth standard cumulants λ_3 and λ_4, is to truncate moderately the tails of a distribution and sample this truncated distribution. The problem then is to estimate the location and scale parameters from a random sample $x_1, x_2, \ldots, x_n$ of size n from a truncated distribution of the type

$$\frac{1}{\sigma}\frac{f((x-\mu)/\sigma)}{F(u'')-F(u')} \qquad u' \leq x \leq u''$$

where $F(u) = (1/\sigma)\int_a^u f((x-\mu)/\sigma)\,dx$, a being the natural lower bound of x.

Truncated distributions also arise naturally in numerous experimental situations as follows (Cohen, 1957, 1959):

EXAMPLE 3.7.1 Right Truncation. To ensure meeting a maximum weight specification of 12 oz on a certain radio component for an aircraft installation, all production of this component is weighed and those units that exceed the specified maximum weight are discarded. For a random sample of 50 units selected from the accepted production, $\bar{x} = 9.35$ oz and $s^2 = 1.1494$; $n = 50$ and $u'' = 12$.

EXAMPLE 3.7.2 Left Truncation. To ensure meeting a lower specification limit of 0.1215 in. on the thickness of a certain insulating washer, all production of this component is sorted through go/no go gauges and all of thickness less than this value are discarded. For a random sample of 100 washers selected from the screened production, it is found that $\bar{x} = 0.12462$ and $s^2 = 2.1319 \times 10^{-6}$; $n = 100$ and $u' = 0.1215$.

EXAMPLE 3.7.3 Double Truncation. The entire production of a certain bushing is sorted through go/no go gauges with the result that items of diameter in excess of 0.6015 in. and those less than 0.5985 in. are discarded. For a random sample of 75 bushings selected from the screened production, $\bar{x} = 0.60015$ and $s^2 = 0.3762 \times 10^{-6}$; $n = 75$, $u' = 0.5985$, and $u'' = 0.6015$.

ML Estimators

Assuming that the underlying distribution is normal, and writing $z_i = (x_i - \mu)/\sigma$, $z' = (u' - \mu)/\sigma$, and $z'' = (u'' - \mu)/\sigma$, the ML equations for estimating μ and σ are given by

$$\frac{\partial \log L}{\partial \mu} = \frac{n}{\sigma}\left[\frac{1}{n}\sum_{i=1}^{n} z_i - g_1(z') + g_2(z'')\right] = 0 \tag{3.7.1}$$

and

$$\frac{\partial \log L}{\partial \sigma} = \frac{n}{\sigma}\left[-1 + \frac{1}{n}\sum_{i=1}^{n} z_i^2 - z'g_1(z') + z''g_2(z'')\right] = 0 \tag{3.7.2}$$

where $g_1(z') = f(z')/[F(z'') - F(z')]$ and $g_2(z'') = f(z'')/[F(z'') - F(z')]$; $f(z) = \exp(-\frac{1}{2}z^2)/\sqrt{2\pi}$ and $F(z) = \int_{-\infty}^{z} f(z)\,dz$.

Pearson and Lee (1908), Fisher (1931), Cohen (1955, 1963), Cohen and Woodward (1953), Halperin (1952), Samford and Taylor (1959), and Taylor (1973) devised procedures to solve these equations. However, these procedures involve some complicated functions and are essentially iterative in nature and, therefore, too laborious and time consuming; for a brief description of these procedures, one may refer to Tiku and Stewart (1977, pp. 1486-1489).

MML Estimators

From a plot it becomes clear that the values of $g_1(z')$ and $g_2(z'')$ when plotted against z' and z'' lie almost on a straight line over small intervals of z' and z''. We therefore seek approximations of the type

$$g_1(z') \simeq \alpha_1 - \beta_1 z' \quad \text{and} \quad g_2(z'') \simeq \alpha_2 + \beta_2 z''$$

These approximations when incorporated in (3.7.1) and (3.7.2) give the following MML equations:

$$\frac{\partial \log L}{\partial \mu} \simeq \frac{\partial \log L^*}{\partial \mu} = \frac{n}{\sigma}\left[\frac{1}{n}\sum_{i=1}^{n} z_i - (\alpha_1 - \beta_1 z') + (\alpha_2 + \beta_2 z'')\right] = 0 \tag{3.7.3}$$

and

$$\frac{\partial \log L}{\partial \sigma} \simeq \frac{\partial \log L^*}{\partial \sigma} = \frac{n}{\sigma}\left[-1 + \frac{1}{n}\sum_{i=1}^{n} z_i^2 - z'(\alpha_1 - \beta_1 z') + z''(\alpha_2 + \beta_2 z'')\right] = 0 \tag{3.7.4}$$

To determine (α_1, β_1) and (α_2, β_2), the difficulty is that $z' = (u' - \mu)/\sigma$ and $z'' = (u'' - \mu)/\sigma$ are not known, although u' and u'' are. For large n, however, z' is likely to be covered by the interval $[(u' - h)/s, (u' - k)/s]$, where

$$h = \bar{x} + \frac{s}{\sqrt{n}} \quad \text{and} \quad k = \bar{x} - \frac{s}{\sqrt{n}}$$

and similarly for z''; $\bar{x} = \Sigma_{i=1}^{n} x_i/n$ and $s^2 = \Sigma_{i=1}^{n} (x_i - \bar{x})^2/(n - 1)$. Of course, $\bar{x}$ and s^2 are biased estimators of μ and σ^2, respectively, but we are only interested in z' being covered by the interval above, not that it should necessarily be its midpoint. Thus

$$\beta_1 = \frac{s}{k - h}\left[\frac{f\left(\frac{u' - k}{s}\right)}{D(k)} - \frac{f\left(\frac{u' - h}{s}\right)}{D(h)}\right] \quad \text{and} \quad \alpha_1 = \frac{f\left(\frac{u' - h}{s}\right)}{D(h)} + \beta_1\left(\frac{u' - h}{s}\right) \tag{3.7.5}$$

and, similarly,

$$\beta_2 = \frac{s}{h - k}\left[\frac{f\left(\frac{u'' - k}{s}\right)}{D(k)} - \frac{f\left(\frac{u'' - h}{s}\right)}{D(h)}\right] \quad \text{and} \quad \alpha_2 = \frac{f\left(\frac{u'' - h}{s}\right)}{D(h)} - \beta_2\left(\frac{u'' - h}{s}\right) \tag{3.7.6}$$

where

$$D(c) = F\left(\frac{u'' - c}{s}\right) - F\left(\frac{u' - c}{s}\right)$$

If there is no truncation on the left, then $\alpha_1 = \beta_1 = 0$, and if there is no truncation on the right, then $\alpha_2 = \beta_2 = 0$.

As n tends to infinity, $(u' - h)/s$ and $(u' - k)/s$ both tend to $z^* = [u' - E(\bar{x})]/E(s)$ and, consequently, β_1 and α_1 tend to β_1^* and α_1^*, respectively, where

$$\beta_1^* = -\left[\frac{d}{dz} g_1(z)\right]_{z=z^*} = \frac{z^* f(z^*)}{F(z^{**}) - F(z^*)} \quad \text{and} \quad \alpha_1^* = \frac{f(z^*)}{F(z^{**}) - F(z^*)} + \beta_1^* z^*$$

and similarly for β_2^* and α_2^*; $z^{**} = [u'' - E(\bar{x})]/E(s)$. Here (Johnson and Kotz, 1970, p. 81)

$$z^* = \frac{z' - [g_1(z') - g_2(z'')]}{\{1 + z' g_1(z') - z'' g_2(z'') - [g_1(z') - g_2(z'')]^2\}^{\frac{1}{2}}}$$

and similarly for z^{**}.

The values of $g_1(z')$ and $g_2(z'')$ and the differences (1) $g_1(z') - (\alpha_1^* - \beta_1^* z')$ and (2) $g_2(z'') - (\alpha_2^* + \beta_2^* z'')$ for numerous values of the truncated points z' and z'' are as follows:

z'	z''	F(z'') - F(z')	$g_1(z')$	(1)	$g_2(z'')$	(2)
$-\infty$	2.32	0.99	0	0	0.0273	0.001
	1.96	0.975	0	0	0.0569	-0.000
	1.64	0.95	0	0	0.1095	0.012
	1.28	0.90	0	0	0.1954	0.032
	0.84	0.80	0	0	0.3506	0.084
	0.00	0.50	0	0	0.7978	0.292
-2.32	2.32	0.98	0.0276	0.001	0.0276	0.001
	1.64	0.94	0.0278	0.005	0.1107	0.015
	0.84	0.79	0.0342	0.014	0.3551	0.096
-1.64	1.28	0.85	0.1225	0.034	0.2070	0.055

It is clear that differences (1) and (2) are negligible if the truncated area is small, say $T(z', z'') = 1 - [F(z'') - F(z')] \leq 0.20$. Consequently, for large n and $T(z', z'') \leq 0.20$, the MML estimators are almost identical with the ML estimators.

If the truncated area $T(z', z'') > 0.20$, the MML estimates should be obtained iteratively; in the first calculation one uses the estimates $\bar{x}$ and s in (3.7.5) and (3.7.6) and calculates the MML estimates $\hat{\mu}$ and $\hat{\sigma}$ given by (3.7.3) and (3.7.4) as

$$\hat{\mu} = K + D\sigma \quad (\sigma \text{ to be replaced by } \hat{\sigma}) \tag{3.7.7}$$

and $\hat{\sigma}$ is the positive root of the equation

$$\sigma^2 - [\alpha_2(u'' - \mu) - \alpha_1(u' - \mu)]\sigma - \left[\frac{1}{n}\sum_{i=1}^{n} (x_i - \mu)^2 + \beta_1(u' - \mu)^2 + \beta_2(u'' - \mu)^2\right] = 0 \tag{3.7.8}$$

or

$$\sigma^2 - [(\alpha_2 u'' - \alpha_1 u') - (1 + \beta_1 + \beta_2)KD]\sigma - \left[\frac{1}{n}\sum_{i=1}^{n} x_i^2 + \beta_1 u'^2 + \beta_2 u''^2 - (1 + \beta_1 + \beta_2)K^2\right] = 0 \tag{3.7.9}$$

according as μ is known or unknown. Here

$$K = \frac{\bar{x} + \beta_1 u' + \beta_2 u''}{1 + \beta_1 + \beta_2} \quad \text{and} \quad D = \frac{\alpha_2 - \alpha_1}{1 + \beta_1 + \beta_2} \tag{3.7.10}$$

In the first iteration one uses the estimates $\hat{\mu}$ and $\hat{\sigma}$ in place of $\bar{x}$ and s in (3.7.5) and (3.7.6) and repeats the calculations above, and so on. The estimates stabilize very quickly in one or two iterations, as we have already illustrated in Section 3.4.

In practice, one might not know whether the truncated area T(z', z'') is small or not; therefore, it is advisable to do one or two iterations always. These iterations can be carried out very easily since $\hat{\mu}$ and $\hat{\sigma}$ are explicit functions of sample observations. Note that these iterated MML estimates $\hat{\mu}$ and $\hat{\sigma}$ are asymptotically identical with the ML estimates $\hat{\hat{\mu}}$ and $\hat{\hat{\sigma}}$ (if one endeavors to calculate them) since the differences (1) and (2) above reduce to zero quickly in one or two iterations.

The approximate bias in $\hat{\mu}$ and $\hat{\sigma}$ and their approximate variances and covariance are given by the expected values of the first and second derivatives of log L* with respect to μ and σ; these are particular cases, obtained by putting $k = 1$ in the more general expressions given by Tiku and Stewart (1977).

Returning to Examples 3.7.1 to 3.7.3, we calculated the MML estimates $\hat{\mu}$ and $\hat{\sigma}$ directly from the values of $\bar{x}$ and s substituted in (3.7.5) and (3.7.6) and without doing any further iterations as suggested above and obtained the following:

In Example 3.7.1, $n = 50$, $\bar{x} = 9.35$, $s = 1.072101$; $u'' = 12$; $\alpha_1 = 0$, $\beta_1 = 0$; $\alpha_2 = 0.12872$, and $\beta_2 = -0.04443$. Substituting these values in (3.7.7) and (3.7.9), we obtain the MML estimates

$$\hat{\mu} = 9.3737 \quad \text{and} \quad \hat{\sigma} = 1.0905$$

with

$$V(\hat{\mu}) \simeq 0.02515 \qquad \text{cov}\,(\hat{\mu}, \hat{\sigma}) \simeq 0.00192 \qquad V(\hat{\sigma}) \simeq 0.01422$$

evaluated at $\mu = 9.3737$ and $\sigma = 1.0905$.

These may be compared with Cohen's (1959) ML estimates

$$\hat{\hat{\mu}} = 9.38 \quad \text{and} \quad \hat{\hat{\sigma}} = 1.091$$

with

$$V(\hat{\hat{\mu}}) \simeq 0.027 \qquad \text{cov}\,(\hat{\hat{\mu}}, \hat{\hat{\sigma}}) \simeq 0.004 \qquad V(\hat{\hat{\sigma}}) \simeq 0.016$$

evaluated at $\mu = 9.38$ and $\sigma = 1.091$.

In Example 3.7.2, $n = 100$; $\bar{x} = 0.12462$, $s = 1.4601 \times 10^{-3}$; $u' = 0.1215$; $\alpha_1 = 0.23424$, $\beta_1 = -0.08989$; $\alpha_2 = 0$, and $\beta_2 = 0$. The MML estimates are

$$\hat{\mu} = 0.12455 \quad \text{and} \quad \hat{\sigma} = 0.00148$$

with

$$V(\hat{\mu}) \simeq 2.5063 \times 10^{-8} \qquad \text{cov}\,(\hat{\mu}, \hat{\sigma}) \simeq -0.059 \times 10^{-8} \qquad V(\hat{\sigma}) \simeq 1.5034 \times 10^{-8}$$

Cohen's (1959) ML estimates are

$$\hat{\hat{\mu}} = 0.1245 \quad \text{and} \quad \hat{\hat{\sigma}} = 0.00155$$

with

$$V(\hat{\hat{\mu}}) \simeq 2.98 \times 10^{-8} \qquad \text{cov}\,(\hat{\hat{\mu}}, \hat{\hat{\sigma}}) \simeq -0.065 \times 10^{-8} \qquad V(\hat{\hat{\sigma}}) \simeq 1.85 \times 10^{-8}$$

In Example 3.7.3, $n = 75$; $\bar{x} = 0.6015$, $s = 0.6133 \times 10^{-3}$; $u' = 0.5985$, $u'' = 0.6015$; $\alpha_1 = 0.08974$, $\beta_1 = -0.02918$; $\alpha_2 = 0.21326$, and $\beta_2 = -0.08021$. The MML estimates are

$$\hat{\mu} = 0.60018 \quad \text{and} \quad \hat{\sigma} = 0.00068$$

with

$$V(\hat{\mu}) \simeq 0.6922 \times 10^{-8} \qquad \text{cov}(\hat{\mu}, \hat{\sigma}) \simeq 0.063 \times 10^{-8} \qquad V(\hat{\sigma}) \simeq 0.452 \times 10^{-8}$$

Cohen's (1957) ML estimates are

$$\hat{\hat{\mu}} = 0.60018 \quad \text{and} \quad \hat{\hat{\sigma}} = 0.00066$$

with

$$V(\hat{\hat{\mu}}) \simeq 0.6819 \times 10^{-8} \qquad \text{cov}(\hat{\hat{\mu}}, \hat{\hat{\sigma}}) \simeq 0.099 \times 10^{-8} \qquad V(\hat{\hat{\sigma}}) \simeq 0.5471 \times 10^{-8}$$

The agreement between the two methods is remarkably close, as expected.

Note that in all the examples above

$$\begin{vmatrix} V(\hat{\sigma}) & \text{cov}(\hat{\mu}, \hat{\sigma}) \\ \text{cov}(\hat{\mu}, \hat{\sigma}) & V(\hat{\sigma}) \end{vmatrix} < \begin{vmatrix} V(\hat{\hat{\sigma}}) & \text{cov}(\hat{\hat{\mu}}, \hat{\hat{\sigma}}) \\ \text{cov}(\hat{\hat{\mu}}, \hat{\hat{\sigma}}) & V(\hat{\hat{\sigma}}) \end{vmatrix}$$

and this kind of phenomenon is not new; it occurred before in Smith et al. evaluations for the log-normal distribution (see Table 3.2.1), indicating that MMLE can be somewhat more efficient than the corresponding MLE for small samples.

The MMLE for truncated distributions also has some interesting applications in experimental design models; for details, see Tiku and Stewart (1977). Note that the MMLE obtained in this chapter can be used to develop hypothesis-testing procedures which are expected to be robust to departures from the respective distribution; this is discussed in the next chapter with particular reference to departures from a normal distribution.

It may be noted that all the MML equations above are asymptotically identical with the corresponding ML equations. Moreover, they are exactly of the same form as the MML equations (2.7.8) and (2.7.9) based on type II censored normal samples. It follows, therefore, that the MML estimators obtained in this chapter have the same asymptotic properties as the MML estimators $\hat{\mu}_c$ and $\hat{\sigma}_c$ given in Chapter 2. In particular, they are asymptotically unbiased and minimum variance estimators.

3.8 CONCLUSIONS AND SUMMARY

In this chapter the problem of estimation of the parameters of the log-normal, logistic, exponential, and Rayleigh distributions is discussed. It is shown that the ML equations based on type II censored samples do not admit explicit solutions and solving them iteratively can become problematic; MML equations are therefore defined and the MMLEs are obtained as explicit functions of sample observations. These MMLEs are asymptotically identical with the MLEs, and are for small samples very close to the MLEs numerically. Examples are given to illustrate the closeness of the ML and MML estimates. The MMLEs given in this chapter may be used in constructing robust hypothesis-testing procedures exactly on the same lines as in the next two chapters. It is hoped that these hypothesis-testing procedures will turn out to be robust to departures from the respective distribution; however, more research is needed to establish this point. The robustness properties of various hypothesis-testing procedures based on the MMLEs calculated from type II censored normal samples are discussed in the next two chapters; these procedures are shown to be remarkably robust to departures from normality.

EXERCISES

3.1 Let $x_{(i)}$, $i = 1, 2, \ldots, n$, be the order statistics of a random sample of size n from the logistic distribution, $z = \pi(x - \mu)/\sigma\sqrt{3}$,

$$p(z) = \frac{e^{-z}}{(1 + e^{-z})^2} \qquad -\infty < z < \infty$$

Then $(a = r_1 + 1,\ b = n - r_2)$

$$\lim_{n \to \infty} \frac{1}{n} E \sum_{i=a}^{b} z_{(i)} = \int_h^k z\,p(z)\,dz$$

and

$$\lim_{n \to \infty} \frac{1}{n} E \sum_{i=a}^{b} z_{(i)}^2 = \int_h^k z^2 p(z)\,dz$$

where h and k are determined by the equations $(q_1 = r_1/n,\ q_2 = r_2/n)$

$$F(h) = q_1 \quad \text{and} \quad F(k) = 1 - q_2 \qquad F(z) = \int_{-\infty}^{z} p(z)\,dz$$

(a) Show that

$$\int_h^k z\,p(z)\,dz = h(1 - q_1) - kq_2 + \log\left(\frac{1 - q_2}{q_1}\right) \qquad (q_1 > 0)$$

and

$$\int_h^k z^2 p(z)\,dz = k^2(1 - q_2) - h^2 q_1 - 2\left[\int_0^k z(1 + e^{-z})^{-1}\,dz + \int_h^0 z(1 + e^{-z})^{-1}\,dz\right]$$

(b) Show that the integral $\int_0^t z(1 + e^{-z})^{-1} dz$ can be expressed as a rapidly convergent series as follows:

$$\int_0^t z(1+e^{-z})^{-1}dz = \frac{1}{2}t^2 + t\left(e^{-t} - \frac{1}{2}e^{-2t} + \frac{1}{3}e^{-3t} - \cdots\right) + \left(e^{-t} - \frac{1}{4}e^{-2t} + \frac{1}{9}e^{-3t} - \cdots\right) - \left(1 - \frac{1}{4} + \frac{1}{9} - \cdots\right) \quad \text{for } t > 0$$

$$= 0 \quad \text{for } t = 0$$

$$= t\left(e^{t} - \frac{1}{2}e^{2t} + \frac{1}{3}e^{3t} - \cdots\right) - \left(e^{t} - \frac{1}{4}e^{2t} + \frac{1}{9}e^{3t} - \cdots\right) - \left(1 - \frac{1}{4} + \frac{1}{9} - \cdots\right) \quad \text{for } t < 0$$

where $(1 - 1/4 + 1/9 - \cdots) = 0.8225\ldots = \pi^2/12$.

(c) Calculate the values of the integrals in part (a) for $q_1 = q_2 = 0.1$ and thereby show that the MML10 estimators $\hat{\mu}$ and $\hat{\sigma}$ given by (3.3.5) and (3.3.6) are asymptotically unbiased.

3.2 Consider a part of the data given in the insulating fluid example in Nelson (1982, p. 462):

Sample 1:	0.00	0.18	0.55	0.66	0.71	1.30	1.63	2.17	2.75	···
Sample 2:	0.02	0.06	0.50	0.70	1.17	2.80	3.57	3.70	3.72	···
Sample 3:	0.20	0.60	0.78	1.08	1.13	2.44	3.17	5.55	6.63	···

with the largest observation censored in each sample. These data are assumed to come from an exponential distribution $E(0, \sigma)$.

(a) Calculate the BLUE and the MMLE of σ and show that they are exactly the same.

(b) Suppose that one smallest observation in each of the three samples was also censored. Calculate the BLUE and the MMLE of σ and the standard errors of these estimates. Show that the MMLE has smaller standard error than the BLUE.

3.3 In the data of Example 3.2.1, suppose that the two largest observations were censored. Compute the MMLEs of the parameters μ, σ, and τ of a log-normal distribution $LN(\mu, \sigma, \tau)$. Do they seem to be seriously biased?

3.4 To ensure meeting a maximum weight specification of 12 oz on a certain radio component for an aircraft installation, all production of this component is weighed and those units that exceed the specified maximum weight are discarded. Two machines are used in the production and for random samples each of 50 units selected from the two accepted productions,

Machine I: $\bar{x}_1 = 9.3500$ and $s_1^2 = 1.1493$

Machine II: $\bar{x}_2 = 8.9800$ and $s_2^2 = 1.1512$

(Tiku and Stewart, 1977). Assuming the linear model for the production, namely,

$$x_{ij} = \mu + g_i + e_{ij} \qquad i = 1, 2, \; j = 1, 2, \ldots, n \qquad (n = 50)$$

where e_{ij} are independently distributed as truncated normals

$$\frac{(\sqrt{2\pi}\sigma)^{-1} \exp[-(x_{ij} - \mu - g_i)^2/2\sigma^2]}{F(z_i'')} \qquad x_{ij} < u_i'' \qquad (u_i'' = 12.0)$$

$z_i'' = (u_i'' - \mu - g_i)/\sigma$, estimate g_1 and g_2, and test the null hypothesis $H_0: g_1 = g_2$. Show that at 5% significance level, H_0 would be rejected against $H_1: g_1 > g_2$; therefore, machine I needs correcting.

4
Robust Tests

4.1 INTRODUCTION

Although it is very convenient to assume normality in many practical situations (Scheffé, 1959, Chap. 10), the fact is that the underlying distribution is hardly ever exactly normal. This fact was recognized about 53 years ago by Pearson and Adyanthaya (1929) and Pearson (1932). To quote Geary (1947) again: "Normality is a myth; there never was, and never will be, a normal distribution." Accepting the fact that the underlying distribution is usually nonnormal, the difficulty is that we do not know its functional form exactly and there is no way of locating this form with certainty through formal (goodness-of-fit tests, for example; Chapter 6) or informal [plotting techniques, for example (Lawless, 1982, Sec. 2.4)] procedures. It is only realistic, therefore, to assume that the underlying distribution is one of a reasonably wide class of distributions, which one exactly is not known; see also Sprott (1982) for similar remarks. Thus in most practical situations where we find it expedient to assume normality, what we should perhaps assume instead is that the underlying distribution no doubt has finite mean and variance, but is (1) either a symmetric distribution with shorter or longer tails than a normal, or (2) is a somewhat skewed distribution. Very short-tailed distributions like the uniform (which in fact has no tails) or very long-tailed distributions like the Cauchy (which has infinite mean and infinite variance) or very skewed distributions like the exponential might not be too realistic as alternatives to normality. Other realistic possibilities are (3) that while the bulk of observations in a sample are "good," a small number of them are "bad" observations (e.g., outliers). What we are in search of are inference procedures for location (mean) and scale (standard deviation) which are optimal or nearly optimal for the normal distribution but maintain high efficiency under alternatives (1), (2), and (3) to normality and normal samples. We also require that these inference procedures, or simple variants of them, maintain high efficiency for distributions other than (1) and (2), for example, very long-tailed or very skewed distributions, or when the sample contains a rather large number of outliers. In this chapter we illustrate how such inference procedures can be obtained by using the MMLE based on type II censored normal samples. We discuss the robustness properties of these procedures and the corresponding procedures based on other robust estimators, namely, Huber estimators BS82, W24, and so on.

4.2 ROBUST STUDENTIZED ONE-SAMPLE TESTS

Let $x_1, x_2, \ldots, x_n$ be a random sample from a distribution with mean μ and variance σ^2. The problem is to test the null hypothesis $H_0: \mu = 0$. In Chapter 1 we have shown that the classical t test for testing H_0 is asymptotically robust with respect to departures from normality. For finite n, the t test does possess type I robustness (Chapter 1) if the underlying distribution is symmetric and departs only moderately from the normal distribution. In this section we develop a robust t test which has substantially the same power as the classical t test for the normal distribution but has considerably higher power than the classical t test for symmetric long-tailed distributions. Such a test is constructed by using the MMLE based on type II censored normal samples. This purports to reflect the following premise of Tiku (1980a, 1982a), which has already been substantiated in Chapter 2

> Non-normality essentially comes from the tails and once the extreme observations (representing these tails) are censored, there is hardly any difference between a normal sample and a non-normal sample and in that situation whatever is good for normal is automatically good for non-normal samples.

This statement is of course in the spirit of the following colorful remark due to Edgeworth (1887):

> The method of least squares is seen to be our best course when we have thrown overboard a certain portion of our data—a sort of sacrifice which has often to be made by those who sail upon the stormy seas of probability.

Consider now a symmetric type II censored sample $x_{(r+1)} < \cdots < x_{(n-r)}$ and denote the MMLE matching pair by $(\hat{\mu}, \hat{\sigma})$; the expressions for $\hat{\mu}$ and $\hat{\sigma}$ are given in Chapter 2 [equations (2.7.18) and (2.7.19)]. Analogous to the classical Student t statistic, we define

$$t_c = \sqrt{m}\,\frac{\hat{\mu}}{\hat{\sigma}} \qquad (4.2.1)$$

[$m = n - 2r + 2r\beta$ (see Chapter 2)]. One may note that if $r = 0$, then t_c reduces to the classical t statistic which has, under the assumption of normality, a central t distribution with $n - 1$ degrees of freedom if, of course, the null hypothesis H_0 is true.

Choice of r

In calculating the statistic t_c and the subsequent Studentized statistics based on the MMLE $\hat{\mu}$ and $\hat{\sigma}$, the best choice of r would be the following (Tiku, 1980a, 1982a):

1. $r = [0.5 + 0.1n]$ for plausible alternatives (1), (2), and (3), mentioned in the beginning of this chapter, to normality and normal samples (see also Stigler, 1977, and Spjøtvall and Aastvett, 1980).
2. $r = 0$ for extreme short-tailed distributions.
3. $r = [0.5 + 0.3n]$ for "disaster" distributions.
4. For extreme skew distributions, a more appropriate procedure is obtained by censoring a sample on only one side, in the direction of the long tail; this is discussed in Section 4.4.

It can be argued that in routine data analysis, one cannot perhaps be expected to exercise a judgment as to which family (1 to 4) the underlying distribution belongs to. In such a situation, choose $r = [0.5 + 0.1n]$ always. Thus, in what follows, we will take $r = [0.5 + 0.1n]$ always, unless stated otherwise, and refer our MMLE as MML10 to indicate 10% censoring on each side. This notation was in fact used by Dunnett (1982).

Using the t_c statistic (4.2.1), one then rejects H_0 in favor of H_1: $\mu \neq 0$ if $|t_c| \geq C_{\alpha^*/2}(0)$ for some constant $C_{\alpha^*/2}(0)$. One rejects H_0 in favor of H_1: $\mu > 0$ if $t_c > C_{\alpha^*}(R)$ for some constant $C_{\alpha^*}(R)$; similarly, one rejects H_0 in favor of H_1: $\mu < 0$ if $t_c \leq C_{\alpha^*}(L)$ for some constant $C_{\alpha^*}(L)$. The constants $C_{\alpha^*/2}(0)$, $C_{\alpha^*}(R)$, and $C_{\alpha^*}(L)$ are determined by a preassigned significance level α^* and the null distribution of t_c, which is given by Theorem 4.2.1. Before we state and prove this theorem, we take note of the following lemma. For a discussion and proof of this lemma, the reader is referred to David (1981, Sec. 9.6).

LEMMA 4.2.1 Let $T = \Sigma_j c_j x_{j:n}$ be a linear function of order statistics $x_{j:n}$ of a random sample of size n. Then, under some very general regularity conditions, T converges to normal in distribution as $n \to \infty$.

In the lemma, the regularity conditions are usually the existence of the variance of T and c_j's have to be bounded uniformly. For the MMLE, these conditions are expected to be satisfied unless, of course, the underlying distribution has a discontinuity toward the middle. The distributions of the latter type are, however, very rare in practice and should not cause much concern. In Chapters 4 and 5 the phrase "under very general regularity conditions" will mean conditions for which Lemma 4.2.1 holds, unless stated otherwise.

THEOREM 4.2.1 If the underlying distribution is normal, the asymptotic null distribution of t_c is a central t distribution with A - 1 degrees of freedom, where $A = n - 2r$ (r is fixed or $q = r/n$ is fixed), and the asymptotic nonnull distribution of t_c is a noncentral t distribution with A - 1 degrees of freedom and noncentrality parameter $\Delta_c^2 = m\mu^2/\sigma^2$.

Proof: Theorem 4.2.1 follows immediately from Lemma 2.7.2.

One would expect Theorem 4.2.1 to hold as a close approximation for small n (say, $n \geq 10$). This is due to the fact that the likelihood function L is closely approximated by L* (Chapter 2).

THEOREM 4.2.2 The asymptotic distribution of t_c is normal, under some very general regularity conditions.

Proof: As n tends to infinity (r or q fixed), $\hat{\sigma}$ tends to its expected value $E(\hat{\sigma})$ in probability, in which case the normality of $\sqrt{m}\,\hat{\mu}/\hat{\sigma}$, a linear function of order statistics, follows from Lemma 4.2.1.

The theorems above, and all the subsequent theorems, are true in both the situations when r is fixed or $q = r/n$ is fixed. In the context of robustness, however, $q = r/n$ is fixed, in which case $A = n(1 - 2q)$ and q is usually taken to be equal to 0.1, as explained earlier.

It follows from Theorem 4.2.1 that under the assumption of normality, the constants $C_{\alpha^*/2}(0)$, $C_{\alpha^*}(R)$, and $C_{\alpha^*}(L)$ are, for large A, equal to $t_{\alpha^*/2}(\nu)$, $t_{\alpha^*}(\nu)$, and $-t_{\alpha^*}(\nu)$ ($\nu = A - 1$), respectively, where $t_{\alpha^*}(\nu)$ is the $100(1 - \alpha^*)\%$ point of a central t distribution with ν degrees of freedom; these provide accurate approximations to the true $100(1 - \alpha^*)\%$ points even for small A, as indicated by results given by Tiku (1980a, 1982a) and Tan (1984a). This is not surprising since L* approximates L very closely (Tan, 1984a). From Theorem 4.2.2 it follows that the asymptotic power function of the t_c test for testing H_0 versus $H_1: \mu \neq 0$ and for testing H_0 versus $H_1: \mu > 0$ are given, respectively, by $P(|Z + \Delta_c| \geq Z_{\alpha^*/2})$ and $P(Z \geq Z_{\alpha^*} - \Delta_c)$, where $\Delta_c = \mu/\sqrt{\text{var}(\hat{\mu})}$; Z is a standard normal variate and Z_{α^*} is its $100(1 - \alpha^*)\%$ point. Note that under the assumption of normality, Δ_c is for large samples exactly equal to $\mu\sqrt{m}/\sigma = 0.983$, since $\text{var}(\hat{\mu}) = \sigma^2/m$. In comparing the t_c test with other tests, in what follows we will consider only the alternatives $H_1: \mu > 0$ for conciseness. The results for the other two alternatives are exactly similar. Now, under the assumption of normality, the power function of the t_c test for large samples is given by

$$1 - \beta^* = P(Z \geq Z_{\alpha^*} - \Delta_c) \quad (4.2.2)$$

where $\Delta_c = \mu\sqrt{m}/\sigma$.

Comparison of the t_c Test with the Classical t Test

The power function of Student's t test is given by

$$P(Z \geq Z_{\alpha*} - \Delta) \qquad \Delta = \frac{\mu\sqrt{n}}{\sigma} \tag{4.2.3}$$

obtained simply by putting r = 0 in (4.2.2).

Using (4.2.2) and (4.2.3), one may compare the power values of the t_c test with those of the t test for testing H_0 versus $H_1: \mu > 0$.

Normal Distribution. For a normal distribution, we give in Table 4.2.1 the values of the power calculated from (4.2.2) and (4.2.3). It is clear from Table 4.2.1 that under the assumption of normality, the t_c test is almost as powerful as the optimal t test, as expected.

Symmetric Nonnormal Distributions. For symmetric nonnormal distributions, $E(\hat{\mu}) = \mu$, and the power function of t_c for testing H_0 versus H_1: $\mu > 0$ is for large A (hence large n) given by

$$1 - \beta* = P(Z \geq Z_{\alpha*} - \Delta_c) \qquad \Delta_c = \frac{\mu}{\sqrt{\text{var}(\hat{\mu})}} \simeq \frac{\mu\sqrt{m}}{E(\hat{\sigma})} \tag{4.2.4}$$

since $\sigma(t_c) = \sqrt{m\ \text{var}(\hat{\mu})}/E(\hat{\sigma})$ is almost 1 (see Table 4.2.2). The corresponding power function of Student's t is given by (4.2.3), for all populations with finite mean and finite variance. Comparing (4.2.3) with (4.2.4), the t_c test will therefore be more powerful than the t test if $\Delta_c > \Delta$. This condition is satisfied by most long-tailed symmetric distributions. To serve as an example, we consider the t family given by (2.3.1). For this family the $E(\hat{\sigma})$ values can readily be computed by using the tables of Tiku and Kumra (1981). We give in Table 4.2.2 the $E(\hat{\sigma})$ values for this family for sample size n = 20.

Table 4.2.1 Values of the Power for the t and t_c Tests

Δ:	0	0.5	1	1.5	2	2.5	3	3.5	4
t	0.05	0.13	0.26	0.44	0.64	0.80	0.91	0.97	0.99
t_c	0.05	0.12	0.25	0.43	0.63	0.79	0.90	0.96	0.99
t	0.01	0.03	0.09	0.20	0.37	0.57	0.75	0.88	0.95
t_c	0.01	0.03	0.09	0.20	0.36	0.55	0.73	0.87	0.95

Table 4.2.2 Values of $E(\hat{\sigma})$ for the Family (2.3.1); n = 20

	p: 2	2.5	3.5	4.5	5	∞^a
			q = 0.1			
$E(\hat{\sigma})/\sigma$	0.732	0.838	0.909	0.936	0.944	0.995
$\sigma(t_c)$	1.06	1.05	1.03	1.02	1.02	1.01
			q = 0.2			
$E(\hat{\sigma})/\sigma$	0.673	0.790	0.874	0.908	0.919	0.988
$\sigma(t_c)$	1.06	1.05	1.03	1.03	1.03	1.02

[a]This is in fact a normal distribution.

Using results from Table 4.2.2, we have the following values of Δ_c/Δ for n = 20 for the family (2.3.1); q = r/n = 0.1:

	p: 2	2.5	3.5	4.5	5	∞
Δ_c/Δ	1.34	1.17	1.08	1.05	1.04	0.988

It is clear that for long-tailed symmetric distributions such as t_3 (central t with 3 degrees of freedom), the t_c test is considerably more powerful than the t test. Note that if Δ_c and Δ are zero (i.e., $\mu = 0$), the power $1 - \beta^*$ reduces to the significance level α^*; Δ_c^2 and Δ^2 are usually called noncentrality parameters.

Skew Nonnormal Populations. For skew populations, $\hat{\sigma}$ also tends to its expected value $E(\hat{\sigma})$ in probability as A becomes large (r or q fixed) in which case $\hat{\mu}\sqrt{m}/\hat{\sigma}$, being a linear function of order statistics, is normally distributed; its noncentrality parameter is given by

$$\Delta_c \simeq \frac{(\mu + \text{bias})\sqrt{m}}{E(\hat{\sigma})} \tag{4.2.5}$$

For distributions that have only a small amount of skewness, the bias is negligible (Tiku, 1982a, Table I) and the noncentrality parameter reduces to $\Delta_c \simeq \mu\sqrt{m}/E(\hat{\sigma})$. The noncentrality parameter of Student's t is Δ^2, where as usual $\Delta = \mu\sqrt{n}/\sigma$ since $\bar{x}$ is unbiased and s tends to σ as n becomes large. The ratio Δ_c/Δ is always less than 1 for most skew distributions and therefore the t_c test is at least as powerful as the t test. As an example, we consider the family of gamma distributions given by (2.3.2). For this family, the values of $E(\hat{\sigma})$ and $\sigma(t_c) = \sqrt{mV(\hat{\mu})}/E(\hat{\sigma})$ can be calculated from the values of $E(Z_i)$ and $E(Z_iZ_j)$ $[Z_i = (X_{(i)} - \theta)/\delta]$ (see Appendix B). For n = 20 and q = r/n = 0.1, for example, we have the following values (see also Tiku, 1982a, Table I):

a:	4	3	2	1	0.5
$E(\hat{\sigma})/\delta$	1.93	1.66	1.33	0.885	0.561
$\sigma(t_c)$	1.01	1.02	1.02	1.05	1.09

Thus we obtain the following values of Δ_c/Δ, which indicate that the t_c test is at least as powerful as the t test:

	n		
a	20	40	80
4	1.02	1.02	1.02
3	1.03	1.03	1.03

Comparison of the t_c Test with the Test Based on Trimmed Sample Estimators

Consider the trimmed matching pair of estimators $(\hat{\mu}_T, \hat{\sigma}_T)$ based on a type II symmetrically censored sample of size n - 2r with r = [0.5 + 0.1n]. The studentized statistic based on this pair is given by

$$t_{trim} = \frac{\hat{\mu}_T\sqrt{n - 2r}}{\hat{\sigma}_T} \tag{4.2.6}$$

As A = n - 2r (r or q fixed) becomes large, $\hat{\sigma}_T$ tends to its expected value $E(\hat{\sigma}_T)$ and the distribution of t_{trim} tends to normal by Lemma 4.2.1; its noncentrality parameter for symmetric populations is given by Δ^2_{trim}, where

$$\Delta_{trim} \simeq \frac{\mu\sqrt{n - 2r}}{E(\hat{\sigma}_T)} \tag{4.2.7}$$

For large A, $E(\hat{\sigma}_T) \simeq \sqrt{E(\hat{\sigma}^2_T)}$ and $E(\hat{\sigma}^2_T)$ can be calculated from the formula for $\hat{\sigma}^2_T$ given in Section 2.8. For the normal population, for example, the ratio $\Delta_{trim}/\Delta = 0.971$ for n = 20 (q = r/n = 0.1); the t_{trim} test is, therefore, somewhat less powerful than the t_c test for the normal population, as expected; see also Table 4.2.3.

Table 4.2.3 Simulated Values of the Power; n = 10 (q = r/n = 0.1)

μ:	0.0	0.3	0.5	1.0	2.0	0.0	0.3	0.5	1.0	2.0
	Normal					Logistic				
t	0.050	0.22	0.42	0.90	1.00	0.051	0.14	0.22	0.50	0.94
t_{trim}	0.051	0.21	0.40	0.88	1.00	0.051	0.13	0.22	0.50	0.94
t_c	0.051	0.22	0.42	0.89	1.00	0.051	0.14	0.23	0.51	0.94
	(n - 1)N(0, 1) and 1N(0, 10)					Student's t (df 2)				
t	0.024	0.09	0.17	0.44	0.73	0.041	0.14	0.22	0.54	0.84
t_{trim}	0.049	0.18	0.33	0.77	1.00	0.045	0.15	0.27	0.65	0.94
t_c	0.054	0.19	0.34	0.78	1.00	0.052	0.17	0.29	0.67	0.94

Using tables of Tiku and Kumra (1981) for the expected values and variances and covariances of order statistics, $E(\hat{\sigma}_T)$ can be calculated for the family (2.3.1) of t distributions and the values of Δ_{trim}/Δ evaluated; for n = 20, for example, these values turn out to be slightly smaller than the corresponding values of Δ_c/Δ. The conclusion to be drawn from this is that the t_{trim} test may be somewhat less powerful than the t_c test, at any rate for the family (2.3.1); see also Table 4.2.3.

Note that $E(\hat{\sigma})/\sigma$ and $E(\hat{\sigma}_{trim})/\sigma$ assume different values for different distributions but $E(s)/\sigma \simeq 1$ always (for large samples). Consequently, the power functions of t_c and t_{trim} tests are sensitive to changes in the underlying distributions but not the power function of the classical t test. Moreover, the null distributions of t_c and t_{trim} are fairly insensitive (robustness of validity) to changes in the underlying distributions. The t_c and t_{trim} tests have, therefore, robustness of validity as well as robustness of efficiency (Box and Tiao, 1973; Sprott, 1982).

For small samples, Tiku (1980a) simulated (from 10,000 Monte Carlo runs) the values of the power $1 - \beta^*$ for numerous distributions and always found the t_{trim} test to trail behind the t_c test. We reproduce some of these values in Table 4.2.3 for illustration. One may note that in view of Lemma 4.2.1 and the implied robustness, the null distribution of t_c is always approximated by Student's t distribution with n - 2r - 1 degrees of freedom; the null distribution of t_{trim} is also approximated by a Student's t distribution with n - 2r - 1 degrees of freedom (Dixon and Tukey, 1968). The null distribution of t is approximated by Student's t distribution with n - 1 degrees of freedom. Note that for $\mu = 0$, the power $1 - \beta^*$ is equal to the significance level α^* and this is remarkably stable (robust) from distribution to distribution for the t_c test (see also Dunnett, 1982).

Comparison of t_c Test with Tests Based on w24 and BS82

The Studentized statistics based on the estimators w24 and BS82 are given by

$$t_w = \frac{\sqrt{n_w}\,\hat{\mu}_w}{\hat{\sigma}_w} \quad \text{and} \quad t_B = \frac{\sqrt{n_B}\,\hat{\mu}_B}{\hat{\sigma}_B} \tag{4.2.8}$$

respectively. Asymptotically, t_W and t_B are normally distributed (Huber, 1981). It is not clear, however, what the values of n_W and n_B should be, although asymptotically $n_W = n_B = n$. It is generally believed that n_W and n_B should both be less than or equal to $n - 1$ (Gross, 1976).

For symmetric distributions, $\hat{\mu}_W$ and $\hat{\mu}_B$ are unbiased estimators of μ, and therefore the noncentrality parameters of t_W and t_B are for large n given by Δ_w^2 and Δ_B^2, respectively, where

$$\Delta_w \le \frac{\sqrt{n-1}\,\mu}{E(\hat{\sigma}_w)} \quad \text{and} \quad \Delta_B \le \frac{\sqrt{n-1}\,\mu}{E(\hat{\sigma}_B)}$$

For $n = 20$, the simulated values of $E(\hat{\sigma}_W)$ and $E(\hat{\sigma}_B)$ are given in Table 2.8.6.

For most symmetric populations with finite mean and finite variance, the noncentrality parameter Δ_w^2 and Δ_B^2 do not seem to be any larger than Δ_c^2. For the t distribution family (2.3.1), for example, we have the following values:

p:	∞	5	2.5	2
$\Delta_W/\Delta \le$	0.99	1.05	1.17	1.37
$\Delta_c/\Delta =$	0.98	1.04	1.17	1.34

Consequently, the t_W and t_B tests do not appear to be any more powerful than the t_c test for the family of long-tailed symmetric distributions (2.3.1); see also Dunnett (1982, Table IV), who arrives at similar conclusions in a pairwise multiple comparisons procedure.

For short-tailed symmetric distributions and skew distributions, the t_W and t_B tests do not even possess robustness of validity (e.g., stability of type I error); see also Dunnett (1982, Table III) for similar conclusions.

For plausible alternatives to normality, therefore, the t_W and t_B tests do not seem to have any advantage over the t_c test, the latter being even much easier to compute.

The t_W and t_B tests are, however, more powerful than the t_c test (based on the MML10 estimators) for "disaster" distributions; see also Dunnett (1982, Table IV). Such distributions, however, rarely occur in practice and when they do, can easily be detected by informal plotting techniques (Lawless, 1982) or by formal goodness-of-fit tests (Chapter 6). One can then base the t_c

test on the MML30 estimators, and this t_C test is more powerful than the t_W and t_B tests, as expected (Tiku, 1980a); see also Table 2.8.7.

4.3 TESTING EQUALITY OF TWO LOCATION PARAMETERS WHEN SCALE PARAMETERS ARE EQUAL

Let $x_{i1}, x_{i2}, \ldots, x_{in_i}$ (i = 1, 2) be independent random samples from populations with density $(1/\sigma)f((x - \mu_i)/\sigma)$, respectively; the functional form f is not known exactly but may be normal. In many practical situations, the problem of interest is to test the null hypothesis $H_0: \mu_1 = \mu_2$.

If the parent distribution is normal, the UMPU test for H_0 versus $H_1: \mu_1 \neq \mu_2$ and the UMP test for H_0 versus $H_1: \mu_1 > \mu_2$ are based on Student's two-sample t statistic

$$t = \frac{\bar{d}}{s(1/n_1 + 1/n_2)^{\frac{1}{2}}}$$

where $\bar{d} = \bar{x}_1 - \bar{x}_2$ and $s^2 = [(n_1 - 1)s_1^2 + (n_2 - 1)s_2^2]/(n_1 + n_2 - 2)$, $\bar{x}_i$ and s_i^2 being the sample mean and sample variance of the ith sample (see Lehmann, 1959). It is well known that under the normality assumption the null distribution of t is a central t distribution with $n_1 + n_2 - 2$ degrees of freedom, while the nonnull distribution of t is a noncentral t distribution with $n_1 + n_2 - 2$ degrees of freedom and noncentrality parameter $\Delta^2 = d^2/[\sigma^2(1/n_1 + 1/n_2)]$, where $d = \mu_1 - \mu_2$. Thus, under the normality assumption, one rejects H_0 at level α^* in favor of $H_1: \mu_1 \neq \mu_2$ if $|t| \geq t_{\alpha^*/2}(\nu)$, $\nu = n_1 + n_2 - 2$; one rejects H_0 at level α^* in favor of $H_1: \mu_1 > \mu_2$ if $t \geq t_{\alpha^*}(\nu)$. As in the one-sample case, again, it can be shown that the t test is asymptotically robust, and for finite n_1 and n_2 it possesses type I error robustness if $n_1 = n_2$ or if the distribution is symmetric; if $n_1 \neq n_2$ and the distribution is skew, the effects of departure from normality may be considerable (Chapter 1). Since the distribution is not normal in most practical problems (Geary, 1947), it is advantageous to use a robust test which has considerably higher power than the t test under nonnormal distributions but has substantially the same power as the t test under normal distributions. By using the MML10, based on type II censored samples, we now proceed to derive such a test and compare it with the t test and many other tests, including some of the prominent nonparametric tests.

Consider a type II symmetrically censored sample from the ith population

$$x_{i(a_i)} < \cdots < x_{i(b_i)} \qquad a_i = r_i + 1 \quad b_i = n_i - r_i \qquad (i = 1, 2)$$

where $r_i = [0.5 + 0.1n_i]$. Let $(\hat{\mu}_i, \hat{\sigma}_i)$ (i = 1, 2) be the MML10 estimators calculated from the two samples, respectively. Then the combined estimator of σ^2 is

$$\hat{\sigma}^2 = \frac{(A_1 - 1)\hat{\sigma}_1^2 + (A_2 - 1)\hat{\sigma}_2^2}{A_1 + A_2 - 2}$$

where $A_i = n_i - 2r_i$. Thus the MML10 for $(d = \mu_1 - \mu_2, \sigma)$ is $(\hat{d} = \hat{\mu}_1 - \hat{\mu}_2, \hat{\sigma})$. Analogous to the two-sample Student's t, we define the statistic T as

$$T = \frac{\hat{d}}{\hat{\sigma}(1/m_1 + 1/m_2)^{\frac{1}{2}}} \tag{4.3.1}$$

$m_i = n_i - 2r_i + 2r_i\beta_i$ $(i = 1, 2)$. We reject H_0 in favor of H_1: $\mu_1 \neq \mu_2$ if $|T| \geq C_{\alpha^*/2}(0)$ for some constant $C_{\alpha^*/2}(0)$ and reject H_0 in favor of H_1: $\mu_1 > \mu_2$ if $T \geq C_{\alpha^*}(R)$ for some constant $C_{\alpha^*}(R)$. The constants $C_{\alpha^*/2}(0)$ and $C_{\alpha^*}(R)$ are determined by a significance level α^* and the null distribution of T, which is given by Theorems 4.3.1 and 4.3.2.

THEOREM 4.3.1 Under the assumption of normality, the null distribution of T for large samples is a central t distribution with $A_1 + A_2 - 2$ degrees of freedom, while the nonnull distribution of T for large samples is a noncentral t distribution with $A_1 + A_2 - 2$ degrees of freedom and noncentrality parameter $\Delta_c^2 = d^2/[\sigma^2(1/m_1 + 1/m_2)]$, $d = \mu_1 - \mu_2$.

Proof: The results follow from the fact that for large samples, $\hat{d}$ is normal $N(d, \sigma\sqrt{1/m_1 + 1/m_2})$ and $(A_1 + A_2 - 2)\hat{\sigma}^2/\sigma^2$ is a central chi-square with $A_1 + A_2 - 2$ degrees of freedom independently of $\hat{d}$ (Lemma 2.7.2).

One would expect the results of Theorem 4.3.1 to work well even for small sample sizes, since L* approximates L very closely (Chapter 2).

THEOREM 4.3.2 Under some very general regularity conditions, the asymptotic distribution of T is normal with mean $E(\hat{d})/[E(\hat{\sigma})\sqrt{1/m_1 + 1/m_2}]$ and variance 1 (or almost 1).

Proof: Theorem 4.3.2 follows from the fact that $\hat{\sigma}$ converges to $E(\hat{\sigma})$ in probability, so that $\sigma(T)$ is 1 or almost 1. It follows that asymptotically T is a linear function of order statistics. The results then follow from Lemma 4.2.1.

If the distributions are symmetric, then $E(\hat{d}) = d$. If further the distributions are normal, then for large A, $E(\hat{\sigma}) = \sigma$ and $\sigma(T) = 1$, so that the noncentrality parameter is asymptotically equal to $\Delta_c^2 = d^2/\sigma^2(1/m_1 + 1/m_2)$. It follows from Theorem 4.3.2, therefore, that the asymptotic power functions of T for testing H_0 versus H_1: $\mu_1 \neq \mu_2$ and for testing H_0 versus H_1: $\mu_1 > \mu_2$ are given, respectively, by $P(|Z + \Delta_c| \geq Z_{\alpha^*/2})$ and $P(Z \geq Z_{\alpha^*} - \Delta_c)$, where Z is a standard normal variate and Z_{α^*} is its $100(1 - \alpha^*)\%$ point. As in the one-sample situation, in what follows, we again restrict ourselves to $H_1: \mu_1 > \mu_2$ unless stated otherwise; note, however, that the results for $H_1 : \mu_1 \neq \mu_2$ are exactly similar.

Comparisons of the Power Functions

The asymptotic power function of the T test is given by

$$P(Z \geq Z_{\alpha*} - \Delta_c) \qquad \Delta_c \simeq \frac{E(\hat{d})}{E(\hat{\sigma})(1/m_1 + 1/m_2)^{\frac{1}{2}}} \tag{4.3.2}$$

The corresponding asymptotic power function of Student's t test is given by

$$P(Z \geq Z_{\alpha*} - \Delta) \qquad \Delta = \frac{d}{\sigma(1/n_1 + 1/n_2)^{\frac{1}{2}}} \tag{4.3.3}$$

If the distributions are normal, then $\Delta_c = d/[\sigma(1/m_1 + 1/m_2)^{\frac{1}{2}}]$, so that $\Delta \geq \Delta_c$. However, since $m_i = n_i - 2r_i + 2r_i\beta_i$ is only slightly smaller than n_i, Δ_c is only slightly smaller than Δ; for example, if $n_1 = n_2$, then $\Delta_c/\Delta = 0.983$. For normal populations, therefore, the loss of power resulting from the use of the statistic T in place of the uniformly most powerful Student's t is only slight (Tiku, 1980a, p. 138).

For symmetric long-tailed distributions and skew distributions, $E(\hat{\sigma})$ is usually smaller than σ [e.g., $E(\hat{\sigma})/\sigma < 1$]. Since m_i is only slightly smaller than n_i, the result $E(\hat{\sigma})/\sigma < 1$ usually leads to $\Delta_c > \Delta$. Therefore, the T test is more powerful than the t test for all such distributions. For the families of distributions (2.3.1) and (2.3.2), for example, we have the following values of Δ_c/Δ; $n_1 = n_2 = 20$:

Family (2.3.1)				Family (2.3.2)				
p:	5	2.5	2	a:	4	3	2	1
	1.04	1.17	1.34		1.02	1.03	1.05	1.12

Analogous to the T test, one may define a Studentized test based on the pair TM10 $(\hat{\mu}_{T_i}, \hat{\sigma}_{T_i})$ (i = 1, 2). However, this test is expected to be somewhat less powerful than the T test, for reasons given in Section 4.4 [see also Tiku (1980a, 1982a) and Dunnett (1982), who give power values of these tests].

Simulations reveal that the two-sample tests based on w24 and BS82 estimators are somewhat less powerful than the T test (based on the MML10 estimators) for most symmetric populations with finite mean and variance (see also Dunnett, 1982, Table IV). For skew populations, w24 and BS82 (and other similar estimators) should not be used; in fact, Huber (1973), Bickel

(1978), and Carroll (1979) have shown that their adaptation leads to quite complicated results in many situations (see also Table 4.5.1). Moreover, Carroll (1979, p. 676) has shown that the trimmed sample estimators TM10 perform better than Huber and Hampel estimators for skew populations (see also Doksum and Wong, 1983). One commonly suggested remedy would be a transformation to symmetry, perhaps using the power transformation of Box and Cox (1964, 1982). However, the results of Bickel and Doksum (1981) and Doksum and Wong (1983) indicate that the Box and Cox transformation methods based on normal-theory likelihood should be used with caution. Other important works in this area are Carroll (1980, 1982), Hernandez and Johnson (1980), Hinkley (1975), and Hinkley and Runger (1984).

Dunnett (1982) considered numerous symmetric distributions and used the MML10 and TM10 estimators and the classical estimators ($\bar{x}$, s), and also the w24 and BS82 estimators, in constructing confidence intervals for all the possible differences $\mu_i - \mu_j$ of k means in a multiple comparisons procedure. He found that the confidence intervals based on the MML10 estimators are the shortest on the whole, at any rate for symmetric distributions which constitute plausible alternatives to normality. Note, however, that his out(10) is in fact the mixture normal distribution $0.9N(0, 1) + 0.1N(0, 10)$.

Comparison of the T Test with Some Prominent Nonparametric Tests

For testing the null hypothesis $H_0: \mu_1 = \mu_2$, nonparametric methods have often been used, particularly if the normality assumption is suspect. An important feature of the nonparametric tests is the stability of their type I error from distribution to distribution. We consider here the three most prominent nonparametric tests: the Wilcoxon test, the Terry-Hoeffding normal-score test, and the Smirnov-Kolmogorov-Massey test (Milton, 1970, pp. 23, 25, 27). Since a vast literature and numerous textbooks are available on nonparametric statistical methods (see, e.g., Lehmann, 1975), we introduce these tests only briefly but emphasize the power comparisons with the T test based on the MML10 estimators. To avoid the problem of ties, we will restrict ourselves to continuous distributions with density functions of the type $(1/\sigma)f((x - \mu)/\sigma)$.

Wilcoxon Test. Combine the two random samples $x_{i1}, x_{i2}, \ldots, x_{in_i}$ ($i = 1, 2$) into a single sample of size $N = n_1 + n_2$. Arrange these N observations in ascending order of magnitude but note which observation comes from which sample. Then the Wilcoxon rank sum statistic is defined as

$$W = \sum_{i=1}^{N} i z_i \qquad (N = n_1 + n_2)$$

where

$$z_i = \begin{cases} 1 & \text{if the ith observation in the ordered sample is an } x_1 \text{ observation} \\ 0 & \text{otherwise} \end{cases}$$

In other words, W is the sum of the ranks of the x_1 observations in the combined sample. One then rejects $H_0: \mu_1 = \mu_2$ in favor of $H_1: \mu_1 > \mu_2$ for large values of W. If n_1 and n_2 are small, the type I error probability can be conveniently computed by a recursive formula given by Mann and Whitney (1947). (In fact, Mann and Whitney table tabulated the type I error probabilities for $n_1 \le 8$, $n_2 \le 8$.) For large n_1 and n_2, a normal approximation to the null distribution of W can be used to compute the type I error probability of W, namely, the null distribution of $[W - E(W)]/\sqrt{\text{var}(W)}$ is referred to a standard normal distribution, where $E(W) = n_1(N+1)/2$ and $\text{var}(W) = n_1 n_2(N+1)/12$ (Lehmann, 1975, p. 14).

The asymptotic power function of W is given by (Lehmann, 1975, p. 72)

$$P\left[Z \ge Z_{\alpha*} - \frac{\sqrt{12 n_1 n_2}}{\sqrt{N+1}} f^*(0)\frac{d}{\sigma}\right] \qquad d = \mu_1 - \mu_2 \tag{4.3.4}$$

where Z and $Z_{\alpha*}$ are the same as (4.3.3); $f^*(0)$ is the density at zero of the difference $u_1 - u_2$ of two independent random variates each having the probability density function $f(u)$, $u = (x - \mu)/\sigma$. For $n_1 = n_2$, (4.3.4) reduces to (since n_1 and n_2 both tend to infinity)

$$P[Z \ge Z_{\alpha*} - \sqrt{12}\, f^*(0)\Delta] \qquad \Delta = \frac{d\sqrt{n}}{\sigma\sqrt{2}} \qquad n = n_1 = n_2 \tag{4.3.5}$$

If f(u) is the normal N(0, 1) distribution, $u_1 - u_2$ is normally distributed with mean zero and variance 2 and hence $f^*(0) = 1/2\sqrt{\pi}$. Therefore, the noncentrality parameter in the power function (4.3.5) is Δ_W^2, where $\Delta_W = \sqrt{12}\, f^*(0)\Delta = 0.977\Delta$. Since the corresponding noncentrality parameter in the power function (4.3.2) is Δ_c^2, where $\Delta_c = 0.983\Delta$ for normal distributions, the T test based on the MML10 is asymptotically more powerful than the Wilcoxon W test for normal distributions.

For the t family of distributions given by (2.3.1), by making the transformation ($y = u_1 - u_2$, $z = u_2$) in the joint density of $u_1 = (x_1 - \mu_1)/\sigma$ and $u_2 = (x_2 - \mu_2)/\sigma$ and integrating out z, we obtain (Tiku, 1980a)

$$f^*(y) = \frac{2}{k[B(1/2, p - 1/2)]^2} \int_0^\infty \left\{ \left[1 + \frac{(y+z)^2}{k}\right]\left[1 + \frac{z^2}{k}\right] \right\}^{-p} dz$$

so that

$$f^*(0) = \frac{2}{k[B(1/2, p - 1/2)]^2} \int_0^\infty \left(1 + \frac{z^2}{k}\right)^{-2p} dz$$

$$= \frac{1}{\sqrt{k}[B(1/2, p - 1/2)]^2} \int_0^\infty \left(\frac{z^2}{k}\right)^{(1/2)-1} \left(1 + \frac{z^2}{k}\right)^{-2p} d\left(\frac{z^2}{k}\right)$$

$$= \frac{B(1/2, 2p - 1/2)}{\sqrt{k}[B(1/2, p - 1/2)]^2} \qquad k = 2p - 3$$

Now $\Delta_W = \sqrt{12n_1n_2/(N + 1)}\, f^*(0)(d/\sigma) = n\sqrt{12/(2n + 1)}\, f^*(0)(d/\sigma)$ for $n_1 = n_2 = n$. For $n_1 = n_2$, the noncentrality parameters Δ_c^2 and Δ^2 for the MML10 T test and for the classical t test are given, respectively, by

$$\Delta_c \simeq \frac{d}{E(\hat{\sigma})\sqrt{1/m_1 + 1/m_2}} = \frac{md}{E(\hat{\sigma})\sqrt{2m}} = \frac{\sqrt{n(1 - 2q + 2q\beta)}\, d}{E(\hat{\sigma})\sqrt{2}}$$

and

$$\Delta = \frac{d}{\sigma\sqrt{1/n_1 + 1/n_2}} = \frac{nd}{\sigma\sqrt{2n}} = \frac{\sqrt{n}\, d}{\sigma\sqrt{2}}$$

It follows that for large n,

$$\frac{\Delta_c}{\Delta} \simeq \frac{\sqrt{1 - 2q + 2q\beta}\,\sigma}{E(\hat{\sigma})}$$

$$= \frac{0.983\,\sigma}{E(\hat{\sigma})} \qquad \left(q = \frac{r}{n} = 0.1\right) \tag{4.3.6}$$

and

$$\frac{\Delta_W}{\Delta} \simeq \sqrt{12}\, f^*(0)$$

Let $Z_j = (x_{(j)} - \mu)/\sigma$, where $x_{(j)}$ is the jth order statistic in a random sample of size n from the t family given by (2.3.1). Now, as $n \to \infty$, Z_{n-r} and Z_{r+1} tend to ξ_q and $-\xi_q$, respectively, where $\int_{-\infty}^{\xi_q} f(z;p)\, dz = 1 - q$; $z = (x - \mu)/\sigma$. Now

$$\lim_{n\to\infty} \frac{1}{n} \sum_{i=r+1}^{n-r} Z_i^2 = \int_{-\xi_q}^{\xi_q} z^2 f(z;p)\, dz$$

$$= \frac{2}{\sqrt{k}\, B(1/2, p - 1/2)} \int_0^{\xi_q} z^2\left(1 + \frac{z^2}{k}\right)^{-p} dz$$

$$= \frac{k}{B(1/2, p - 1/2)} \int_0^{\xi_q} \left(\frac{z^2}{k}\right)^{\frac{1}{2}} \left(1 + \frac{z^2}{k}\right)^{-p} d\left(\frac{z^2}{k}\right)$$

$$= I_{1-u_q}\left(\frac{3}{2},\ p - \frac{3}{2}\right) \tag{4.3.7}$$

where $u_q = (1 + \xi_q^2/k)^{-1}$ (Harter and Moore, 1966; Tiku, 1967, 1980a), and

$$\lim_{n\to\infty} \frac{1}{n}E(B) = \sigma q\alpha E(Z_{n-r} - Z_{r+1}) = 2\sigma q\alpha\xi_q \tag{4.3.8}$$

and

$$\frac{1}{n}E(C) = E\left[\frac{\sigma^2}{n}\sum_{i=r+1}^{n-r} Z_i^2 + \sigma^2 q\beta(Z_{n-r}^2 + Z_{r+1}^2) - \frac{m}{n}(\hat{\mu} - \mu)^2\right]$$

$$\simeq \sigma^2\left[\int_{-\xi_q}^{\xi_q} z^2 f(z;p)\,dz + 2q\beta\xi_\beta^2\right] \tag{4.3.9}$$

[Note that $E(\hat{\mu} - \mu)^2 = \mathrm{var}(\hat{\mu}) \le \sigma^2/m$ and tends to zero as $n \to \infty$.] For $n_1 = n_2 = n$, $A_i = A = n - 2r = n(1 - 2q)$, $E(B_i) = E(B)$, and $E(C_i) = E(C)$, so that the asymptotic value of $E(\hat{\sigma})$ is given by

$$E(\hat{\sigma}) = \frac{E(B) + \sqrt{E(B^2) + 4AE(C)}}{2\sqrt{A(A-1)}}$$

which can be computed from (4.3.8) and (4.3.9) and substituted in (4.3.6) to obtain the value of the ratio Δ_C/Δ; for the t family (2.3.1), Δ_C/Δ turns out to be slightly greater than $\sqrt{12}\,f^*(0)$ and, therefore, the T test should be slightly more powerful than the W test. For $p = 5/2$ and $7/2$, for example, we have the following asymptotic values ($n_1 = n_2$):

p:	5/2	7/2	∞
$(1/\sigma)E(\hat{\sigma})$	0.824	0.902	1.000
Δ_C/Δ	1.193	1.089	0.983
$\sqrt{12}\,f^*(0)$	1.184	1.080	0.977

It is interesting to note that the asymptotic values of $E(\hat{\sigma})/\sigma$ above are very close to the corresponding values for $n = 20$ ($q = r/n = 0.1$) given in Table 4.2.2.

<u>The Terry-Hoeffding Normal Score Test and the Smirnov-Kolmogorov-Massey Test.</u> The statistics for the Terry-Hoeffding normal score test and the Smirnov-Kolmogorov-Massey test are given, respectively, by

$$C_1 = \sum_{i=1}^{N} z_i E(V_i) \tag{4.3.10}$$

and

$$D^+ = \max_{1,2,\ldots,N} \left\{ \sum_{i=1}^{j} [n_2(1 - z_i) - n_1 z_i] \right\} \qquad (4.3.11)$$

where $E(V_i)$ is the expected value of the ith order statistic in a random sample of size N from the normal N(0, 1) distribution (Owen, 1962). Large values of C_1 and D^+ lead to the rejection of H_0 (d = 0) in favor of H_1 (d > 0).

The asymptotic null distribution of C_1 is normal; for small samples, the accuracy of a normal approximation and an alternative t approximation to the null distribution of C_1 has been studied by Klotz (1964). The null distribution of D^+ assumes a neat form if $n_1 = n_2 = n$ (Lehmann, 1975), namely

$$P_{H_0}(D^+ \geq c) = \frac{\binom{2n}{n-a} - \binom{2n}{n-2a} + \cdots}{\binom{2n}{n}} \qquad \left(c = \frac{a}{n}\right) \qquad (4.3.12)$$

The series continues as long as $n - ia \geq 0$, i = 1, 2, 3, ... The asymptotic null distribution of D^+ is not normal, however, even if $n_1 = n_2$. Under H_0, the probability $P(\sqrt{n_1 n_2/N}\, D^+ \geq \xi)$ $(N = n_1 + n_2)$ is approximated by $K(\xi)/2$, where $K(\xi)$ is given in Table F of Lehmann's (1975) book. Extensive tables of the cumulative probabilities of the null distributions of W, C_1, and D^+ are available (Wilcoxon et al., 1973; Klotz, 1964; Kim and Jennrich, 1973).

Numerical Comparison of the T Test with the W, C_1, and D^+ Tests. Using the tables mentioned above, Tiku (1980a) simulated (from 4000 Monte Carlo runs) the values of the power of W, C_1, D^+, and T (based on MML10 estimators) for testing H_0 (d = 0) against H_1 (d > 0), for sample sizes (n_1, n_2) = (8, 6), (8, 8), (10, 8), and (10, 10). To attain a significance level close to a preassigned value for W, C_1, and D^+ tests, randomization between the two appropriate cumulative probabilities was performed. (Note that the W, C_1, and D^+ are discrete variables.) He found that the T test is in general more powerful than the nonparametric tests based on W, C_1, and D^+, except for some very short-tailed distributions (uniform, for example) and some extremely skew distributions (chi-square with 1 df or log-normal, for example) in which case the statistics C_1 and D^+, respectively, were found to be considerably more powerful than the T test. For very short-tailed distributions like the uniform and "disaster" distributions like the Cauchy, if one uses T based on the MML0 (no censoring) and MML30 (30% censoring on each side of the ordered samples) estimators, respectively, the power of the T test is substantially enhanced and is higher than that of the C_1 and W tests, as expected. For extremely skew distributions, a more powerful T test (based on samples censored only on one side) is developed in Section 4.4. For illustration, we reproduce Tiku's (1980a) values in Table 4.3.1, for the following distributions:

(1) normal N(0, 1), (2) Dixon's outlier model: (n - 1)N(0, 1) and 1N(0, 10), (3) Tiku's outlier model with k = 1 and $\lambda = 2$ [under this model the k

Table 4.3.1 Values of the Power for the 1% Significance Level; $n_1 = n_2 = 8$

d:	0.0	1.0	2.0	3.0	0.0	1.0	2.0	3.0	0.0	1.0	2.0	3.0
	(1)				(2)				(3)			
D^+	0.010	0.24	0.81	0.98	0.010	0.17	0.60	0.92	0.007	0.17	0.67	0.97
C_1	0.010	0.28	0.88	1.00	0.010	0.15	0.43	0.59	0.004	0.08	0.38	0.78
W	0.010	0.29	0.86	1.00	0.010	0.18	0.55	0.81	0.006	0.10	0.46	0.85
T	0.011	0.28	0.85	1.00	0.010	0.19	0.69	0.97	0.012	0.27	0.85	1.00
	(4)				(5)[a]				(6)			
D^+	0.010	0.16	0.55	0.83	0.010	0.10	0.34	0.58	0.010	0.50	0.94	1.00
C_1	0.010	0.14	0.45	0.67	0.010	0.08	0.23	0.36	0.010	0.44	0.83	0.95
W	0.010	0.15	0.53	0.78	0.010	0.10	0.28	0.48	0.010	0.48	0.89	0.98
T	0.008	0.17	0.55	0.84	0.006	0.08	0.29	0.54	0.007	0.42	0.89	0.99

[a]The corresponding values of the power of T based on MML30 are 0.006, 0.10, 0.36, and 0.64.

smallest and largest observations in a random sample of size n from the normal N(0, 1) are multiplied by a constant λ (Chapter 7)], (4) Student's t with 2 degrees of freedom, (5) Cauchy, (6) exponential

As applications of these two-sample statistics, we consider the following data.

EXAMPLE 4.3.1 (Lehmann, 1975, p. 37) Two drugs A and B are assigned randomly to 16 patients, with 8 receiving drug A and the other 8 drug B. The problem is to compare the effect of these two drugs on the relief of postoperative pain. Following are the numbers of hours of relief observed:

A: 6.8 3.1 5.8 4.5 3.3 4.7 4.2 4.9

B: 4.4 2.5 2.8 2.1 6.6 0.0 4.8 2.3

One wants to test the null hypothesis $H_0: \mu_A = \mu_B$ against the alternative hypotheses $H_1: \mu_A > \mu_B$.

To calculate the Wilcoxon rank sum statistics W, we combine the two samples to form a single sample of size 16. From an ordering of these 16 observations, we obtain the following rankings:

A:	6	7	8	10	11	13	14	16
B:	1	2	3	4	5	9	12	15

Thus

$$W = 6 + 7 + 8 + \cdots = 85$$

and the probability (Lehmann, 1975, Table B)

$$P_{H_0}(W \geq 85) = P_{H_0}(W_{XY} \leq 15) = 0.042$$

To calculate the statistic D^+, we have the following:

j:	1	2	3	4	5	6	7	8	9	10	11	12	13	14	15	16
$\sum_{i=1}^{j} (1 - z_i)$	1	2	3	4	5	5	5	5	6	6	6	7	7	7	8	8
$\sum_{i=1}^{j} z_i$	0	0	0	0	0	1	2	3	3	4	5	5	6	7	7	8

Therefore,

$$D^+ = 8(5) = 40$$

Since $a = 40/8 = 5$,

$$P_{H_0}(D^+ \geq 40) = \frac{\binom{16}{3}}{\binom{16}{8}} = 0.044$$

To calculate the statistic T, we have the following:

$r_1 = r_2 = r = [0.5 + 0.1(8)] = 1 \qquad \alpha_1 = \alpha_2 = \alpha = 0.670 \qquad \beta_1 = \beta_2 = \beta = 0.848$

(Table B.1)

$$\hat{\mu}_1 = 4.563 \quad \hat{\sigma}_1 = 1.268 \quad \hat{\mu}_2 = 3.215 \quad \hat{\sigma}_2 = 1.586$$

$$\hat{\sigma} = \frac{\sqrt{1.608 + 2.514}}{\sqrt{2}} = 1.436$$

$$T = \frac{4.563 - 3.215}{1.436\sqrt{2/7.656}} = 1.837 \qquad (\text{df} = 10)$$

and

$$P_{H_0}(T \geq 1.837) = 0.047$$

For Student's two-sample t,

$\bar{x}_1 = 4.662 \quad s_1 = 1.222 \quad \bar{x}_2 = 3.188 \quad s_2 = 2.017$

$$s = \frac{\sqrt{1.404 + 4.067}}{\sqrt{2}} = 1.668$$

$$t = \frac{4.662 - 3.188}{1.668\sqrt{2/8}} = 1.769 \quad (\text{df} = 14)$$

and

$$P_{H_0}(t \geq 1.769) = 0.051$$

At the 5% significance level, the W, D^+, and T tests reject H_0 in favor of H_1, but the t test just fails to do so. The failure of the t test may be due to the observations 0.0 and 6.6 in the second sample, which are apparently outliers (see Chapter 7 and Special Appendix C).

4.4 TESTING EQUALITY OF LOCATION PARAMETERS OF EXTREME SKEW DISTRIBUTIONS

In Section 2.3 we illustrated that if x_1 and x_2 are independent and both have the truncated gamma distribution (2.3.10), truncated moderately on the right-hand side, then $x = x_1 - x_2$ has a symmetric distribution with kurtosis $\beta_2^* = \mu_4/\mu_2^2$ close to 3. We will exploit this phenomenon here to develop robust procedures for testing assumed values of the difference of location parameters of extreme skew distributions. Our procedure is to censor the samples in relation to the skewness of the distributions, so that the type II censored samples from skew and normal populations are much alike. Then, by assuming normality for the censored samples, we calculate the MML estimators and using these estimates, we construct t-type tests as in Section 4.3.

To illustrate this, we let $x_{i1}, x_{i2}, \ldots, x_{in_i}$ $(i = 1, 2)$ be two independent random samples from skew distributions of the type $(1/\sigma)f((x - \mu_i)/\sigma)$. Without any loss of generality, we assume that f has a long tail on the right-hand side. Since this long tail is clearly represented by the largest observations in a sample, we censor (give zero weights to) a few largest observations and consider the type II censored samples (Tiku, 1982b)

$$x_{(1,1)}, x_{(1,2)}, \ldots, x_{(1,n_1-r_1)}$$

and

$$x_{(2,1)}, x_{(2,2)}, \ldots, x_{(2,n_2-r_2)}$$

with $r_1 = [0.5 + 0.2n_1]$ and $r_2 = [0.5 + 0.2n_2]$. Note that r_1 and r_2 are now the number of observations censored on the right-hand side of the two ordered samples. Let $\hat{\sigma}_{ic}$ be the MMLE of σ calculated from the ith sample [equation (2.7.11)], so that the MMLE of (μ_1, μ_2, σ) is $(\hat{\mu}_{1c}, \hat{\mu}_{2c}, \hat{\sigma}_c)$, where

$$\hat{\mu}_{ic} = K_{ic} + D_i \hat{\sigma}_c \qquad (i = 1, 2) \tag{4.4.1}$$

and

$$\hat{\sigma}_c^2 = \frac{(A_1 - 1)\hat{\sigma}_{1c}^2 + (A_2 - 1)\hat{\sigma}_{2c}^2}{A_1 + A_2 - 2}$$

with $A_i = n_i - r_i$; K_{ic} and D_i are calculated from the ith sample and are the same as K and D, respectively, given by (2.7.8). Note that in (4.4.1), D_1 and D_2 are pure constants and depend only on $p_1 = r_1/n_1$ and $p_2 = r_2/n_2$ (the proportions of observations censored on the right-hand side of the two samples). If $n_1 = n_2$, then $p_1 = p_2$ and $D_1 = D_2$; consequently, $\hat{d}_c = \hat{\mu}_{1c} - \hat{\mu}_{2c} = K_{1c} - K_{2c}$, which is an unbiased estimator of $d = \mu_1 - \mu_2$ since the bias in K_{1c} and K_{2c} cancels out. It is shown by Tiku (1982b) that for skew distributions, such as the gamma distribution family (2.3.2), $\hat{d}_c = \hat{\mu}_{1c} - \hat{\mu}_{2c}$ is considerably more efficient than $\bar{d} = \bar{x}_1 - \bar{x}_2$ as an estimator of $d = \mu_1 - \mu_2$. To test H_0: $d = \mu_1 - \mu_2 = 0$, one may then use $\hat{d}_c$ and $\hat{\sigma}_c$ and proceed as in Section 4.3 to construct a Studentized statistic t_S, namely,

$$t_S = \frac{\hat{d}_c}{\hat{\sigma}_c(1/m_1 + 1/m_2)^{\frac{1}{2}}} \tag{4.4.2}$$

where $m_i = n_i - r_i + r_i\beta_i$; β_i is the value of β given in Table B.1 with $r_i = r$ and $n_i = n$ $(i = 1, 2)$.

One then rejects H_0: $d = 0$ in favor of H_1: $d > 0$ at significance level α^* if

$$t_S \geq C_{\alpha^*}(R)$$

As $A_i = n_i - r_i$ $(i = 1, 2)$ become large, $\hat{\sigma}_c$ converges to $E(\hat{\sigma}_c)$ in probability. Consequently, the asymptotic distribution of t_S is normal with mean Δ_S and standard deviation $\sigma(t_S)$, where

$$\Delta_S = \frac{d}{E(\hat{\sigma}_c)(1/m_1 + 1/m_2)^{\frac{1}{2}}} \quad \text{and} \quad \sigma(t_S) = \frac{[V(\hat{d}_c)]^{\frac{1}{2}}}{E(\hat{\sigma})(1/m_1 + 1/m_2)^{\frac{1}{2}}}$$

One may note that $\sigma(t_S)$ turns out to be considerably greater than 1 for extreme skew distributions (see Table 4.4.1); see also Tiku (1982b, Table I). For large samples, the power function of t_S is given by

$$P(Z \geq Z_{\alpha^*} - \Delta_S) \qquad \Delta_S = \frac{d}{\sqrt{V(\hat{d}_c)}}$$

Table 4.4.1 Expected Values and Variances for the Gamma Family (2.3.2); $A = n(1 - q)$ and $q = r/n = 0.2$

n = 20				n = 40				n = 80			
(1)[a]	(2)	(3)	(4)	(1)	(2)	(3)	(4)	(1)	(2)	(3)	(4)
					a = 4						
3.60	0.181	1.71	1.09	3.60	0.091	1.71	1.09	3.60	0.046	1.71	1.09
					a = 3						
2.64	0.132	1.44	1.10	2.64	0.066	1.44	1.11	2.64	0.033	1.44	1.11
					a = 2						
1.70	0.084	1.12	1.13	1.70	0.042	1.11	1.14	1.69	0.021	1.11	1.14
					a = 1						
0.767	0.036	0.695	1.20	0.765	0.018	0.689	1.21	0.764	0.009	0.686	1.22
					a = 0.5						
0.322	0.014	0.399	1.31	0.320	0.0071	0.390	1.34	0.319	0.0036	0.385	1.36

[a]Values of (1) $[E(K_c) - \theta]/\delta$, (2) $V(K_c)/\delta^2$, (3) $E(\hat{\sigma}_c)/\delta$, and (4) $\sigma(t_S) = \sqrt{mV(K_c)}/E(\hat{\sigma}_c)$.

where Z is a standard normal variate and $Z_{\alpha*}$ its $100(1 - \alpha^*)\%$ point. The corresponding power function of the Student's two-sample statistic is given by (4.2.3). For $n_1 = n_2 = n$, for example, we have the following values of the ratio of the positive root of the two noncentrality parameters Δ_S^2 and Δ^2, calculated from the expected values and variances given in Table 4.4.1. Note that $m_1 = m_2 = n[1 - 0.2 + 0.2(0.7818)] = 0.9564n$.

	a				
n	4	3	2	1	0.5
20	1.04	1.06	1.10	1.17	1.32
80	1.04	1.06	1.10	1.17	1.32

It is clear that the t_S test is considerably more powerful than Student's two-sample t test, for extreme skew distributions. In practice the underlying distribution will not be known exactly but we have assumed here that it belongs to a family of extreme skew distributions. The values of $\sigma(t_S)$ may be worked out on the basis of exponentiality from (2.5.8); the percentage points of t_S are determined in terms of these values of $\sigma(t_S)$ and the percentage points of the standard normal distribution. These percentage points provide satisfactory approximations to the true percentage points for most skew distributions (Tiku, 1982b). This is otherwise obvious from the fact that $\sigma(t_S)$ does not change a great deal from distribution to distribution (Table 4.4.1). For $n_1 = n_2 = n \geq 20$, for example, $\sigma(t_S)$ could be taken to be equal to 1.20 (Table 4.4.1).

To approximate the percentage points of t_S for small samples, let

$$t = \frac{t_S}{\rho} \tag{4.4.3}$$

where t has a Student's t distribution with $A_1 + A_2 - 2$ degrees of freedom. Equating the variances on both sides of (4.4.3), we obtain

$$\rho = \sigma(t_S) \frac{\sqrt{A_1 + A_2 - 4}}{\sqrt{A_1 + A_2 - 2}} \tag{4.4.4}$$

If t_{α^*} is the $100(1 - \alpha^*)\%$ point of t, then the approximate $100(1 - \alpha^*)\%$ point of t_S is given by ρt_{α^*}. For $n_1 = n_2 = 8$ ($r_1 = r_2 = r = 2$; $\alpha = 0.7719$, $\beta = 0.7680$), for example,

$$V(\hat{d}_c) = 0.174 \quad E(\hat{\sigma}_c) = 0.66 \quad \sigma(t_S) = 1.22 \quad \rho = 1.09$$

(on the basis of exponentiality); the 99% point of t_S is, therefore, 1.09(2.764) = 3.01. The simulated 99% point of t_S is 2.90 (Tiku, 1982b, p. 2547), a value very close to 3.01.

The simulated values (based on 4000 Monte Carlo runs) of the probabilities $P(t_S \geq 2.90 \mid H_0)$, given by Tiku (1982b, Table II) for $n_1 = n_2 = 8$ and $r_1 = r_2 = 2$, are as follows:

Normal	χ_4^2	χ_2^2	χ_1^2	Log-normal ($\sigma = 1$)
0.010	0.012	0.010	0.009	0.012

The robustness of the type I error of the t_S test is obvious. Similar results are of course true for $n_1 \neq n_2$, since $r_1 = [0.5 + 0.2n_1]$ and $r_2 = [0.5 + 0.2n_2]$ and therefore $p_1 = r_1/n_1$ and $p_2 = r_2/n_2$ are almost equal; consequently, $\hat{d}_c$ is almost unbiased.

The values of the power of the t_S test are compared with the corresponding values of numerous other tests by Tiku (1982b, Table II) and it is shown that the t_S test is more powerful than even the Smirnov-Kolmogorov-Massey D^+ test, which is particularly powerful for extreme skew distributions.

4.5 COMPARISON OF TWO LOCATION PARAMETERS WHEN THE SCALE PARAMETERS ARE UNEQUAL

Let $x_{i1}, \ldots, x_{in_i}$ $(i = 1, 2)$ be two independent random samples from populations with densities $(1/\sigma_i)f((x - \mu_i)/\sigma_i)$, respectively. For testing H_0: $\mu_1 = \mu_2$, it is conveniently assumed that $\sigma_1 = \sigma_2$. In numerous practical situations, however, one often has $\sigma_1 \neq \sigma_2$. The problem of testing H_0: $\mu_1 = \mu_2$ under the normality assumption but with $\sigma_1 \neq \sigma_2$ (heteroscedastic variances) is referred to as the Behren-Fisher problem. One approach to solve this problem was proposed by Welch (1937), who used the following statistic:

$$t^* = \frac{\bar{x}_1 - \bar{x}_2}{\sqrt{s_1^2/n_1 + s_2^2/n_2}} \qquad (4.5.1)$$

where $\bar{x}_i$ and s_i^2 are the sample mean and the sample variance of the ith sample $(i = 1, 2)$. Using t^*, one then rejects H_0 in favor of H_1: $\mu_1 \neq \mu_2$ if $|t^*| \geq C_{\alpha^*/2}$ for some constant $C_{\alpha^*/2}$, where $C_{\alpha^*/2}$ is determined by the level α^* and the null distribution of t^*. To obtain the null and nonnull distributions of t^* under the assumption of normality, one realizes that $\bar{x}_1 - \bar{x}_2$ is distributed as normal with mean $\mu_1 - \mu_2$ and variance $(\sigma_1^2/n_1) + (\sigma_2^2/n_2)$ while $(s_1^2/n_1) + (s_2^2/n_2)$ is distributed as $a_1\chi^2_{f_1} + a_2\chi^2_{f_2}$, a linear combination of independent central chi-square variates, independently of $\bar{x}_1 - \bar{x}_2$, where $a_i = \sigma_i^2/n_i(n_i - 1)$ and $f_i = n_i - 1$. It is difficult to work out the exact distribution of a linear combination of two chi-square variates, and therefore the distribution of t^* is intractable. To obtain an approximation to the null distribution of t^*, Welch (1937) approximated the distribution of $s_1^2/n_1 + s_2^2/n_2$ by a scaled chi-square variate χ_f^2/ρ by equating the mean and variance on both sides of $a_1\chi^2_{f_1} + a_2\chi^2_{f_2} = \chi^2_f/\rho$. Thus he obtained

$$a_1 f_1 + a_2 f_2 = \frac{f}{\rho}$$

and

$$a_1^2 f_1 + a_2^2 f_2 = \frac{f}{\rho^2}$$

This gives

$$f = \frac{(a_1 f_1 + a_2 f_2)^2}{a_1^2 f_1 + a_2^2 f_2}$$

$$= \frac{(\sigma_1^2/n_1 + \sigma_2^2/n_2)^2}{[\sigma_1^4/n_1^2(n_1 - 1)] + [\sigma_2^4/n_2^2(n_2 - 1)]} \tag{4.5.2}$$

and

$$\rho^{-1} = \frac{\sigma_1^2/n_1 + \sigma_2^2/n_2}{f}$$

so that

$$t^* \simeq \frac{\bar{x}_1 - \bar{x}_2}{[(\sigma_1^2/n_1 + \sigma_2^2/n_2)\chi_f^2/f]^{\frac{1}{2}}}$$

$$\simeq \frac{Z}{(\chi_f^2/f)^{\frac{1}{2}}} \quad \text{(a Student's t variate)}$$

Thus we have the following theorem.

THEOREM 4.5.1 Under the assumption of normality, the nonnull distribution of t^* is approximately a noncentral t-distribution with f degrees of freedom and noncentrality parameter Δ^2, where

$$\Delta = \frac{\mu_1 - \mu_2}{(\sigma_1^2/n_1 + \sigma_2^2/n_2)^{\frac{1}{2}}}$$

The null distribution of t^* is approximately a central Student's t distribution with f degrees of freedom.

Using Theorem 4.5.1, $C_{\alpha^*/2}$ is then approximated by $t_{\alpha^*/2}(f)$. For large sample sizes n_1 and n_2, σ_1 and σ_2 can justifiably be replaced by s_1 and s_2, respectively, in which case an estimate $\hat{f}$ of f is given by

$$\hat{f} = \frac{(s_1^2/n_1 + s_2^2/n_2)^2}{[s_1^4/n_1^2(n_1 - 1)] + [s_2^4/n_2^2(n_2 - 1)]} \tag{4.5.3}$$

Under the assumption of normality, the Student's t distribution above is known to provide remarkably accurate approximations to the percentage points of t^*, even for small samples (Welch, 1937; Yuen, 1974). Unfortunately, these percentage points are quite sensitive to departures from normality. One would, therefore, look for a test that has good robustness properties, for example, has robustness of validity (stability of type I error from distribution to distribution) and also has higher power overall than the t^* test for nonnormal populations but has almost the same power for normal populations. Such a test is obtained by replacing the sample means $\bar{x}_1$ and the sample standard deviations s_i by the corresponding MML10 estimators $\hat{\mu}_i$ and $\hat{\sigma}_i$ and by replacing n_i by m_i. A robust analog of t* is then defined as

$$t_c^* = \frac{\hat{\mu}_1 - \hat{\mu}_2}{(\hat{\sigma}_1^2/m_1 + \hat{\sigma}_2^2/m_2)^{\frac{1}{2}}} \tag{4.5.4}$$

One rejects H_0 at level α^* in favor of $H_1: \mu_1 \neq \mu_2$ if $|t_c^*| \geq t_{\alpha^*/2}(\hat{\nu}_1)$ where $\hat{\nu}_1$ is determined by the equations

$$\frac{1}{\hat{\nu}_1} = \frac{h_1^2}{A_1 - 1} + \frac{(1 - h_1)^2}{A_2 - 1}$$

and

$$h_1 = \frac{\hat{\sigma}_1^2/m_1}{\hat{\sigma}_1^2/m_1 + \hat{\sigma}_2^2/m_2} \tag{4.5.5}$$

THEOREM 4.5.2 Under the assumption of normality, the null distribution of t_c^* for large samples is a central t distribution (approximately) with $\hat{\nu}_1$ degrees of freedom, while the nonnull distribution of t_c^* for large samples is a noncentral t distribution (approximately) with $\hat{\nu}_1$ degrees of freedom and noncentrality parameter Δ_c^2, where

$$\Delta_c^2 = \frac{(\mu_1 - \mu_2)^2}{\sigma_1^2/m_1 + \sigma_2^2/m_2}$$

Proof: The results follow immediately from Lemma 2.7.2 and Theorem 4.5.1.

Since the noncentral t with ν degrees of freedom and noncentrality parameter Δ^2 converges to normal $N(\Delta, 1)$ distribution as $\nu \to \infty$, the asymptotic power functions of the t* and t_c^* tests are given respectively by

$$P(|Z + \Delta| \geq Z_{\alpha^*/2}) \quad \text{and} \quad P(|Z + 0.983\Delta| \geq Z_{\alpha^*/2}) \tag{4.5.6}$$

where $\Delta = (\mu_1 - \mu_2)/(\sigma_1^2/n_1 + \sigma_2^2/n_2)^{\frac{1}{2}}$; Z is a standard normal variate and Z_{α^*} its $100(1 - \alpha^*)\%$ point. Note that Δ reduces to $\Delta = (\mu_1 - \mu_2)\sqrt{n}/\sigma\sqrt{1+\phi}$ ($\sigma_1^2 = \phi\sigma^2$, $\sigma_2^2 = \sigma^2$) if $n_1 = n_2 = n$.

The values of the power of t* and t_c^* tests, calculated from (4.5.6), are given below. It is clear that for normal samples, t_c^* is only slightly less powerful than t*:

Δ:	0.0	1.0	2.0	3.0	4.0
t*	0.01	0.06	0.28	0.66	0.92
t_c^*	0.01	0.05	0.27	0.65	0.91

The asymptotic power function of the analogous statistic based on the trimmed sample estimators TM10,

$$t^*_{trim} = \frac{\hat{\mu}_{1trim} - \hat{\mu}_{2trim}}{\{[\hat{\sigma}^2_{1trim}/(n_1 - 2r_1) + \hat{\sigma}^2_{2trim}/(n_2 - 2r_2)]\}^{\frac{1}{2}}} \tag{4.5.7}$$

is given by $P(|Z + 0.971\Delta| \geq Z_{\alpha^*/2})$ $(n_1 = n_2)$, and therefore the t^*_{trim} test is slightly less powerful than the t^*_c test, for normal distributions.

Since $\hat{\mu}_1 - \hat{\mu}_2$ is unbiased (or almost unbiased), and $\sigma(t^*_c)$ is 1 (or almost 1) for large samples, we have the following theorem:

THEOREM 4.5.3 Under some very general regularity conditions, the asymptotic power function of the t^*_c test for testing H_0 versus H_1 at level α^* is given by

$$P(|Z + \Delta_c| \geq Z_{\alpha^*/2}) \tag{4.5.8}$$

where $\Delta_c \simeq (\mu_1 - \mu_2)/[E(\hat{\sigma}_2)(\phi_c/m_1 + 1/m_2)^{\frac{1}{2}}]$ $[\phi_c = E(\hat{\sigma}_1^2)/E(\hat{\sigma}_2^2)]$; Δ_c reduces to $0.983(\mu_1 - \mu_2)\sqrt{n}/[E(\hat{\sigma}_2)(1 + \phi_c)^{\frac{1}{2}}]$ if $n_1 = n_2 = n$.

Proof: This follows from the fact that $(\hat{\sigma}_1^2/m_1 + \hat{\sigma}_2^2/m_2)$ converges in probability to $[E(\hat{\sigma}_2)]^2(\phi_c/m_1 + 1/m_2)$ as n_1 and n_2 tend to infinity, in which case t^*_c is a linear combination of order statistics.

For comparing the asymptotic power functions of t^* and t^*_c tests, one may note that the asymptotic power function of the t^* test is always given by the first equation in (4.5.6).

For the family of distributions (2.3.1), for example, $E(\hat{\sigma}) = h\sigma$ $(h < 1)$, $\phi_c = \phi h\sigma/h\sigma = \phi$ and $\Delta_c/\Delta > 1$ for $n_1 = n_2$. Therefore, the t^*_c test is asymptotically more powerful than the t^* test for $n_1 = n_2$, for the family (2.3.1). The same result is of course true for $n_1 \neq n_2$.

To investigate the robustness of validity of the tests based on t^*, t^*_c, and t^*_{trim}, and the test based on the Huber estimators (say BS82), namely,

$$t^*_B = \frac{\hat{\mu}_{1B} - \hat{\mu}_{2B}}{(\hat{\sigma}^2_{1B}/n_1 + \hat{\sigma}^2_{2B}/n_2)^{\frac{1}{2}}} \tag{4.5.9}$$

Tiku and Singh (1981) simulated (from 4000 Monte Carlo runs) the values of the probabilities $P(|t^*| \geq a_1)$, $P(|t^*_c| \geq a_2)$, $P(|t^*_{trim}| \geq a_3)$, and $P(|t^*_B| \geq a_4)$ for numerous symmetric and skew distributions; a_1 and a_2 are the $100(1 - \alpha^*/2)\%$ points of Student's t distributions with ν and ν_1 degrees of freedom given by (4.5.3) and (4.5.5), respectively, and a_3 and a_4 are the

Table 4.5.1 Simulated Values of the Probability[a] $P(|t| \geq a \mid H_0)$

	$n_1 = n_2 = 10$				$n_1 = 10,\ n_2 = 20$			
α^*	t^*	t^*_c	t^*_{trim}	t^*_B	t^*	t^*_c	t^*_{trim}	t^*_B
				Uniform				
0.05	0.054	0.057	0.060	0.096	0.047	0.042	0.049	0.068
0.01	0.012	0.016	0.016	0.040	0.010	0.010	0.011	0.020
				Normal				
0.05	0.050	0.056	0.054	0.076	0.049	0.054	0.054	0.066
0.01	0.009	0.011	0.010	0.022	0.011	0.012	0.012	0.017
				(n - 1)N(0, 1) and 1N(0, 10)				
0.05	0.013	0.055	0.049	0.070	0.018	0.056	0.050	0.060
0.01	0.001	0.010	0.009	0.019	0.001	0.012	0.011	0.016
				Student's t (df = 2)				
0.05	0.034	0.050	0.041	0.061	0.036	0.056	0.042	0.058
0.01	0.005	0.008	0.006	0.015	0.005	0.011	0.008	0.013
				Weibull: c = 2 ($\beta_1^* = 0.40$)				
0.05	0.055	0.055	0.055	0.092	0.052	0.052	0.053	0.077
0.01	0.010	0.013	0.012	0.029	0.010	0.013	0.012	0.022

[a] t^*_B is based on the BS82 estimators, which is one of the most efficient pair of Huber's M estimators (Gross, 1976, pp. 411-416).

$100(1 - \alpha^*/2)\%$ points of Student's t distributions with ν_2 and ν_3 degrees of freedom, respectively:

$$\frac{1}{\nu_2} = \frac{h_2^2}{A_1 - 1} + \frac{(1 - h_2)^2}{A_2 - 1}$$

and (4.5.10)

$$h_2 = \frac{\hat{\sigma}^2_{1trim}/A_1}{\hat{\sigma}^2_{1trim}/A_1 + \hat{\sigma}^2_{2trim}/A_2}$$

$$\frac{1}{\nu_3} = \frac{h_3^2}{n_1} + \frac{(1 - h_3)^2}{n_2}$$

and (4.5.11)

$$h_3 = \frac{\hat{\sigma}_{1B}^2/n_1}{\hat{\sigma}_{1B}^2/n_1 + \hat{\sigma}_{2B}^2/n_2}$$

In (4.5.11), the divisors should perhaps be smaller than n_1 and n_2, but it is not clear what their values should be when sample sizes are finite; n_1 and n_2 are, however, the correct divisors asymptotically (Boos, 1980; Gross, 1976).

The values of these probabilities are given in Table 4.5.1 for a few representative distributions. It is clear that only the t_c^* and t_{trim}^* tests have robustness of validity (stability of type I error over numerous distributions). Note here that in Table 4.5.1, the Weibull distribution is given by

$$\text{constant} \times e^{-x^c} x^{c-1} \qquad 0 < x < \infty$$

Tiku and Singh (1981a) determined the values of the power of the t_c^* and t_{trim}^* tests and found the latter trailing behind always; Dunnett (1982) reports similar results. For example, Tiku and Singh (1981a) report the following values of the power $1 - \beta^*$; $d = \mu_1 - \mu_2$, $\sigma_1^2/\sigma_2^2 = 4$, $n_1 = n_2 = 10$, and $\alpha^* = 0.01$:

	d = 1.0		d = 2.0		d = 3.0		d = 4.0	
Distribution	t_c^*	t_{trim}^*	t_c^*	t_{trim}^*	t_c^*	t_{trim}^*	t_c^*	t_{trim}^*
Uniform	0.04	0.04	0.12	0.11	0.31	0.30	0.60	0.58
Normal	0.09	0.08	0.40	0.39	0.81	0.79	0.98	0.98
Student's t (df = 2)	0.05	0.04	0.22	0.20	0.49	0.47	0.74	0.73
Weibull (c = 2)	0.02	0.02	0.11	0.11	0.33	0.31	0.63	0.61

Note that the t_c^* and t_{trim}^* tests, but not the t^* and t_B^* tests, can also be used if censoring of extreme observations is imposed by constraints of the experiment rather than applied for the purpose of achieving robustness (see Example 5.2.1).

EXAMPLE 4.5.1 Consider the data of Example 4.3.1. Here

$$t_c^* = \frac{4.563 - 3.215}{(0.210 + 0.328)^{\frac{1}{2}}}$$

$$= 1.837$$

$$h_1 = \frac{0.210}{0.538} = 0.390 \quad \text{and} \quad \frac{1}{\nu_1} = 0.030 + 0.074 = 0.104$$

The degrees of freedom ν of t* is given by

$$\nu = \frac{1}{0.086} = 11.53$$

The probability

$$P_{H_0}(t^* \geq 1.769) > 0.051$$

The t* test, like the t test, fails to reject H_0 at 5% significance level; the effect of the potential outliers 0.0 and 6.6 in the second sample seems persistent.

For the t_c^* test,

$$t^* = \frac{4.662 - 3.188}{(1.494/8 + 4.067/8)^{\frac{1}{2}}}$$

$$= 1.769$$

$$h = \frac{0.187}{0.695} = 0.269 \quad \text{and} \quad \frac{1}{\nu} = 0.010 + 0.076 = 0.086$$

The degrees of freedom ν_1 of t_c^* is given by

$$\nu_1 = \frac{1}{0.104} = 9.5$$

The probability

$$P_{H_0}(t_c^* \geq 1.837) < 0.05$$

The t_c^* test, unlike the t* test, rejects the null hypothesis H_0 at the 5% significance level.

Note that the degrees of freedom of t and t* are 14 and 11.53, respectively, the difference between these degrees of freedom being 2.47. The degrees of freedom of t_c and t_c^* are 10 and 9.54, respectively, the difference between these two degrees of freedom being 0.46 only. This indicates that the t test is more sensitive than the t_c test to differences between the population variances resulting from potential outliers, as expected.

4.6 COMPARISON OF TWO VARIANCES

Let $x_{i1}, x_{i2}, \ldots, x_{in_i}$ (i = 1, 2) be two independent random samples from populations of the type $(1/\sigma_i)f((x - \mu_i)/\sigma_i)$. To test $H_0: \phi = \sigma_1^2/\sigma_2^2 = 1$ against $H_1: \phi > 1$ under the assumption of normality, the UMP test is based on the statistic F, which is the ratio of the two sample variances, namely,

$$F = \frac{s_1^2}{s_2^2} \tag{4.6.1}$$

One rejects H_0 in favor of H_1 at level α^* if $F \geq F_{\alpha^*}(n_1 - 1, n_2 - 1)$, where $F_{\alpha^*}(n_1 - 1, n_2 - 1)$ is the $100(1 - \alpha^*)\%$ point of a central F distribution with $(n_1 - 1, n_2 - 1)$ degrees of freedom. This follows from the fact that under the assumption of normality, the sampling distribution of F is a central F distribution with $(n_1 - 1, n_2 - 1)$ degrees of freedom.

We have seen in Chapter 1 that the type I error of the above F test is very sensitive to changes in the underlying distribution (particularly the kurtosis μ_4/μ_2^2). Therefore, the classical F test does not have robustness of validity. It is possible to develop a robust and powerful test for testing the null hypothesis $H_0: \phi/\phi_0 = 1$ against the alternative hypotheses $H_1: \phi/\phi_0 > 1$, where ϕ_0 is any specified value between 1 and the true value ϕ $(1 \leq \phi_0 \leq \phi)$. For describing how such tests can be derived, we begin by distinguishing between the following three cases:

1. μ_1 and μ_2 are known, in which case define $y_{1i} = (x_{1i} - \mu_1)^2$, i = 1, 2, ..., n_1, and $y_{2i} = (x_{2i} - \mu_2)^2$, i = 1, 2, ..., n_2; note that $E(y_{1i}) = \sigma_1^2$ and $E(y_{2i}) = \sigma_2^2$.
2. μ_1 and μ_2 are unknown but equal and equal to μ, say, in which case define $y_{1i} = x_{1i}^2$, i = 1, 2, ..., n_1, and $y_{2i} = x_{2i}^2$, i = 1, 2, ..., n_2; note that $E(y_{1i}) = \mu^2 + \sigma_1^2$ and $E(y_{2i}) = \mu^2 + \sigma_2^2$.
3. μ_1 and μ_2 are unknown and unequal, in which case we assume that the sample sizes are large and define $y_{1i} = (x_{1i} - \bar{x}_1)^2$, i = 1, 2, ..., n_1, and $y_{2i} = (x_{2i} - \bar{x}_2)^2$, i = 1, 2, ..., n_2; note that $E(y_{1i}) = \sigma_1^2(1 - 1/n_1) \simeq \sigma_1^2$ for large n_1, and $E(y_{2i}) = \sigma_2^2(1 - 1/n_2) \simeq \sigma_2^2$ for large n_2. It may also be noted that $\text{cov}(x_{1i} - \bar{x}_1, x_{1j} - \bar{x}_1) = -\sigma_1^2/n_1 \simeq 0$ for large n_1, and $\text{cov}(x_{2i} - \bar{x}_2, x_{2j} - \bar{x}_2) = -\sigma_2^2/n_2 \simeq 0$ for large n_2.

From the definitions of y_{1i} and y_{2j}, it is clear that the distributions of y_{1i} and y_{2j} will usually be positively skew. Thus by using results given in Section 4.4, robust and powerful tests for testing H_0 versus H_1 can be developed by using the MMLE based on the samples y_{1i}, i = 1, 2, ..., n_1, and y_{2j}, j = 1, 2, ..., n_2. We then consider the following censored samples:

$$y_{(1,1)}, y_{(1,2)}, \ldots, y_{(1,n_1-r_1)} \tag{4.6.2}$$

and

$$y_{(2,1)}, y_{(2,2)}, \ldots, y_{(2,n_2-r_2)} \tag{4.6.3}$$

where $r_1 = [0.5+0.2n_1]$ and $r_2 = [0.5+0.2n_2]$ are the number of largest observations censored in the two ordered samples, respectively. Let K_{1c} and $\hat{\sigma}_{1c}$, and K_{2c} and $\hat{\sigma}_{2c}$, be the estimators (4.4.1) calculated from the censored samples (4.6.2) and (4.6.3), respectively. Define

$$\sigma_c^* = \left[\frac{(A_1 - 1)(\hat{\sigma}_{1c}/\phi)^2 + (A_2 - 1)\hat{\sigma}_{2c}^2}{A_1 + A_2 - 2}\right]^{\frac{1}{2}} \tag{4.6.4}$$

To test $H_0: \phi/\phi_0 = 1$, the proposed statistic is (Tiku, 1982b)

$$t_S^* = \frac{K_{1c} - K_{2c}}{\sigma_c^*(1/m_1 + 1/m_2)^{\frac{1}{2}}} \tag{4.6.5}$$

large values of t_S^* lead to the rejection of H_0 in favor of $H_1: \phi/\phi_0 > 1$. In the expression (4.6.4), ϕ is not known but it is replaced by ϕ_0, $1 \le \phi_0 \le \phi$.

It is clear that if the underlying distributions are normal, then the null distribution of t_S^* is exactly the same as that of t_S (based on samples from χ_1^2 populations). The t_S^* test has reasonable type I robustness (Tiku, 1982b). For $n_1 = n_2 = 10$, for example, Tiku (1982b) gave the following simulated (based on 4000 Monte Carlo runs) values of the probability $P(t_S^* \ge c \mid H_0)$ for the case (3), which is the most important one from a practical point of view:

c	Normal	Logistic	Double Exponential	Student's t_4	Uniform
2.10	0.049	0.051	0.051	0.049	0.056
2.90	0.009	0.006	0.007	0.010	0.013

For large samples, the power function of t_S^* is given by

$$P(Z \ge Z_{\alpha*} - \Delta_S^*) \qquad \Delta_S^* = \frac{E(d_S^*)}{[V(d_S^*)]^{\frac{1}{2}}} \tag{4.6.6}$$

$d_S^* = K_{1c} - K_{2c}$; Z is as usual a standard normal variate and $Z_{\alpha*}$ is its $100(1 - \alpha*)\%$ point.

Under the assumption of normality, the power function of the statistic F is given by

$$P\left(F \geq \frac{1}{\phi} F_{\alpha*}\right) \tag{4.6.7}$$

where F has the F distribution with $(n_1 - 1, n_2 - 1)$ degrees of freedom and $F_{\alpha*}$ is its $100(1 - \alpha*)$% point.

The values of (4.6.6) can be calculated from the appropriate values given in Table 4.4.1 (values for $a = 0.5$). For $n_1 = n_2 = 80$, for example, we have for normal populations

$$\Delta_S^* = \frac{0.319(\phi - 1)}{\{0.0036[(\phi/\phi_0)^2 + 1]\}^{\frac{1}{2}}}$$

and consequently, we have the following values of the power; $\phi_0 = 1$:

$\sqrt{\phi}$:	1.0	1.1	1.2	1.3	1.4	1.5	1.6
F	0.05	0.21	0.51	0.75	0.91	0.97	0.99
t_S^*	0.05	0.18	0.38	0.59	0.75	0.86	0.92

It is clear that the t_S^* test has satisfactory power compared to the F test, even under the assumption of normality, in which case the F test is UMP (see also Tiku, 1982b; Tiku and Balakrishnan, 1984a).

The F test should not be used, particularly for symmetric nonnormal populations, but there is no such restriction on the t_S^* test. The t_S^* test has in fact very good power properties for nonnormal populations (Tiku, 1982b; Tiku and Balakrishnan, 1984a).

EXAMPLE 4.6.1 Consider the following 20 values of $y_{1i} = (x_{1i} - \bar{x}_1)^2$ and $y_{2i} = (x_{2i} - \bar{x}_2)^2$, arranged in ascending order of magnitude (Tiku, 1982b):

y_1: 0.0001 0.0002 0.0018 0.0270 0.0415 0.1912 0.1927 0.2089
0.2430 0.3289 0.5191 0.7829 1.2303 1.5136 1.5565 1.8547
2.5946 2.9363 3.8233 5.4535

y_2: 0.0132 0.0176 0.0803 0.0859 0.0992 0.1122 0.2364 0.2711
0.2877 0.3259 0.4002 0.5173 0.5334 0.6864 0.8225 0.9606
1.0467 1.0807 1.1664 1.8295

One wants to test the null hypothesis $H_0: \sigma_1^2/\sigma_2^2 = 1$; $\sigma_1^2 = V(x_1)$ and $\sigma_2^2 = V(x_2)$.

To calculate the statistic t_S^*, we censor the $r_1 = r_2 = [0.5 + 0.2(20)] = 4$ largest observations in each of the two ordered samples and obtain the following; here $n_1 = n_2 = 20$, $r_1/n_1 = r_2/n_2 = 0.2$, $\alpha_1 = \alpha_2 = 0.7454$, and $\beta_1 = \beta_2 = 0.7859$ (Table B.1), and

$K_{1c} = 0.7577$ $\quad K_{2c} = 0.4420$ $\quad B_1 = 3.2551$ $\quad B_2 = 1.5389$ $\quad C_1 = 10.6133$

$C_2 = 2.2870$ $\quad \hat{\sigma}_{1c} = 0.9528$ $\quad \hat{\sigma}_{2c} = 0.4433$ $\quad \hat{\sigma}_c = 0.7430$ $(\phi_0 = 1)$

$t_S^* = 1.314$

The approximate 95% point of t_S^* is 2.15 [based on χ_1^2 and (4.4.4)]; the simulated 95% point is 2.10. [In fact, Tiku and Balakrishnan (1984a, Table I) give values of $E(K)$, $V(K)$, and $E(\hat{\sigma})$ for sample sizes $n = 10(1)40(5)100$ which can be used to obtain the approximate percentage points of t_S^* from (4.4.4).] Since the computed value of t_S^* is smaller than the 95% point, there is no evidence against H_0 at 5% significance level.

If one adopts the classical approach, then $F = s_1^2/s_2^2 = 2.223$, which is larger than the 95% point of a central F distribution with (19, 19) degrees of freedom and hence H_0 would be rejected. This is, however, erroneous since x_{1i} and x_{2i} are two independent random samples coming from the same logistic distribution.

Comparison with Other Tests

A number of robust tests for testing the equality of two (in fact, any number k) population variances is available. The prominent ones are the following (see Brown and Forsythe, 1974; Conover et al., 1981).

Levene (1960). Let $z_{ij} = |x_{ij} - \bar{x}_i|$; $i = 1, 2$, $j = 1, 2, \ldots, n_i$; then Levene's test is based on the ANOVA-type statistic

$$W_0 = \frac{n_1 n_2 (\bar{z}_1 - \bar{z}_2)^2/(n_1 + n_2)}{[\Sigma_{j=1}^{n_1} (z_{1j} - \bar{z}_1)^2 + \Sigma_{j=1}^{n_2} (z_{2j} - \bar{z}_2)^2]/(n_1 + n_2 - 2)} \tag{4.6.8}$$

For large samples, the null distribution of W_0 is approximately an F distribution with $(1, n_1 + n_2 - 2)$ degrees of freedom.

W_{50} test: In Levene's test, the sample means $\bar{x}_1$ and $\bar{x}_2$ are replaced by the sample medians $\tilde{x}_1$ and $\tilde{x}_2$, respectively.

W_{10} test: In Levene's test, the sample means are replaced by the 10% trimmed sample means; a 10% trimmed mean is the mean of the $n - 2r$ observations obtained by censoring $r = [0.5 + 0.1n]$ smallest and largest observations in a random sample of size n.

Layard (1973). Layard's test is based on the statistic

$$S = \frac{(n_1 - 1)(n_2 - 1)[\log s_1^2 - \log s_2^2]^2/(n_1 + n_2 - 2)}{\hat{\tau}^2} \qquad (4.6.9)$$

where s_1^2 and s_2^2 are the sample variances and

$$\hat{\tau}^2 = 2 + \left(1 - \frac{2}{n_1 + n_2}\right)\left\{\frac{(n_1 + n_2)\Sigma_{j=1}^{n_1}(x_{1j} - \bar{x}_1)^4 + \Sigma_{j=1}^{n_2}(x_{2j} - \bar{x}_2)^4}{[\Sigma_{j=1}^{n_1}(x_{1j} - \bar{x}_1)^2 + \Sigma_{j=1}^{n_2}(x_{2j} - \bar{x}_2)^2]^2} - 3\right\}$$

For large samples, the null distribution of S is approximately a chi-square with 1 degree of freedom.

Miller (1968). Miller's jackknife statistic, and its null distribution, is exactly the same as Levene's W_0 but with z_{ij} replaced by

$$u_{ij} = n_i \log s_i^2 - (n_i - 1)\log s_{i(j)}^2 \qquad i = 1, 2, \quad j = 1, 2, \ldots, n_i$$

$$s_{i(j)}^2 = \frac{1}{n_i - 2}\sum_{\substack{k=1 \\ k \neq j}}^{n_i}[x_{ik} - \bar{x}_{i(j)}]^2 \qquad (4.6.10)$$

where $\bar{x}_{i(j)}$ is the mean of the ith sample with jth observation jackknifed.

Fligner and Killeen (1976). The modified version of the Fligner-Killeen test, as suggested by Conover et al. (1981), is to make use of the ranks of $|x_{ij} - \tilde{x}_i|$ and assign increasing normal scores

$$a_{N,i} = \Phi^{-1}\left[\frac{1}{2} + \frac{i}{2(n + 1)}\right] \qquad N = n_1 + n_2$$

based on those ranks. The test is then based on the statistic

$$\chi^2 = \frac{n_1 n_2(\bar{A}_1 - \bar{A}_2)^2}{(n_1 + n_2)V^2} \qquad (4.6.11)$$

where

$$V^2 = \sum_{i=1}^{N}\frac{(a_{N,i} - \bar{a})^2}{N - 1} \qquad \bar{a} = \frac{n_1\bar{A}_1 + n_2\bar{A}_2}{n_1 + n_2}$$

For large samples, the null distribution of χ^2 is approximately a chi-square with 1 degree of freedom.

Fligner and Killeen (1976) proposed another test,

$$\frac{(N - 2)\chi^2}{N - 1 + \chi^2}$$

but this has essentially the same power properties as the χ^2 test above and will not therefore be considered further. Large values of these statistics lead to the rejection of H_0 in favor of H_1.

To compare the t_S^* test (and an analogous t* test which is based on symmetric type II censored samples) with the tests above, Tiku and Balakrishnan (1984a) simulated the power values for sample sizes $(n_1, n_2) = (30, 30)$, $(40, 20)$, and $(40, 40)$. The distributions of $(x_1 - \mu_1)/\sigma_1$ and $(x_2 - \mu_2)/\sigma_2$, μ_1 and μ_2 unknown and unequal, they considered are the following symmetric distributions:

(1) normal, (2) logistic, (3) double exponential, (4) Student's t with 6 df, (5) Student's t with 4 df, (6) mixture model: $0.90N(0,1) + 0.10N(0,4)$, (7) outlier model: $(n-j)N(0,1)$ and $jN(0,4)$ if $n_1 = n_2 = n$, $j = [n/10]$

Table 4.6.1 Simulated Values of the Power for Testing $\phi = 1$ Against $\phi \neq 1$; $n_1 = n_2 = 40$ and Type I Error $\alpha^* = 0.10$

Distribution	$\phi = \sigma_1^2/\sigma_2^2$	F	Le	W_{50}	W_{10}	La	J	F - K	$\lvert t_S^* \rvert$	$\lvert t^* \rvert$
(1)	1	0.10	0.11	0.09	0.11	0.09	0.10	0.10	0.09	0.09
	2	0.65	0.50	0.47	0.49	0.54	0.50	0.45	0.50	0.58
	4	1.00	0.98	0.98	0.98	0.99	0.99	0.96	0.97	0.99
(2)	1	0.17	0.09	0.09	0.08	0.09	0.10	0.10	0.10	0.11
	2	—	0.43	0.41	0.43	0.43	0.42	0.38	0.48	0.53
	4	—	0.96	0.95	0.96	0.95	0.91	0.93	0.96	0.98
(3)	1	0.25	0.09	0.09	0.10	0.10	0.11	0.12	0.12	0.12
	2	—	0.33	0.30	0.32	0.33	0.30	0.29	0.46	0.49
	4	—	0.87	0.85	0.86	0.83	0.78	0.82	0.90	0.91
(4)	1	0.23	0.10	0.10	0.12	0.11	0.11	0.11	0.12	0.13
	2	—	0.42	0.38	0.40	0.40	0.40	0.39	0.47	0.54
	4	—	0.93	0.91	0.92	0.89	0.87	0.90	0.93	0.95
(5)	1	0.31	0.09	0.08	0.07	0.09	0.12	0.10	0.09	0.11
	2	—	0.32	0.29	0.31	0.31	0.33	0.31	0.45	0.52
	4	—	0.86	0.84	0.85	0.79	0.75	0.86	0.93	0.95
(6)	1	0.51	0.10	0.09	0.10	0.09	0.13	0.09	0.11	0.12
	2	—	0.32	0.27	0.28	0.26	0.29	0.31	0.51	0.54
	4	—	0.76	0.75	0.75	0.57	0.53	0.84	0.94	0.94
(7)	1	0.43	0.06	0.07	0.05	0.06	0.10	0.10	0.10	0.10
	2	—	0.24	0.21	0.22	0.15	0.21	0.30	0.47	0.52
	4	—	0.79	0.76	0.78	0.54	0.50	0.87	0.96	0.97

Table 4.6.2 Values of the Power for Half-Normal; $(n_1, n_2) = (40,40)$ and the Significance Level $\alpha^* = 0.10$

ϕ	F	Le	W_{50}	W_{10}	La	J	F - K	$\lvert t_S^* \rvert$	$\lvert t^* \rvert$
1	0.15	0.13	0.09	0.10	0.10	0.10	0.11	0.12	0.15
2	0.65	0.50	0.41	0.45	0.46	0.43	0.42	0.60	0.65
4	0.99	0.96	0.92	0.95	0.95	0.92	0.92	0.97	0.99

These power values (based on 1000 Monte Carlo runs) for $(n_1, n_2) = (40, 40)$ are given in Table 4.6.1 for significance level 0.10. It is clear from these values that both the t_S^* and t^* tests are more powerful than all the other robust tests. Comparing the t_S^* test with the t_S test, Tiku and Balakrishnan (1984a) concluded that although the former is somewhat less powerful for symmetric distributions but has better type I error robustness for skew distributions; see Table 4.6.2, which gives the simulated values of the power when the distributions of $(x_1 - \mu_1)/\sigma_1$ and $(x_2 - \mu_2)/\sigma_2$ are half-normal, namely,

$$f(z) = \text{constant} \times e^{-z^2/2} \qquad 0 < z < \infty \tag{4.6.12}$$

The recommendation is therefore that one should use the t_S test if the underlying distribution is symmetric, at any rate not too skew, but otherwise one should use the t_S^* test. The percentage points of both t_S and t_S^* can easily be calculated from the tables of E(K), var (K), and $E(\hat{\sigma})$ given by Tiku and Balakrishnan (1984a).

4.7 SPROTT APPROACH TO ROBUST TESTS

Let $f_\lambda(x;\theta)$ be a probability density function, θ being a location parameter; λ is a "shape" parameter and is a quantitative measure of deviation from normality. Define the relative likelihood function

$$R = \frac{f_\lambda(x;\theta)}{f_\lambda(x;\hat{\theta})} \tag{4.7.1}$$

where $\hat{\theta} = \hat{\theta}(x, \lambda)$ is the maximum likelihood estimate of θ based on the observed x, for a specified λ. To find the approximate distribution of R, expand $\log R(\theta)$ as a Taylor series (around $\hat{\theta}$):

$$\log R(\theta) \simeq -\frac{1}{2} u^2\left(1 + \frac{1}{3} uF_3 - \frac{1}{12} u^2 F_4\right) \tag{4.7.2}$$

where

$$u = (\hat{\theta} - \theta)\sqrt{I_{\hat{\theta}}}$$

and

$$I_{\hat{\theta}} = I_{\hat{\theta}}(x) = -\frac{\partial^2 \log R}{\partial \hat{\theta}^2} = -\left[\frac{\partial^2}{\partial \theta^2} \log R(\theta)\right]_{\theta=\hat{\theta}}$$

is the "observed information" and

$$F_3 = \frac{\partial^3 \log R}{\partial \hat{\theta}^3} I_{\hat{\theta}}^{-3/2}$$
$$F_4 = \frac{\partial^4 \log R}{\partial \hat{\theta}^4} I_{\hat{\theta}}^{-2} \tag{4.7.3}$$

are the third and fourth standardized derivatives of log R; F_3 and F_4 estimate, respectively, the skewness and kurtosis of the distribution of R (Bartlett, 1953). Suppose that F_3 is zero (almost) and let

$$F_4(\hat{\theta}) = \frac{6}{M+1} \quad \text{and} \quad u\left(\frac{M}{M+1}\right)^{\frac{1}{2}} = t_M \tag{4.7.4}$$

Then (4.7.2) becomes

$$-\frac{1}{2}\frac{M+1}{M}t_M^2\left(1 - \frac{1}{12}\frac{M+1}{M}\frac{6}{M+1}t_M^2\right) = -\frac{M+1}{2}\left(\frac{1}{M}t_M^2 - \frac{1}{2M^2}t_M^4\right)$$
$$\simeq -\frac{M+1}{2}\log\left(1 + \frac{1}{M}t_M^2\right) \tag{4.7.5}$$

which is the log-likelihood of Student's t distribution with M degrees of freedom. Thus the distribution of

$$t_M = \frac{(\hat{\theta} - \theta)\sqrt{MI_{\hat{\theta}}}}{\sqrt{M+1}} \tag{4.7.6}$$

is approximately Student's t distribution with M degrees of freedom. If we write $I_{\hat{\theta}} = 1/\hat{\sigma}_\theta$, where $\hat{\sigma}_\theta$ is the standard error of $\hat{\theta}$, then

$$t_M = \frac{(\hat{\theta} - \theta)\sqrt{M}}{\hat{\sigma}_\theta\sqrt{M+1}} \tag{4.7.7}$$

has approximately Student's t distribution with M degrees of freedom for all values of λ, and therefore the t_M test for testing H_0: $\theta = 0$, say, will have criterion robustness (stability of type I error for different values of λ).

For a location-scale family $f_\lambda(x;\theta,\sigma)$, $I_{\hat{\theta}}$ will involve σ, but σ is replaced by its maximum likelihood estimate $\hat{\sigma}$. Realize that $\hat{\theta}$ and $\hat{\sigma}$ are both functions of λ and therefore a knowledge of λ is needed to compute the statistic t_M. To locate plausible value(s) of λ, one may examine the relative likelihood (Fraser, 1976)

$$R_M(\lambda) = \frac{f_\lambda[x;\hat{\theta}(\lambda), \hat{\sigma}(\lambda)]}{\sup_\lambda f[x;\hat{\theta}(\lambda), \hat{\sigma}(\lambda)]} \tag{4.7.8}$$

and map out ranges of plausible values of λ, for example, those values of λ for which $R_M(\lambda)$ is appreciably greater than 0 (Sprott, 1982).

Estimators which are formulated in such a way that they can utilize the information provided by the sample about the underlying distribution are called "adaptive" estimators (Sprott, 1982; Hogg, 1982). Consider the family of distributions [which is essentially the t distribution family (2.3.1)] considered by Sprott (1982), namely,

$$f_\lambda\left(\frac{x-\theta}{\sigma}\right) = \frac{1}{\sigma}\frac{\Gamma((\lambda+1)/2)}{\sqrt{\pi\lambda}\,\Gamma(\lambda/2)}\left[1+\frac{1}{\lambda}\left(\frac{x-\theta}{\sigma}\right)^2\right]^{-(\lambda+1)/2} \qquad -\infty < x < \infty \tag{4.7.9}$$

For a specified value of λ, the maximum likelihood estimates $\hat{\theta}$ and $\hat{\sigma}$ are obtained iteratively as solutions of the equations

$$\frac{\partial \log L}{\partial \theta} = 0 \quad \text{and} \quad \frac{\partial \log L}{\partial \sigma} = 0$$

where the log-likelihood is given by

$$\log L = \text{constant} - n\log\sigma - \frac{1}{2}(\lambda+1)\sum_{i=1}^{n}\log\left[1+\frac{1}{\lambda}\left(\frac{x_i-\theta}{\sigma}\right)^2\right]$$

Noting that $F_3(\hat{\theta}) \simeq 0$, one may work out the value of $F_4(\hat{\theta})$ with $\sigma = \hat{\sigma}$ from the second and fourth derivatives of $\log R$ and obtain the value of M from the equation $F_4(\hat{\theta}) = 6/(M+1)$. The value of t_M is then obtained from (4.7.7) under H_0 and referred to Student's t tables.

EXAMPLE 4.7.1 Consider the data of Example 2.7.1 (Darwin's data; Fisher, 1966). For these data, Sprott (1982) gives the following values [assuming that the underlying distribution is (4.7.9)]:

λ	$\hat{\theta}$	$\hat{\sigma}_\theta$	$F_3(\hat{\theta})$	$F_4(\hat{\theta})$	M	$\hat{\sigma}_\theta\sqrt{(M+1)/M}$	$t_M(\theta = 0)$	P
2	26.6543	6.9883	0.16	0.4154	13	7.2521	3.68	0.005
5	25.9804	8.1758	-0.13	0.4627	12	8.5097	3.05	0.01
10	24.0966	8.8991	-0.12	0.3877	14	9.2114	2.62	0.02

From (4.7.8), the plausible values of λ are seen to be small, perhaps $\lambda \leq 2$ (Sprott, 1978); H_0: $\theta = 0$ will therefore be rejected in favor of H_1: $\theta \neq 0$, at any significance level greater than 0.5%. Realize that the Sprott procedure is adaptive since it utilizes the information about λ provided by the sample.

For the MML10 procedure, we have $\hat{\theta} = 27.713$ and $\hat{\sigma} = 25.498$; therefore,

$$t_c = \frac{(27.713)\sqrt{14.3}}{25.498} = 4.11 \quad (df = 10)$$

The probability $P(|t_0| \geq 4.11) = 0.002$; H_0 is rejected at any significance level greater than 0.2%. For the classical t-test, $\bar{x} = 20.933$ and $s = 37.7228$. Therefore

$$t = \frac{20.933\sqrt{15}}{37.7228} = 2.15 \quad (df = 14)$$

The probability $P(|t| \geq 2.15) = 0.049$; the classical test does not reject H_0 at any significance level less than 4.9%. This typically illustrates the insensitivity of the classical t test to potential outliers in the data. On the other hand, the t_M and t_c tests, while having type I robustness, are quite sensitive (robustness of efficiency) to these outliers. The computation of t_M is, however, quite involved, but t_c is easy to compute.

EXAMPLE 4.7.2 Consider the following data (the Cushney-Peebles sleep data; Fisher, 1970), which represent the differences in the number of hours of sleep gained using two drugs:

1.2 2.4 1.3 1.3 0 1.0 1.8 0.8 4.6 1.4

Assuming that the underlying distribution is (4.7.9), Sprott (1982) gives the following values:

λ	$\hat{\theta}$	$\hat{\sigma}_\theta$	$F_3(\hat{\theta})$	$F_4(\hat{\theta})$	M	$t_M(\theta = 0)$	$P[t_M \geq t_\alpha(M)] = \alpha$
2	1.2805	0.18659	0.25	1.299	4	6.16	< 0.005
5	1.3522	0.18659	0.26	0.8206	6	4.65	
10	1.4321	0.32013	0.23	0.6548	8	4.27	

From (4.7.8), the plausible values of λ are seen to be small; in fact, $R_M(\lambda)$ has its maximum at $\lambda = 1$ (Sprott, 1982). The null hypothesis H_0 is rejected.

For the t_c test, we have

$$\hat{\theta} = 1.4344 \quad \text{and} \quad \hat{\sigma} = 0.7171 \quad (m = 9.6626)$$

$$t_c = \frac{1.4344\sqrt{9.6626}}{0.7171} = 6.207 \quad (df = 7)$$

The probability $P(|t_c| \geq 6.207)$ being very small, the null hypothesis H_0 is rejected.

For the classical t test, $\bar{x} = 1.58$ and $s = 1.230$;

$$t = \frac{1.58\sqrt{10}}{1.230} = 4.062 \quad (\text{df} = 9)$$

The probability $P(|t| \geq 4.062)$ being very small, the null hypothesis is rejected.

In all the examples above, the t_M and t_c procedures are in agreement but the latter is much easier to compute. The t_c procedure also readily generalizes to multisample and multivariate situations, as we will see later.

To test assumed values of a location parameter θ, Hogg (1982) proposed test procedures based on estimators of the type

$$\tilde{\theta} = \sum_{i=1}^{k} w_i \tilde{\theta}_i \qquad \left(\sum_{i=1}^{k} w_i = 1\right) \tag{4.7.10}$$

$\tilde{\theta}_1, \tilde{\theta}_2, \ldots, \tilde{\theta}_k$ are unbiased estimators of θ and $w_i \geq 0$ are weights which are some special functions of sample observations. The weights w_i, in fact, are quantitative measurements of the tail lengths and/or skewness of the postulated distributions; Harter et al. (1979) proposed estimators which are similar in nature. These proposals seem promising but clearly need more study.

4.8 ROBUST CONFIDENCE INTERVALS

The $100(1 - \alpha^*)\%$ confidence interval for μ, based on the MML10 estimators $\hat{\mu}$ and $\hat{\sigma}$, is approximately

$$\left\{ \hat{\mu} - t_{(1/2)\alpha^*}(\nu) \frac{\hat{\sigma}}{\sqrt{m}},\ \hat{\mu} + t_{(1/2)\alpha^*}(\nu) \frac{\hat{\sigma}}{\sqrt{m}} \right\} \qquad \nu = A - 1$$

since (4.8.1)

$$P\left[-t_{(1/2)\alpha^*}(\nu) \leq \frac{(\hat{\mu} - \mu)\sqrt{m}}{\hat{\sigma}} \leq t_{(1/2)\alpha^*}(\nu)\right] \simeq 1 - \alpha^*$$

$t_{(1/2)\alpha^*}(\nu)$ is as usual the $100(1 - \frac{1}{2}\alpha^*)\%$ point of a Student's t distribution with ν degrees of freedom.

The corresponding confidence intervals based on the trimmed estimators TM10 and the classical estimators $(\bar{x}, s)$ are given by

$$\left\{ \hat{\mu}_{\text{trim}} - t_{(1/2)\alpha^*}(\nu) \frac{\hat{\sigma}_{\text{trim}}}{\sqrt{n - 2r}},\ \hat{\mu}_{\text{trim}} + t_{(1/2)\alpha^*}(\nu) \frac{\hat{\sigma}_{\text{trim}}}{\sqrt{n - 2r}} \right\} \qquad \nu = A - 1 \tag{4.8.2}$$

and

$$\left\{ \bar{x} - t_{(1/2)\alpha^*}(\nu) \frac{s}{\sqrt{n}},\ \bar{x} + t_{(1/2)\alpha^*}(\nu) \frac{s}{\sqrt{n}} \right\} \qquad \nu = n - 1 \tag{4.8.3}$$

respectively, and similarly for the Studentized tests based on the BS82 and w24 estimators. The confidence interval (4.8.3) is, however, expected to be much wider than (4.8.1) and (4.8.2) for long-tailed symmetric distributions.

For the data of Example 2.7.1, we have the following confidence intervals ($\alpha^* = 0.05$):

ML10:	(12.696, 42.730),	width = 30.034
TM10:	(11.398, 42.056),	width = 30.658
Sprott ($\lambda = 2$):	(11.560, 41.749),	width = 30.189
($\lambda = 5$):	(8.165, 43.795),	width = 35.630
BS82:	(8.634, 44.518),	width = 35.884
w24:	(8.336, 44.292),	width = 35.956
Classical:	(0.041, 41.825),	width = 41.784

Realize that a correct choice of λ is very crucial for Sprott's method to produce satisfactory confidence intervals.

The exceedingly good performance of the MML10 procedure compared to the Sprott procedure and other procedures can in fact be explained as follows: For the t-distribution family (2.3.1), the minimum variance bound for estimating μ is given by

$$R^2(\mu) = \frac{1}{-E(\partial^2 \log L/\partial\mu^2)}$$

$$= \frac{(p - 1.5)(p + 1)}{np(p - 0.5)} \sigma^2 \qquad (p \geq 2) \tag{4.8.4}$$

If the value of p were known exactly, one would use the Studentized t statistic based on the ML estimators of μ and σ. The asymptotic power function of this test for testing $H_0: \mu = 0$ versus $H_1: \mu > 0$ would be

$$P(Z \geq Z_{\alpha*} - \Delta_{opt}) \tag{4.8.5}$$

where

$$\Delta_{opt} = \mu \frac{\sqrt{np(p - 0.5)}}{\sqrt{(p - 1.5)(p + 1)}} \frac{1}{\sigma} \tag{4.8.6}$$

Z is a standard normal variate and $Z_{\alpha*}$ is its $100(1 - \alpha^*)\%$ point.

The asymptotic power function of the analogous t_c test based on the MML10 estimator $\hat{\mu}$ of μ is given by

$$P(Z \geq Z_{\alpha*} - \Delta^*) \tag{4.8.7}$$

where

$$\Delta^* = \frac{\mu}{\sqrt{V(\hat{\mu})}} \tag{4.8.8}$$

Since for large samples $\sigma(t_c) = \sqrt{mV(\hat{\mu})}/E(\hat{\sigma}) \simeq 1$ (see Table 4.2.2, for example),

$$\Delta^* \simeq \Delta_c = \frac{\mu\sqrt{m}}{E(\hat{\sigma})}$$

$m = n(1 - 2q + 2q\beta)$ ($q = 0.1$, $\beta = 0.8309$). For $n = 20$, for example, we have the following exact values of Δ_{opt}/Δ^* and Δ_{opt}/Δ_c calculated from the Tiku-Kumra tables mentioned earlier:

p:	2.0	2.5	3.5	4.5	5.0	∞
Δ_{opt}/Δ^*	1.12	1.07	1.03	1.02	1.02	1.02
Δ_{opt}/Δ_c	1.06	1.02	1.00	1.00	1.00	1.02

Realizing that Δ_{opt}/Δ^* or Δ_{opt}/Δ_c is very close to 1 and realizing that one cannot do any better than the optimal power function (4.8.5), the MML10 procedure is clearly nearly optimal. Sprott's procedure will often use a value of p which is different from its true value and that will render this procedure less powerful than the optimal (4.8.5) and therefore almost certainly less powerful than the MML10 procedure.

All the test procedures above can readily be generalized to multivariate populations (see Tiku and Singh, 1982, and Tiku and Balakrishnan, 1985).

4.9 THE BAYESIAN APPROACH

Over the years, inference procedures have been developed from the Bayesian point of view essentially on the following lines. Let $L(\underset{\sim}{x};\underset{\sim}{\theta})$ be the likelihood function of a random sample $\underset{\sim}{x}' = (x_1, x_2, \ldots, x_n)$; $\underset{\sim}{\theta}' = (\theta_1, \theta_2, \ldots, \theta_p)$ is a vector of unknown parameter values belonging to a parameter space Ω. Let $P(\underset{\sim}{\theta})$, $\underset{\sim}{\theta} \in \Omega$, be the prior distribution of $\underset{\sim}{\theta}$, representing prior information or prior knowledge about $\underset{\sim}{\theta}$. The Bayesian approach combines the prior distribution $P(\underset{\sim}{\theta})$ with the likelihood function $L(\underset{\sim}{x};\underset{\sim}{\theta})$ to form the posterior distribution $P(\underset{\sim}{\theta}|\underset{\sim}{x})$, which is proportional to the product of $P(\underset{\sim}{\theta})$ and $L(\underset{\sim}{x};\underset{\sim}{\theta})$, namely,

$$P(\underset{\sim}{\theta}|\underset{\sim}{x}) \propto P(\underset{\sim}{\theta})L(\underset{\sim}{x};\underset{\sim}{\theta}) \tag{4.9.1}$$

Inferences about $\underset{\sim}{\theta}$ are then derived from the posterior distribution of $\underset{\sim}{\theta}$. Typically, a Bayesian would use the mode or mean or median of $P(\underset{\sim}{\theta};\underset{\sim}{x})$ as a point estimate of $\underset{\sim}{\theta}$; these are called the posterior mode, posterior mean, and posterior median, respectively.

EXAMPLE 4.9.1 Let $x_1, x_2, \ldots, x_n$ be a random sample from a normal distribution with mean μ and variance σ^2. Suppose that it is known a priori that μ has a normal distribution with mean μ_0 and variance σ_0^2. Assume that σ^2 is known and one wants to estimate μ. The posterior distribution of μ is given by

$$P(\mu|\underset{\sim}{x}) \propto \exp\left[-\frac{1}{2\sigma_0^2}(\mu - \mu_0)^2 - \frac{n}{2\sigma^2}(\bar{x} - \mu)^2\right]$$

$$\propto \exp\left[-\frac{\bar{\sigma}^2}{2}(\mu - \bar{\mu})^2\right] \tag{4.9.2}$$

where

$$\bar{\sigma}^2 = \frac{n}{\sigma^2} + \frac{1}{\sigma_0^2} \quad \text{and} \quad \bar{\mu} = \frac{1}{\bar{\sigma}^2}\left(\frac{n}{\sigma^2}\bar{x} + \frac{1}{\sigma_0^2}\mu_0\right) \tag{4.9.3}$$

Thus the posterior distribution of μ is normal with posterior mean $\bar{\mu}$ and posterior variance $1/\bar{\sigma}^2$. Clearly, $\bar{\mu}$ is the weighted average of the sample mean and the prior mean μ_0, with weights inversely proportional to their variances. If $\sigma_0^2 = 0$, then $\bar{\mu}$ reduces to μ_0, and if $\sigma_0^2 = \infty$, then $\bar{\mu}$ reduces to $\bar{x}$, as expected. Realize that $\sigma_0^2 = 0$ implies that μ_0 is the true value of μ and $\sigma_0^2 = \infty$ implies that no prior knowledge of μ is available. A posterior $100(1 - \alpha^*)\%$ interval for μ is given by

$$\left\{\bar{\mu} - Z_{\alpha^*/2}\frac{\sigma}{\sqrt{n}}\left(1 + \frac{1}{n}\frac{\sigma^2}{\sigma_0^2}\right)^{-\frac{1}{2}},\ \bar{\mu} + Z_{\alpha^*/2}\frac{\sigma}{\sqrt{n}}\left(1 + \frac{1}{n}\frac{\sigma^2}{\sigma_0^2}\right)^{-\frac{1}{2}}\right\} \tag{4.9.4}$$

where as usual $Z_{\alpha^*/2}$ is the $100(1 - \frac{1}{2}\alpha^*)\%$ point of a standard normal distribution. A posterior interval is analogous to a confidence interval. Bayesian would discount all values of μ outside this posterior interval.

Prior Distribution

One of the major difficulties with the Bayesian approach is what to choose for a prior distribution. Several approaches for choosing a prior, including Jeffrey's principle of invariance (Jeffreys, 1961), have been discussed by Zellner (1971) and Box and Tiao (1973). Jeffreys' principle states that a prior distribution $P(\underset{\sim}{\theta})$ of $\underset{\sim}{\theta}$ should be invariant with respect to nonsingular transformations of $\underset{\sim}{\theta}$, which leads to the result that (Box and Tiao, 1973)

$$P(\underset{\sim}{\theta}) \propto |I(\underset{\sim}{\theta})|^{\frac{1}{2}} \tag{4.9.5}$$

where $I(\underset{\sim}{\theta})$ is the Fisher information matrix and $|I(\underset{\sim}{\theta})|$ is its determinant. Alternatively, when the situation is such that little is known about $\underset{\sim}{\theta}$ a priori, one chooses some kind of a uniform prior. For example, one chooses $P(\mu) \propto$ const. for the location parameter μ and $P(\log \sigma) \propto$ const. [or $P(\sigma) \propto \sigma^{-1}$] for the scale parameter σ. Usually, μ and σ are assumed to be independent a priori. [Note that $P(\sigma) \propto \sigma^{-1}$ is consistent with Jeffrey's invariance principle.] This is equivalent to saying that when nothing is known about a parameter that takes values between $-\infty$ and ∞, one should assume that any value on the entire real line is equally likely. This principle has been referred to as the Bayes postulate; see Zellner (1971) and Box and Tiao (1973), for example.

Bayesian Approach to Tiku's Robust Procedures

Tiku's robust procedures presented in the earlier sections of this chapter are based on a type II censored sample $x_{(a)}, x_{(a+1)}, \ldots, x_{(b)}$ $(a = r + 1,$ $b = n - r)$, and assumes normality for this censored sample. This purports to reflect the argument that nonnormality comes essentially from the tails, and once the extreme observations representing these tails are censored, there is not much difference between a normal and a nonnormal sample. Now the likelihood function of this censored sample is given by

$$L(x;\mu,\sigma) \propto \sigma^{-A}\left[\prod_{i=a}^{b} \exp\left(-\frac{1}{2\sigma^2} z_i^2\right)\right][F(Z_a)]^r[1 - F(Z_b)]^r \tag{4.9.6}$$

where $A = n - 2r$, $Z_j = (x_{(j)} - \mu)/\sigma$, and $F(Z)$ is the distribution function of a standard normal $N(0,1)$ variate.

Using (4.9.6), Tan (1984a,b) has developed Bayesian robust procedures for making inference about the unknown parameters. We discuss here the one-sample procedure only; for the two-sample and k-sample procedures, the reader is referred to Tan (1984b).

Assume that a priori our knowledge about (μ,σ) is imprecise. We may then follow Box and Tiao (1964a) and assume that the prior distribution $P(\mu,\sigma)$ of (μ,σ) is given by

$$P(\mu,\sigma) \propto \frac{1}{\sigma} \tag{4.9.7}$$

Combining (4.9.7) with (4.9.6), the posterior distribution $P(\mu,\sigma|\underset{\sim}{x})$ of (μ,σ) is

$$P(\mu,\sigma|\underset{\sim}{x}) \propto \sigma^{-1} L(\underset{\sim}{x};\mu,\sigma) \tag{4.9.8}$$

As in Chapter 2, we now replace $F(Z) = f(Z)g^{-1}(Z)$ by its approximation $F(Z) \simeq f(Z)e^{-\log(\alpha-\beta Z)}$, and similarly for $1 - F(Z)$. Then, after some simplification, the posterior distribution $P(\mu,\sigma|\underset{\sim}{x})$ in (4.9.8) reduces to (Tan, 1984a)

$$P(\mu,\sigma|\underset{\sim}{x}) \propto \sigma^{-(A+1)} \exp\left[-\frac{B}{\sigma} - \frac{C}{2\sigma^2} - \frac{m}{2\sigma^2}(\mu - \hat{\mu})^2\right] \tag{4.9.9}$$

where A, B, C, and m are given in Section 2.7 and $\hat{\mu}$ is the MMLE of μ. Let

$\hat{\sigma}$ be the MMLE of σ and note that for large n, C is considerably larger than B. Then (4.9.9) implies that if n is large, the posterior distribution of $y = (A - 1)\hat{\sigma}^2/\sigma^2$ is closely approximated by a central chi-square distribution with $A - 1$ degrees of freedom. From (4.9.9), conditional on y, the posterior distribution of μ is normal with posterior mean $\hat{\mu}$ and posterior variance $\sigma^2/m = (A - 1)\hat{\sigma}^2/(my)$. Therefore, the conditional distribution of $t = \sqrt{m}(\mu - \hat{\mu})/\hat{\sigma}$ given y is

$$P(t|y,\underset{\sim}{x}) = [2\pi(A - 1)]^{-\frac{1}{2}}\sqrt{y}\exp\left[-\frac{yt^2}{2(A - 1)}\right] \tag{4.9.10}$$

It follows that the posterior distribution of t is

$$P(t|\underset{\sim}{x}) = \int_0^\infty P(y|\underset{\sim}{x})P(t|y,\underset{\sim}{x})\,dy \tag{4.9.11}$$

Approximating $P(y|\underset{\sim}{x})$ by the density of a central chi-square distribution with $A - 1$ degrees of freedom and using (4.9.10), we obtain by integrating out y from (4.9.11) an approximation to the posterior distribution of t, namely,

$$P(t|\underset{\sim}{x}) \simeq \frac{1}{\sqrt{A - 1}\,B(1/2, (A - 1)/2)}\left(1 + \frac{t^2}{A - 1}\right)^{-A/2} \tag{4.9.12}$$

The density in (4.9.12) is equivalent to saying that a close approximation to the posterior distribution of $t = \sqrt{m}(\mu - \hat{\mu})/\hat{\sigma}$ is the Student's t distribution with $\nu = A - 1$ degrees of freedom; in fact, Tan (1984a) has shown that it is indeed a very close approximation even for small n. Using (4.9.12), the $100(1 - \alpha^*)\%$ Bayesian interval for t is therefore given by $|t| \geq t_{\alpha^*/2}(\nu)$; equivalently, the $100(1 - \alpha^*)\%$ Bayesian interval for μ is given by

$$\left[\hat{\mu} - \frac{\hat{\sigma}}{\sqrt{m}}t_{\alpha^*/2}(\nu),\ \hat{\mu} + \frac{\hat{\sigma}}{\sqrt{m}}t_{\alpha^*/2}(\nu)\right] \tag{4.9.13}$$

This is in fact the Box-Tiao highest posterior density interval (see Box and Tiao, 1973). At level α^*, the Bayesian would discount all values of μ not in the interval above. Note that the result (4.9.13) is precisely the same result obtained by Tiku's MMLE method (see Section 4.2), but of course the Bayesian concept is different from the sampling theory concept which we have adopted in all other sections.

4.10 CONCLUSIONS AND SUMMARY

In this chapter, a robust analog t_c (based on the MMLE calculated from type II censored normal samples) of the Student's t statistic is defined. The distribution of t_c for large samples is shown to be a noncentral t distribution. The type I error of the t_c test (for testing that the population mean is zero) is shown to be robust to departures from normality and the test is shown to be remarkably powerful compared to the analogous tests based on other robust estimators discussed in Chapter 3. The method is extended to two-sample situations (assuming equality of the population variances) and the resulting test is compared to several other tests, including some of the most popular nonparametric tests; the test is shown to be remarkably robust and powerful. For testing the equality of two population means when the population variances are unequal (Behren-Fisher problem), a Welch-type robust t statistic (based on the MMLE) is defined; this test is shown to have good robustness and power properties compared to several other tests. The problem of testing the equality of two population variances is considered. A t-type test based on the order statistics of $(x_{1i} - \bar{x}_1)^2$ and $(x_{2i} - \bar{x}_2)^2$ is developed. The test is shown to be robust and powerful; in fact, the test is shown to be superior to any other robust test available in the literature. A relatively recent approach to robust tests of location due to Sprott (1982) is also presented and a comparison is made between the Sprott method and the MML method. The reasons for the somewhat better performance of the MML method are given. Finally, a Bayesian interpretation of the MML method due to Tan (1984a) is given and it is shown that with uniform prior on location and scale, the Bayesian highest posterior density interval for the location parameter is identical to the confidence interval based on the MMLE. Numerous examples are given throughout the chapter to illustrate the applications (and the computations involved) of the various procedures. It must be remembered that another endearing feature of the procedures based on the MMLE is that they are completely analogous to the classical procedures based on sample means and sample variances and their distribution (approximate) theory is completely identical to the classical distribution theory based on the normality assumption.

EXERCISES

4.1 The following data represent the gain (in pounds) in weight of 20 pigs assigned randomly to two feeds A and B:

A: 0.09 1.43 2.79 1.60 1.71 3.37 2.06 2.67 8.42 3.67

B: 1.96 1.79 2.60 1.40 2.22 3.45 1.16 5.71 2.93 1.40

Assume that the population variances σ_A^2 and σ_B^2 are equal.

(a) Calculate the classical and the MML10 estimates of $\mu = \mu_A - \mu_B$ and show that the classical estimate has considerably larger standard error.

(b) Use these estimates and their standard errors to test the null hypothesis $H_0: \mu = 0$ against the alternative hypothesis $H_1: \mu \neq 0$. Show that the classical t test does not reject H_0 at 5% significance level but the MML10 t_c test does.

(c) Examine the data carefully and see whether this disagreement could be due to the existence of outliers in the data.

(d) Calculate the Huber estimates w24 and BS82 of μ and their standard errors and use them to test H_0 against H_1 at the 5% significance level and comment. How easy is it to compute Huber estimates as compared to the MML estimates?

4.2 Assume that the variances σ_A^2 and σ_B^2 are not equal.
(a) Calculate the standard errors of the classical and the MML10 estimates of μ.
(b) Calculate the Welch-type classical t* statistic and the MML10 t_c^* statistic and test H_0 against H_1 at the 5% significance level. Are the conclusions any different here than in part (b) of Exercise 4.1?

4.3 For the data of Exercise 2.4, use the classical method and the MML10 method to test that the linear contrast

$$\mu_A + \mu_B - \mu_C - \mu_D$$

is zero at the 5% significance level. Do the two methods lead to the same conclusions? If not, examine the data carefully and see if this disagreement could be due to the existence of outliers in the data (see Special Appendix C).

4.4 The following independent samples represent the weight of comb (mg):

Harmone A:	57	120	101	137	119	117	104	73	53	68
Harmone B:	89	30	82	50	39	22	57	32	96	31

Use the classical method and the MML method to test the equality of the variances $\sigma_A^2 = \sigma_B^2$ at the 5% significance level. Do the two methods agree in not rejecting this hypothesis?

5
Other Important Robust Procedures

5.1 INTRODUCTION

In Chapters 2 and 3 we have developed robust estimation procedures for estimating the location and scale parameters based on type II censored samples. This leads to the MML matching pair $(\hat{\mu}, \hat{\sigma})$ of estimators. Using the MMLE based on type II censored normal samples, we have illustrated in Chapter 4 how robust and powerful procedures can be developed for testing statistical hypotheses about the location and scale parameters. In this chapter we illustrate how the MMLE can be used to develop robust and powerful inference procedures in such important areas as regression, analysis of variance, classification, and sample survey.

5.2 ROBUST REGRESSION VIA MMLEs

Consider the following linear regression model:

$$x_{ij} = \mu + \phi u_i + e_{ij} \quad (j = 1, 2, \ldots, n_i; i = 1, 2, \ldots, k) \tag{5.2.1}$$

where u_i are nonstochastic independent variables, μ and ϕ are unknown parameters, and e_{ij} are random errors. We assume that e_{ij} are IID with mean $E(e_{ij}) = 0$ and common variance $\text{var}(e_{ij}) = \sigma^2$. For purposes of developing robust regression procedures [or because of constraints of the experiment (see Example 5.2.1)], we consider the type II censored samples ($r_{1i} \geq 0$, $r_{2i} \geq 0$; $i = 1, 2, \ldots, k$)

$$x_{(i,a_i)}, x_{(i,a_i+1)}, \ldots, x_{(i,b_i)} \quad (a_i = r_{1i} + 1, b_i = n_i - r_{2i})$$

obtained by arranging the observations x_{ij}, $j = 1, 2, \ldots, n_i$, in ascending order of magnitude and censoring the r_{1i} smallest and r_{2i} largest observations. For simplicity, we will take $r_{1i} = r_{2i} = r_i$ so that we have symmetric type II censored samples; the method is more general, however, and can be used even if $r_{1i} \neq r_{2i}$ (Tiku, 1973, 1982b). We will assume normality for the above type II censored samples and will illustrate the remarkable efficiency of the MMLE and the type I error robustness and high power of the resulting statistical test procedures employed in regression analysis, for both normal and nonnormal populations.

Estimation

Let A_i, B_i, C_i, K_i, m_i, α_i, and β_i be, respectively, the A, B, C, K, m, α, and β (defined in Section 2.7) values calculated from the ith sample, and write

$$M = \sum_{i=1}^{k} m_i \qquad A = \sum_{i=1}^{k} A_i \qquad B = \sum_{i=1}^{k} B_i$$

$$\bar{K} = \sum_{i=1}^{k} \frac{m_i K_i}{M} \qquad \bar{u} = \sum_{i=1}^{k} \frac{m_i u_i}{M}$$

$$S_{xu} = \sum_{i=1}^{k} m_i(u_i - \bar{u})K_i = \sum_{i=1}^{k} m_i(u_i - \bar{u})(K_i - \bar{K})$$

and

$$S_{uu} = \sum_{i=1}^{k} m_i(u_i - \bar{u})u_i = \sum_{i=1}^{k} m_i(u_i - \bar{u})^2$$

Then the MMLEs of ϕ, μ, and σ are given by

$$\hat{\phi} = \frac{S_{xu}}{S_{uu}}$$

$$\hat{\mu} = \bar{K} - \hat{\phi}\bar{u} \tag{5.2.2}$$

and

$$\hat{\sigma} = \frac{B + (B^2 + 4AC)^{\frac{1}{2}}}{2[A(A - 2)]^{\frac{1}{2}}}$$

where $C = S_e + S_L$, with $S_e = \Sigma_{i=1}^{k} C_i$ and $S_L = \Sigma_{i=1}^{k} m_i[K_i - \bar{K} - \hat{\phi}(u_i - \bar{u})]^2$. As in Chapter 2, these results are obtained by replacing the terms $g_1(e_{(i,a_i)})$ and $g_2(e_{(i,b_i)})$ in the derivatives of the log-likelihood function log L by the approximations

$$g_1(e_{(i,a_i)}) \simeq \alpha_i - \beta_i e_{(i,a_i)} \quad \text{and} \quad g_2(e_{(i,b_i)}) \simeq \alpha_i + \beta_i e_{(i,b_i)}$$

$$e_{(i,j)} = \frac{x_{(i,j)} - \mu - \phi u_i}{\sigma}$$

and equating the resulting derivatives to zero and solving them for ϕ, μ, and σ. In other words, the estimators $\hat{\phi}$, $\hat{\mu}$, and $\hat{\sigma}$ are obtained as solutions of the equations

$$\frac{\partial \log L^*}{\partial \phi} = 0 \qquad \frac{\partial \log L^*}{\partial \mu} = 0 \qquad \frac{\partial \log L^*}{\partial \sigma} = 0$$

where the modified maximum likelihood function L^* is given by

$$L^* = \prod_{i=1}^{k} \left\{ \sigma^{-A_i} \exp\left[-\frac{B_i}{\sigma} - \frac{C_i}{2\sigma^2} - \frac{m_i}{2\sigma^2}(K_i - \mu - \phi u_i)^2 \right] H(\underset{\sim}{x}_i) \right\} \tag{5.2.3}$$

$H(\underset{\sim}{x}_i)$ is a function of $\underset{\sim}{x}_i = (x_{(i,a_i)}, \ldots, x_{(i,b_i)})$ and is free of μ, ϕ, and σ. The function (5.2.3) suggests that based on L^*, the regression model (5.2.1) reduces to the regression model of the form $K_i = \mu + \phi u_i + \epsilon_i$ with $E(K_i) = \mu + \phi u_i$, $\text{var}(K_i) = \sigma^2/m_i$, $i = 1, 2, \ldots, k$, and $\text{cov}(K_i, K_j) = 0$, $i \neq j$. Using (5.2.3), one then sees immediately that based on the normal type II censored samples (and consequently L^*), the MMLE of μ and ϕ are in fact the weighted least-squares estimators which are linear (in the K_i's), unbiased and minimum variance estimators of μ and ϕ, respectively, by the Gauss-Markov theorem (Tan, 1979). Further, based on L^*, $\hat{\sigma}^2 \simeq C/(A - 2)$ is, for large A_i, $i = 1, 2, \ldots, k$, the quadratic, unbiased, minimum variance estimator of σ^2 (Tan, 1971). For normal type II censored samples, the following results follow immediately from Lemma 2.7.2 and standard well-known results of regression theory (see Draper and Smith, 1966).

1. Since $E(K_i) = \mu + \phi u_i$ (see also Chapter 2), it follows that

$$E(\hat{\phi}) = \phi \text{ and } E(\hat{\mu}) = \frac{1}{M} \sum_{i=1}^{k} m_i(\mu + \phi u_i) - E(\hat{\phi})\bar{u} = \mu$$

Further, $\text{cov}(K_i, K_j) = 0$ for $i \neq j$, and for large A_i (hence large n_i), $\text{var}(K_j) \simeq \sigma^2/m_j$. It also follows that

$$\text{var}(\hat{\phi}) \simeq \frac{1}{s_{uu}^2} \sum_{i=1}^{k} \frac{[m_i(u_i - \bar{u})]^2}{m_i} = \frac{\sigma^2}{s_{uu}} \tag{5.2.4}$$

$$\text{cov}(\bar{K}, \hat{\phi}) \simeq \sum_{i=1}^{k} \frac{m_i^2(u_i - \bar{u})(\sigma^2/m_i)}{M} = 0 \tag{5.2.5}$$

$$\operatorname{var}(\hat{\mu}) \simeq \operatorname{var}(\bar{K}) + \bar{u}^2 \operatorname{var}(\hat{\phi})$$

$$= \frac{1}{M^2}\sum_{i=1}^{k} m_i^2 \frac{\sigma^2}{m_i} + \bar{u}^2 \frac{\sigma^2}{s_{uu}}$$

$$= \sigma^2\left(\frac{1}{M} + \frac{\bar{u}^2}{s_{uu}}\right) \tag{5.2.6}$$

and

$$\operatorname{cov}(\hat{\phi}, \hat{\mu}) \simeq \operatorname{cov}(\hat{\phi}, \bar{K}) - \bar{u}\operatorname{var}(\hat{\phi})$$

$$= -\frac{\bar{u}\sigma^2}{s_{uu}} \tag{5.2.7}$$

Since L* provides a very close approximation to L (see Chapter 2), one would expect (5.2.4)-(5.2.7) to provide very close approximations to the respective true values even for small A_i, $A_i \geq 3$ (Tiku, 1981a, 1982b).

2. For large $A_i = n_i - 2r_i$ ($i = 1, 2, \ldots, k$), the estimators $\hat{\phi}$ and $\hat{\mu}$ are normally distributed.

3. For large A_i, the distribution of

$$\frac{1}{\sigma^2} C(\mu, \phi) = \frac{1}{\sigma^2}\sum_{i=1}^{k}\left[\sum_{j=a_i}^{b_i} (x_{(i,j)} - \mu - \phi u_i)^2 + r_i\beta_i(x_{(i,a_i)} - \mu - \phi u_i)^2 + r_i\beta_i(x_{(i,b_i)} - \mu - \phi u_i)^2\right]$$

is a chi-square having $A = \Sigma_{i=1}^{k} A_i$ degrees of freedom.

If $\phi = 0$, then $\hat{\mu} = \bar{K}$ and the distribution of $C(\bar{K}, 0)/\sigma^2$, where

$$C(\bar{K}, 0) = \sum_{i=1}^{k}\left[\sum_{j=a_i}^{b_i} (x_{(i,j)} - \bar{K})^2 + r_i\beta_i(x_{(i,a_i)} - \bar{K})^2 + r_i\beta_i(x_{(i,b_i)} - \bar{K})^2\right]$$

$$= \sum_{i=1}^{k}\left[\sum_{j=a_i}^{b_i} x_{(i,j)}^2 + r_i\beta_i(x_{(i,a_i)}^2 + x_{(i,b_i)}^2)\right] - M\bar{K}^2$$

is a chi-square having $A - 1$ degrees of freedom, for large A_i.

4. The distribution of C/σ^2, where

$$C = C(\bar{K}, \hat{\phi}) = \sum_{i=1}^{k} \left\{ \sum_{j=a_i}^{b_i} [x_{(i,j)} - \bar{K} - \hat{\phi}(u_i - \bar{u})]^2 + r_i\beta_i[x_{(i,a_i)} - \bar{K} - \hat{\phi}(u_i - \bar{u})]^2 + r_i\beta_i[x_{(i,b_i)} - \bar{K} - \hat{\phi}(u_i - \bar{u})]^2 \right\}$$

is for large A_i, a chi-square having $A - 2$ degrees of freedom. As a consequence of this, the estimator $\hat{\sigma}$ is, for large A_i, an unbiased estimator of σ and

$$V(\hat{\sigma}) \simeq \frac{\sigma^2}{2(A - 2)}$$

5. For large A_i ($i = 1, 2, \ldots, k$), the distribution of $\Sigma_{i=1}^{k} C_i/\sigma^2$ is a chi-square having $\Sigma_{i=1}^{k} (A_i - 1) = A - k$ degrees of freedom.

6. Since $e_{(i,j)} = (x_{(i,j)} - \mu - \phi u_i)/\sigma$, K_i can be written as

$$K_i = \left(\frac{\sigma}{m_i}\right)\left[\sum_{j=a_i}^{b_i} e_{(i,j)} + r_i\beta_i(e_{(i,a_i)} + e_{(i,b_i)})\right] + (\mu + \phi u_i)$$

and since for symmetric distributions (normal or nonnormal), $E(e_{(i,j)}) = -E(e_{(i,n_i-j+1)})$, it follows that for symmetric distributions $E(K_i) = \mu + \phi u_i$ so that $E(\hat{\phi}) = \phi$ and $E(\hat{\mu}) = \mu$. For skew distributions and equal sample sizes $n_i = n$, $E(K_i) = \mu + \phi u_i + \text{bias}$, where bias assumes a constant value for all $i = 1, 2, \ldots, k$. Thus again one has $E(\hat{\phi}) = \phi + \text{bias}[\Sigma_{i=1}^{k} m_i(u_i - \bar{u})/\Sigma_{i=1}^{k} m_i] = \phi$ since $\Sigma_{i=1}^{k} m_i(u_i - \bar{u}) = 0$. For skew distributions and unequal sample sizes $n_i \neq n$, however, $E(\hat{\phi}) = \phi + \Sigma_{i=1}^{k} [m_i(u_i - \bar{u}) \text{ bias (i)}]/\Sigma_{i=1}^{k} m_i$, but since bias (i) have the same sign and are almost constant for all $i = 1, 2, \ldots, k$, since $q_i = r_i/n_i$, $r_i = [0.5 + 0.1n_i]$, are almost equal and bias (i) only depends on q_i, it follows that $\Sigma_{i=1}^{k}[m_i(u_i - \bar{u}) \text{ bias (i)}] \simeq 0$, so that $E(\hat{\phi}) \simeq \phi$ and hence $\hat{\phi}$ is almost unbiased. For nonnormal populations, the asymptotic distributions of K_i (being linear functions of order statistics) are independent normals under some very general regularity conditions. Thus even for nonnormal populations (symmetric or skew), property 2 holds.

Testing Statistical Hypotheses

For testing statistical hypotheses about ϕ and μ, first note that for large $A_i = n_i - 2r_i$, the K_j's are independently distributed as normal variables with $E(K_j) = \mu + \phi u_j$ and $\text{var}(K_j) = \sigma^2/m_j$ for normal type II censored samples. Further, it is easy to see that with $\mu^* = \mu + \phi\bar{u}$,

$$C(\mu, \phi) = \bar{C}(\mu^*, \phi) = \sum_{i=1}^{k} \sum_{j=a_i}^{b_i} [x_{(i,j)} - \mu^* - \phi(u_i - \bar{u})]^2 + r_i\beta_i [x_{(i,a_i)} - \mu^*$$

$$- \phi(u_i - \bar{u})]^2 + [x_{(i,b_i)} - \mu^* - \phi(u_i - \bar{u})]^2]$$

$$= M(\bar{K} - \mu^*)^2 + \bar{C}(\bar{K}, \phi)$$

$$\bar{C}(\bar{K}, \phi) = \bar{C}(\bar{K}, \hat{\phi}) + s_{uu}(\hat{\phi} - \phi)^2$$

$$= C(\hat{\mu}, \hat{\phi}) + s_{uu}(\hat{\phi} - \phi)^2$$

and

$$C(\hat{\mu}, \hat{\phi}) = \bar{C}(\bar{K}, \hat{\phi}) = \sum_{i=1}^{k} C_i + \sum_{i=1}^{k} m_i[K_i - \bar{K} - \hat{\phi}(u_i - \bar{u})]^2$$

$$= S_e + S_L$$

Noting that

$$C(0,0) = \sum_{i=1}^{k}\left[\sum_{j=a_i}^{b_i} x^2_{(i,j)} + r_i\beta_i(x^2_{(i,a_i)} + x^2_{(i,b_i)})\right]$$

and $\bar{C}(\bar{K}, 0) = C(\bar{K}, 0) = C(\bar{K})$, one has the following equation:

$$C(\mu, \phi) = M(\bar{K} - \mu^*)^2 + s_{uu}(\hat{\phi} - \phi)^2 + S_L + S_e \tag{5.2.8}$$

Thus

$$S_{total} = C(0,0) = M\bar{K}^2 + s_{uu}\hat{\phi}^2 + S_L + S_e$$

$$= S_{\hat{\mu}} + S_{\hat{\phi}} + S_L + S_e \tag{5.2.9}$$

where $S_{\hat{\mu}} = M\bar{K}^2$ and $S_{\hat{\phi}} = s_{uu}\hat{\phi}^2$.

In (5.2.9), each term is a quadratic form in the $x_{(i,j)}$'s, $j = a_i, \ldots, b_i$, $i = 1, 2, \ldots, k$, so that (5.2.9) corresponds to the partition of the total sum of squares S_{total}; $S_{\hat{\mu}}$ is the sum of squares for μ; $S_{\hat{\phi}}$ is the sum of squares for ϕ; S_L is the lack-of-fit sum of squares, which measures the deviation of K_i from the fitted line; and S_e is the error sum of squares.

In (5.2.9), for normal type II censored samples, the distribution of each term is, for large A_i, $i = 1, 2, \ldots, k$, a σ^2 multiple of central chi-square variables from results 2-5 above. By Cochran's theorem (see Tan, 1984b), $s_{uu}(\hat{\phi} - \phi)^2/\sigma^2$, S_L/σ^2, and S_e/σ^2 are for large A_i independently distributed as central chi-square variables with degrees of freedom 1, k - 2, and A - k,

Table 5.2.1 ANOVA Table for Testing $H_0: \phi = 0$

Sources of variation	Sum of squares	df	Mean square	F ratio
μ^*	$S_{\hat{\mu}}$	1	$M_{\hat{\mu}} = S_{\hat{\mu}}$	
Linear fit	$S_{\hat{\phi}} = s_{uu}\hat{\phi}^2$	1	$M_{\hat{\phi}} = S_{\hat{\phi}}$	$F_1 = M_{\hat{\phi}}/M_e$
Lack of fit	S_L	k - 2	$M_L = S_L/(k-2)$	$F_2 = M_L/M_e$
Error	S_e	A - k	$M_e = S_e/(A-k)$, $A = \sum_{i=1}^{k} A_i$	
Total	S_{total}	A		

respectively, for normal type II censored samples. For large A_i, therefore, $S_{\hat{\phi}}/\sigma^2$ is a noncentral chi-square variable with 1 degree of freedom and noncentrality parameter $\delta^2 = s_{uu}\phi^2/\sigma^2$ for normal type II censored samples.

ANOVA Table. To test the null hypothesis $H_0: \phi = 0$, and to test the linearity of the model (5.2.1), we have an ANOVA table (Table 5.2.1). For type II censored normal samples and large A_i, $i = 1, 2, \ldots, k$, F_1 has under H_0 a central F distribution having (1, A - k) degrees of freedom; F_2 has under H_0^* (e.g., the linear model is an adequate fit) a central F distribution having (k - 2, A - k) degrees of freedom. To verify the accuracy of these results, Tiku (1981a) simulated the probabilities of F_1 and F_2 being greater than or equal to the $100(1 - \alpha^*)\%$ points of the central F distributions having (1, A - k) and (k - 2, A - k) degrees of freedom, respectively. His results are given here in Table 5.2.2 for type II censored normal samples. It is clear that these central F distributions give accurate approximations.

For most of the nonnormal distributions, the null distributions of F_1 and F_2 are again closely approximated by central F distributions with degrees of freedom (1, A-k) and (k-2, A-k), respectively. Given in Table 5.2.3 are the simulated values (based on 4000 Monte Carlo runs) of the null probabilities $P[F_1 \geq F_{\alpha^*}(1, A-k)\ H_0]$ and $P[F_2 \geq F_{\alpha^*}(k-2, A-k)\ H_0^*]$ (see also Tiku, 1981a). For these values, u_i are equispaced on the interval (-9, 9) and $n_i = 5$, $i = 1, 2, \ldots, 10$. This is the Atkinson-Cox model considered in Section 5.2.5. The type I robustness of the F_1 and F_2 tests is obvious.

Table 5.2.2 Simulated Probabilities of Falling to the Right of the $100(1-\alpha^*)\%$ Points of the Approximate[a] Distributions; $n_i = n$ and $r_i = r$, $i = 1, 2, \ldots, k$

n	α^*	$\hat{\phi}$	$\hat{\sigma}^2$	F_1	F_2	$\hat{\phi}$	$\hat{\sigma}^2$	F_1	F_2
		k = 3, r = 1				k = 5, r = 1			
5	0.10	0.102	0.103	0.120	0.112	0.113	0.112	0.115	0.104
	0.05	0.052	0.057	0.064	0.056	0.059	0.060	0.060	0.051
	0.01	0.012	0.015	0.016	0.009	0.014	0.015	0.013	0.008
		k = 3, r = 1				k = 5, r = 1			
10	0.10	0.101	0.102	0.107	0.101	0.103	0.107	0.101	0.104
	0.05	0.051	0.050	0.052	0.054	0.050	0.053	0.052	0.052
	0.01	0.009	0.012	0.009	0.010	0.011	0.012	0.010	0.010
		k = 3, r = 2				k = 5, r = 2			
10	0.10	0.104	0.110	0.108	0.108	0.103	0.106	0.101	0.104
	0.05	0.052	0.057	0.055	0.054	0.050	0.054	0.054	0.053
	0.01	0.010	0.014	0.011	0.010	0.011	0.012	0.010	0.010

[a] $\hat{\phi} \cong$ normal with mean $\phi = 0$ and variance $\sigma^2/\Sigma_{i=1}^{k} m_i(u_i - \bar{u})^2$; $(A-2)\hat{\sigma}^2/\sigma^2 \simeq$ chi-square having A - 2 degrees of freedom; $F_1 \simeq$ central F distribution having (1, A - k) degrees of freedom; $F_2 \simeq$ central F distribution having (k - 2, A - k) degrees of freedom.

Table 5.2.3 Values of the Null Probabilities $P[F_1 \geq F_{\alpha^*}(1, A-k) \mid H_0]$ and $P[F_2 \geq F_{\alpha^*}(k-2, A-k) \mid H_0^*]$; Presumed Level $\alpha^* = 0.05$ and 0.01

Normal	Laplace	Student's t_3	Cauchy	Exponential
		F_1		
0.053	0.055	0.057	0.051	0.061
0.010	0.010	0.011	0.007	0.012
		F_2		
0.049	0.050	0.049	0.038	0.063
0.008	0.009	0.009	0.008	0.015

EXAMPLE 5.2.1 In a time-mortality experiment, five doses of a drug are administered to five different groups of sample specimens. However, a few specimens die before observations begin. The experiment is subsequently terminated with certain specimens remaining alive. The log-survival times x of the specimens are recorded during the full periods of observations as follows; log (10 dose) = u:

Group	u	Log-survival times
1	0.310	..., 1.159, 1.173, 2.142, 2.168, 2.220, 2.479, 2.932, 3.449, ...
2	0.477	..., ..., 1.745, 2.483, 2.553, 2.724, 3.482, 3.542, ...
3	0.602	2.624, 2.672, 3.213, 3.523, 3.609
4	1.000	3.466, 3.533, 3.549, 3.689
5	1.398	..., 3.075,[a] 3.081, 3.320, 3.362, 3.594, 3.642, 3.859, ..., ...

[a]We discard this observation to achieve symmetry in censoring, that is, to make r_{15} and r_{25} equal, and equal to 2. This is done for ease of computation.

Assuming that the log-survival times are normally distributed, we have the following for the data above:

	Groups				
i:	1	2	3	4	5
n_i	10	10	5	4	10
r_i	1	2	0	0	2
α_i	0.6737	0.7493	0	0	0.7493
β_i	0.8611	0.7914	1	1	0.7914
m_i	9.722	9.166	5	4	9.166
K_i	2.231	2.716	3.128	3.559	3.474
u_i	0.301	0.477	0.602	1.000	1.398

$$Q = 86.574 - \frac{(27.123)(108.309)}{37.054} = 7.295$$

$$\sum_i m_i(u_i - \bar{u})^2 = 26.692 - \frac{(27.123)^2}{37.054} = 6.839$$

$$\sum_i m_i(K_i - \bar{K})^2 = 326.250 - \frac{(108.309)^2}{37.054} = 9.660$$

$$\hat{\phi} = \frac{7.295}{6.839} = 1.067$$

$$\hat{\sigma} = \frac{5.402 + \sqrt{29.178 + 1748.108}}{2\sqrt{29(27)}} = 0.8498$$

$$(A - 2)\hat{\sigma}^2 = 19.498$$

The corresponding sums of squares are

$$S_{\hat{\phi}} = 1.067(7.295) = 7.781 \qquad df = 1, \quad F = 10.6 \quad df(1, 24)$$

$$S_L = 9.660 - 7.781 = 1.880 \qquad df = 3, \quad F = 0.85 \quad df(3, 24)$$

$$S_e = 19.498 - 1.880 = 17.618 \quad df = 24$$

The linear model is adequate, and the null hypothesis $H_0: \phi = 0$, is rejected at 1% significance level.

One-Sided Test. The statistic F_1 above is used for testing $H_0: \phi = 0$, against two-sided alternative hypotheses $H_1: \phi \neq 0$. To test H_0 against one-sided alternative hypotheses $H_1: \phi > 0$, say, the test is based on the statistic

$$t_c = \hat{\phi}\,\frac{[\Sigma_{i=1}^k m_i(u_i - \bar{u})^2]^{\frac{1}{2}}}{\hat{\sigma}_e} \qquad \hat{\sigma}_e = \left(\frac{S_e}{A - k}\right)^{\frac{1}{2}}$$

Large values of t_c lead to the rejection of H_0 in favor of H_1. The null distribution of t_c is approximately Student's t distribution having A - k degrees of freedom.

For the data above (Example 5.2.1)

$$t_c = \frac{1.067\sqrt{6.839}}{\sqrt{17.618/24}} = 3.26$$

The upper 1% point of Student's t distribution having 24 degrees of freedom being 2.80, H_0 is rejected in favor of H_1: $\phi > 0$. Remember that the statistic t_c has type I robustness with respect to departures from normality and is remarkably powerful (see Chapter 4).

Comparison of the MMLE $\hat{\phi}$ with the Least-Squares Estimator $\tilde{\phi}$ of ϕ

To compare $\hat{\phi}$ with the least-squares estimator based on complete samples, namely, $\tilde{\phi} = [\Sigma_{i=1}^k n_i(u_i - \bar{u})\bar{x}_i]/[\Sigma_{i=1}^k n_i(u_i - \bar{u})^2]$, we define the relative efficiency of $\hat{\phi}$ as compared to $\tilde{\phi}$ as

$$\mathrm{RE}(\hat{\phi}) = 100\sqrt{\frac{V(\tilde{\phi})}{V(\hat{\phi})}} \tag{5.2.10}$$

where $V(\tilde{\phi})$ = variance of $\tilde{\phi}$, and so on. Realize that $\tilde{\phi}$ is unbiased and has variance

$$V(\tilde{\phi}) = \frac{\sigma^2}{\Sigma_{i=1}^{k} n_i(u_i - \bar{u})^2}$$

for all distributions with finite variance of course. On the other hand, $\hat{\phi}$ is unbiased only for symmetric distributions; for skew distributions, however, it is almost unbiased, as explained earlier (in fact, if $n_i = n$, $i = 1, 2, \ldots, k$, then $\hat{\phi}$ is unbiased for skew distributions also). The approximate variance of $\hat{\phi}$ is

$$V(\hat{\phi}) \simeq \frac{\sigma^2}{\Sigma_{i=1}^{k} m_i(u_i - \bar{u})^2}$$

for normal type II censored samples and it is a close approximation indeed (Chapter 2). For any distribution, however,

$$V(\hat{\phi}) = \sum_{i=1}^{k} a_i^2 \operatorname{var}(K_i) \qquad a_i = \frac{m_i(u_i - \bar{u})}{S_{uu}}$$

and

$$\operatorname{var}(K_i) = \boldsymbol{\ell}'\underset{\sim}{V}_i\boldsymbol{\ell}$$

where $m_i\boldsymbol{\ell}' = (1 + r_i\beta_i, 1, \ldots, 1, 1 + r_i\beta_i)$, a vector of $n_i - 2r_i$ elements, and $\underset{\sim}{V}_i$ is the variance-covariance matrix of $e_{(i,j)} = (x_{(i,j)} - \mu - \phi u_i)/\sigma$, $j = r_i + 1, \ldots, n_i - r_i$. Since the elements of $\underset{\sim}{V}_i$ are available for numerous distributions (Balakrishnan, 1984; Balakrishnan and Kocherlakota, 1985a; Barnett, 1966; David et al., 1977; Tiku and Kumra, 1981; Sarhan and Greenberg, 1962), one can readily compute $\operatorname{var}(K_i)/\sigma^2$ and hence $V(\hat{\phi})/\sigma^2$ and $\mathrm{RE}(\hat{\phi})$. Many numerical results indicate that $\hat{\phi}$ is on the whole considerably more efficient than $\tilde{\phi}$. As an example, consider the linear model of Atkinson and Cox (1977) in which $k = 10$, $n_i = n = 5$, and u_i are equally spaced on $(-9, 9)$. Given in Table 5.2.4 are the exact values of $\mathrm{RE}(\hat{\phi})$ for this model for the following distributions:

(1) uniform, (2) normal, (3) Student's t_9 (this is effectively the logistic distribution; Tiku and Jones, 1971), (4) Student's t_7, (5) Laplace (double exponential, (6) Student's t_4, (7) Student's t_3, (8) outlier model: 4N(0, 1) and 1N(0, 4)., and (9) exponential

Table 5.2.4 Values of the Relative Efficiency $RE(\hat{\phi})$

Distribution								
(1)	(2)	(3)	(4)	(5)	(6)	(7)	(8)	(9)
82.7	94.1	99.2	101.3	109.8	111.5	127.1	152.8	104.8

In the Atkinson-Cox model above, the variance of the BLUE (best linear unbiased estimator) of $\hat{\phi}$ for the Laplace distribution (Govindarajulu, 1966) is $0.0048\sigma^2$, and therefore the efficiency of $\hat{\phi}$ relative to the BLUE (based on complete samples) is 97.7%, which is remarkably high; the efficiency of the classical least-squares estimator $\tilde{\phi}$ relative to the BLUE is only 89%. Note that for distributions (4) to (9), $\text{var}(K_i)$ is smaller (see Chapter 2) than $\text{var}(\bar{x}_i)$ for all $n_i = n \geq 5$, $i = 1, 2, \ldots, k$, and therefore the MML estimator $\hat{\phi}$ is more efficient than the classical estimator $\tilde{\phi}$ for all sorts of design points, not only for the design points considered in Table 5.2.4, namely, u_i equispaced on $(-9, 9)$. For distributions (1) and (2), however, $\hat{\phi}$ is always somewhat less efficient than $\tilde{\phi}$. For distribution (3), $\hat{\phi}$ is almost as efficient as $\tilde{\phi}$ for all $n_i = n \geq 5$ and is in fact slightly more efficient for all $n_i = n \geq 10$, $i = 1, 2, \ldots, k$. Similar results hold for unequal n_i (see also Hamouda and Leone, 1977).

Comparison of F_1 Test with Classical F Test for Testing $\phi = 0$

From results given in Tables 5.2.2 and 5.2.3, it is clear that the F test for testing H_0: $\phi = 0$ based on the MMLE possesses type I robustness. Tiku (1981a) has shown that the F_1 test is considerably more powerful than the classical F test for nonnormal distributions and has substantially the same power as the classical F test for normal distributions. Given in Table 5.2.5 are the values of the power of the F and F_1 tests for the design points u_i considered by Atkinson and Cox (1977) (see also Binkley and Birch, 1984).

Comparison of the MMLE with the Atkinson-Cox Method

Atkinson and Cox (1977) obtained a robust estimator of ϕ via discriminant analysis as follows:

Let $x_{(i,j)}$ $(j = 1, 2, \ldots, n)$ be the order statistics of a random sample of n observations on the variable x at the ith level $u = u_i$ of the explanatory variable u $(i = 1, 2, \ldots, k)$. A general linear estimate of location based on these order statistics is

Table 5.2.5 Values of the Power for Testing $\phi = 0$ Against $\phi \neq 0$ in Case of the Atkinson-Cox Example

$\phi = 0$		$\phi = 0.02$		$\phi = 0.05$		$\phi = 0.10$	
F	F_1	F	F_1	F	F_1	F	F_1
			Normal				
0.049	0.053	0.14	0.14	0.52	0.48	0.98	0.96
			Student's t_3				
0.044	0.051	0.14	0.18	0.60	0.74	0.96	1.00
			4N(0, 1) and 1N(0, 4)				
0.042	0.050	0.07	0.10	0.19	0.36	0.53	0.85
			Exponential[a]				
0.042	0.061	0.14	0.15	0.53	0.58	0.97	0.97

[a]For extreme skew distributions like the exponential, a more powerful test would, of course, be achieved by censoring only the largest $r_i = [0.5 + 0.2n_i]$ $(i = 1, 2, \ldots, k)$ observations in the ith group, as explained in Section 4.4.

$$X_i^{(a)} = \sum_{j=1}^{n} a_j x_{(i,j)} \qquad \sum_{j=1}^{n} a_j = 1$$

where $a_1, a_2, \ldots, a_n$ are suitable constants, the same for all i. Atkinson and Cox propose to choose a_j in such a way that the regression relation is recovered in as strong a form as possible; if one or more outliers in the data would seriously distort the relation, then one hopes to choose a_j to give these values little or no weight. One way to do this is to choose a_j in such a way that the proportion of the corrected sum of squares of the $X_i^{(a)}$, accounted for by linear regression on u_i, is maximized. That is, the coefficients a_j are chosen by linear discriminant analysis. This leads to a robust estimator of the type

$$\phi_R^* = \frac{\Sigma_{i=1}^{k} u_i(\Sigma_{j=1}^{n} c_j x_{(i,j)})}{\Sigma_{i=1}^{k} u_i^2} \qquad \left(\sum_{i=1}^{k} u_i = 0\right)$$

For this method to work at all, it is necessary that $k \geq 3$ and $n \leq k - 2$.

Atkinson and Cox (1977) give values of $V(\phi_R^*)$ for a few distributions; from these values, the efficiency of ϕ_R^* relative to the least-squares estimator $\tilde{\phi}$ can be calculated. The values of the relative efficiency $RE(\phi_R^*) = 100\sqrt{V(\tilde{\phi})/V(\phi_R^*)}$ for the normal and Laplace distributions are 90.3 and 103.34, respectively. The corresponding values for the MML10 estimator $\hat{\phi}$ are 94.1 and 109.8.

For the Cauchy distribution, the variance of the least-squares estimator $\tilde{\phi}$ is infinite; the variance of ϕ_R^* is 14,530, and the variance of the MML30 estimator [i.e., the estimator $\hat{\phi}$ based on samples with $r_i = [0.5 + 0.3n_i]$ smallest and largest observations censored ($i = 1, 2, \ldots, k$)] is 10,078, and this is the exact variance calculated from the variances and covariances of order statistics of random samples from a Cauchy distribution given by Barnett (1966).

It may also be noted that the estimator ϕ_R^* is rather restrictive in the sense that the number of observations n_i in the groups have all to be equal [and there are computational difficulties with ϕ_R^*; see Atkinson and Cox (1977)]. The only desirable restriction on the MML estimator $\hat{\phi}$ is that $n_i \geq 4$ ($i = 1, 2, \ldots, k$); for $n_i < 4$, $\hat{\phi}$ would be based only on the medians and might, therefore, lose efficiency. In certain practical situations, however, there will be only one x observation available for most of the u_i levels; this needs a somewhat different treatment, which is discussed in the next section.

Small Group Sizes

In numerous practical situations, some or all of the groups may have only one observation. In such a situation, one proceeds as follows. Suppose that the regression equation is linear and is given by (5.2.1). Calculate the residuals $\tilde{e}_{ij} = x_{ij} - \tilde{\mu} - \tilde{\phi}u_i$ ($j = 1, 2, \ldots, n_i$; $i = 1, 2, \ldots, k$), where $\tilde{\mu}$ and $\tilde{\phi}$ are the least-squares estimators. Now $E(\tilde{e}_{ij}) = 0$ and the distribution of $\tilde{e}_{ij}^2$ will usually be extremely skew (positively), with a long tail on the right-hand side. Censor $R = [0.5 + 0.2N]$ ($N = \Sigma_{i=1}^k n_i$), of those x_{ij} observations which correspond to the R largest $\tilde{e}_{ij}^2$, for reasons discussed in Section 4.4. Realize that in this process some of the groups may be completely eliminated; some groups might lose only the smallest and/or largest observations. In the latter situation, if need be, censor a few more smallest or largest observations in the groups to achieve symmetry in censoring; this is done primarily for ease of computation of the component sums of squares (see Example 5.2.1) entering in the statistics F_1 and F_2. Calculate the MML estimators K_i and $\hat{\phi}$ and proceed as above to calculate the statistics F_1 and F_2; note that if the ith group is left with only one observation x_i, then $K_i = x_i$.

EXAMPLE 5.2.2 The following data (Draper and Smith, 1966, p. 102) indicate the relationship between the amount of erythroidine in an aqueous solution and the colorimeter reading of the turbidity:

Concentration, u (mg/ml)	Colorimeter reading, x
40	69
50	175
60	272
70	335
80	490
90	415
40	72
60	265
80	492
50	180

Here $n = 10$, $\Sigma u_i = 620$, $\Sigma x_i = 2765$, $\Sigma u_i x_i = 194970$, $\Sigma u_i^2 = 41200$; $\tilde{\phi} = 8.529$ and $\tilde{\mu} = -252.2971$. The fitted regression line is

$$x = -252.2971 + 8.529u$$

The estimated errors $\tilde{e}_i$ are given by

-19.86 0.85 12.56 -9.73 59.98 -100.31 -16.86 5.56 61.98 5.85

and

$$\tilde{s}_e^2 = \frac{\Sigma_{i=1}^{10} \tilde{e}_i^2}{8} = 2312.3$$

is the mean error sum of squares.

The x observations that correspond to $R = [0.5 + 0.2(10)] = 2$ largest $\tilde{e}_i^2$ are $x = 415$ $(u = 90)$ and $x = 492$ $(u = 80)$; we set aside these two groups. From the remaining eight groups, we obtain (note that $K_i = x_i$)

$$\hat{\phi} = 9.908 \quad \text{and} \quad \hat{\mu} = -325.0811$$

The fitted regression line is

$$x = -325.0811 + 9.908u$$

From the estimated errors $\hat{e}_i = x_i - \hat{\mu} - \hat{\phi}u_i$, we obtain

$$\hat{s}_e^2 = \frac{\Sigma_{i=1}^{8} \hat{e}_i^2}{6} = 295.3$$

The reduction in the mean error sum of squares is substantial.

To test $H_0: \phi = 0$, the value of F_1 calculated from all the 10 groups is

$$F_1 = \tilde{\phi} \sum_{i=1}^{10} \frac{(u_i - \bar{u})x_i}{\tilde{s}_e^2} = 86.83 \quad (\text{df} = 8)$$

The corresponding value of F_1 calculated from the eight groups above is ($K_i = x_i$, since there is only one observation in the ith group)

$$F_1 = \hat{\phi} \sum_{i=1}^{8} \frac{(u_i - \bar{u})K_i}{\hat{s}_e^2} = 461.4 \quad (df = 6)$$

Although both F_1 values are significant, the latter F_1 test rejects H_0 with a much higher probability.

5.3 GENERAL LINEAR MODELS BASED ON TYPE II CENSORED SAMPLES

The robust procedure developed in Section 5.2 extends immediately to general linear models of the form

$$x_{ij} = \mu + \phi_1 u_{1i} + \phi_2 u_{2i} + \cdots + \phi_p u_{pi} + e_{ij} \tag{5.3.1}$$

$j = 1, 2, \ldots, n_i$, $i = 1, 2, \ldots, k$, where μ is a constant, $\phi_1, \phi_2, \ldots, \phi_p$ are the p regression coefficients, $u_{1i}, \ldots, u_{pi}$ are the (nonstochastic) independent variables, and e_{ij} are random errors. As usual, we assume that e_{ij}'s are independent and identically distributed with mean $E(e_{ij}) = 0$ and variance $V(e_{ij}) = \sigma^2$. To develop robust procedures for estimating ϕ_i, and to test hypotheses about ϕ_i, we consider symmetrically type II censored samples

$$x_{(i,a_i)} < \cdots < x_{(i,b_i)} \qquad (a_i = r_i + 1,\ b_i = n_i - r_i)$$

and assume normality for these censored samples. Let A_i, A, B_i, C_i, m_i, M, K_i, and K be as given in Section 5.2 and write

$$\bar{u}_j = \frac{\Sigma_{i=1}^{k} m_i u_{ji}}{M} \qquad Q_j = \sum_{i=1}^{k} m_i(u_{ji} - \bar{u}_j)K_i$$

$$S_{j\ell} = \sum_{i=1}^{k} m_i(u_{j\ell} - \bar{u}_j)u_{\ell i} \qquad j, \ell = 1, 2, \ldots, p$$

Then the MMLE $\hat{\mu}$, $\hat{\phi}_j$, $j = 1, 2, \ldots, p$, and $\hat{\sigma}_p$ of μ, ϕ_j, $j = 1, 2, \ldots, p$, and σ are given by

$$\hat{\mu} = \bar{K} - \sum_{j=1}^{p} \bar{u}_j \hat{\phi}_j$$

$(\hat{\phi}_1, \ldots, \hat{\phi}_p)$ are the solutions of the equations

$$\sum_{j=1}^{p} S_{ij}\hat{\phi}_j = Q_i \qquad i = 1, 2, \ldots, p$$

and

$$\hat{\sigma}_p = \sqrt{\frac{A - 2}{A - p - 1}}\,\hat{\sigma}$$

where $\hat{\sigma}$ is the same as in Section 5.2 but with C replaced by

$$C = \sum_{i=1}^{k}\left\{\sum_{j=a_i}^{b_i}\left(x_{(i,j)} - \hat{\mu} - \sum_{\ell=1}^{p}\hat{\phi}_\ell u_{\ell i}\right)^2 + r_i\beta_i\left[\left(x_{(i,a_i)} - \hat{\mu} - \sum_{\ell=1}^{p}\hat{\phi}_\ell u_{\ell i}\right)^2 + \left(x_{(i,b_i)} - \hat{\mu} - \sum_{\ell=1}^{p}\hat{\phi}_\ell u_{\ell i}\right)^2\right]\right\}$$

As in Section 5.2, again one may note that $\hat{\phi}_j$ are unbiased estimators of ϕ_j for symmetric distributions. For skew distributions, they are unbiased if $n_i = n$, $i = 1, 2, \ldots, k$, and are almost unbiased if $n_i \neq n$. Again, one notes that the maximum likelihood function L, based on type II censored normal samples, is very closely approximated by the modified maximum likelihood function L*. Here

$$L^* = \prod_{i=1}^{k}\left\{\sigma^{-A_i}\exp\left[-\frac{B_i}{\sigma} - \frac{C_i}{2\sigma^2} - \frac{m_i}{2\sigma^2}\left(K_i - \mu - \sum_{\ell=1}^{p}\phi_\ell u_{\ell i}\right)^2\right]H(\underset{\sim}{x}_i)\right\} \tag{5.3.2}$$

and $H(\underset{\sim}{x}_i)$ is free of the parameters; (5.3.2) implies that based on type II censored normal samples (and consequently L*), the regression model (5.3.1) reduces to the regression model

$$K_i = \mu + \sum_{\ell=1}^{p}\phi_\ell u_{\ell i} + \epsilon_i \tag{5.3.3}$$

where the ϵ_i's are independently distributed as normal with $E(\epsilon_i) = 0$ and variance $V(K_i) = V(\epsilon_i) = \sigma^2/m_i$. Using (5.3.3), one may proceed as in ordinary regression analysis (see Draper and Smith, 1981) to obtain the weighted least-squares estimators of μ and ϕ_j ($j = 1, 2, \ldots, p$), which are of course exactly the same as the MMLE given above. Using (5.3.3), one also obtains the following partitions of the total sum of squares S_{total},

$$S_{total} = S_{\hat{\mu}} + S_{\hat{\underset{\sim}{\phi}}} + S_L + S_e$$

where

$$S_{total} = \sum_{i=1}^{k}\left[\sum_{j=a_i}^{b_i} x^2_{(i,j)} + r_i\beta_i(x^2_{(i,a_i)} + x^2_{(i,b_i)})\right]$$

$$S_{\hat{\mu}} = M\bar{K}^2 \quad \text{(sum of squares for } \mu)$$

$$S_{\hat{\phi}} = \sum_{j=1}^{p} \hat{\phi}_j Q_j \quad \text{(regression sum of squares)}$$

$$S_L = \sum_{i=1}^{k} m_i\left[K_i - \bar{K} - \sum_{j=1}^{p} \hat{\phi}_j(u_{ji} - \bar{u}_j)\right]^2 \quad \text{(lack of fit sum of squares)}$$

$$S_e = \sum_{i=1}^{k} C_i \quad \text{(error sum of squares)}$$

The degrees of freedom for $S_{\hat{\mu}}$, $S_{\hat{\phi}}$, S_L, and S_e are 1, p, k - p - 1, and A - k, respectively.

The F statistic for testing the null hypothesis H_0: $\phi_1 = \cdots = \phi_p = 0$ is given by

$$F_1 = \frac{s^2_{\hat{\phi}}}{s^2_e} \qquad s^2_{\hat{\phi}} = \frac{S_{\hat{\phi}}}{p} \qquad s^2_e = \frac{S_e}{A - k}$$

The F statistic for testing H_0^* (the linear model is an adequate fit) is given by

$$F_2 = \frac{s^2_L}{s^2_e} \qquad s^2_L = \frac{S_L}{k - p - 1}$$

Under the assumption of normality, the null distributions of F_1 and F_2, for large A_i, are central F distributions with (p, A - k) and (k - p - 1, A - k) degrees of freedom, respectively. These F statistics have exactly the same type I robustness and power properties as the F statistics considered in Section 5.2 (see Tables 5.2.3 and 5.2.5).

Of course, there are situations in practice when in the model (5.3.1), $n_i = 1$ for all i = 1, 2, ..., k, for example, a single x observation for each $(u_1, u_2, \ldots, u_p)$ point, or n_i is less than 4 for some i = 1, 2, ..., k. In such situations, one proceeds as in Section 5.2; that is, one obtains the least-squares estimates $\tilde{\mu}$ and $\tilde{\phi}_j$, j = 1, 2, ..., p, by minimizing the error sum of squares and works out the estimated residuals

$$\tilde{e}_{ij} = x_{ij} - \tilde{\mu} - \sum_{i=1}^{p} \tilde{\phi}_i u_{ij} \qquad j = 1, 2, \ldots, n_i, \; i = 1, 2, \ldots, k$$

One then sets aside all those x_{ij} which correspond to the largest $R = [0.5 + 0.2N]$, $N = \Sigma_{i=1}^{k} n_i$, $\tilde{e}_{ij}^2$. One then works out the MMLE $\hat{\mu}$, $\hat{\phi}_j$, $j = 1, 2, \ldots, p$, and $\hat{\sigma}_p$ from the reduced data and constructs the F_1 and F_2 statistics in exactly the same way as in Section 5.2. Note that if there is only one observation x_i in the ith group, then $K_i = x_i$, as explained in Section 5.2.

EXAMPLE 5.3.1 The following data have been considered by several authors (Brownlee, 1965, Sec. 13.12; Daniel and Wood, 1971, Chap. 5; Andrews, 1974); x = stock loss, and the explanatory variables are u_1 = airflow, u_2 = cooling water temperature, u_3 = acid concentration:

Group	u_1	u_2	u_3	x	Group	u_1	u_2	u_3	x
1	80	27	89	42	11	58	18	89	14
2	80	27	88	37	12	58	17	88	13
3	75	25	90	37	13	58	18	82	11
4	62	24	87	28	14	58	19	93	12
5	62	22	87	18	15	50	18	89	8
6	62	23	87	18	16	50	18	86	7
7	62	24	93	19	17	50	19	72	8
8	62	24	93	19	18	50	19	79	8
9	58	23	87	15	19	50	20	80	9
10	58	18	80	14	20	56	20	82	15
					21	70	20	91	15

Assuming the linear model

$$x = \mu + \phi_1 u_1 + \phi_2 u_2 + \phi_3 u_3 + e$$

the least-squares estimates are (Brownlee, 1965, p. 460)

$$\tilde{\mu} = -39.9 \qquad \tilde{\phi}_1 = 0.72 \qquad \tilde{\phi}_2 = 1.29 \qquad \tilde{\phi}_3 = -0.15$$

The fitted regression equation is

$$x = -39.9 + 0.72u_1 + 1.29u_2 - 0.15u_3$$

The mean error sum of squares is

$$\tilde{s}_e^2 = \sum_{i=1}^{21} \frac{\tilde{e}_i^2}{17} = 10.52$$

The x observations that correspond to the $R = [0.5 + 0.2(21)] = 4$ largest $\tilde{e}_i^2$ are x = 42 (group 1), x = 37 (group 3), x = 28 (group 4), and x = 15 (group 21). Setting aside these four groups, we obtain the following MMLEs, which are in fact the least-square estimates calculated from the reduced data since $K_i = x_i$:

$$\hat{\mu} = -37.6 \qquad \hat{\phi}_1 = 0.80 \qquad \hat{\phi}_2 = 0.58 \qquad \hat{\phi}_3 = -0.07$$

The fitted regression equation is

$$x = -37.6 + 0.80u_1 + 0.58u_2 - 0.07u_3$$

The mean error sum of squares is

$$\hat{s}_e^2 = \sum_{i=1}^{17} \frac{\hat{e}_i^2}{13} = 1.57$$

Again, the reduction in the error sum of squares is substantial.

In their analysis of the data above, Daniel and Wood also set aside the four groups (1, 3, 4, and 21) and discuss the unusual behavior of these points; see also Cook and Weisberg (1982) and Belsley et al. (1980), who describe in detail various diagnostic techniques for identifying data points which might have an unusual behavior with respect to an assumed regression model. They then fit the regression equation

$$x = -15.4 - 0.07u_1 + 0.53u_2 + 0.0068u_1^2$$

to the remaining points, resulting in a mean error sum of squares 1.25.

5.4 TESTING DATA ANOMALY IN LINEAR MODELS

Assuming that a linear model is the correct one for the data of Examples 5.2.1 and 5.3.1, the anomaly of the data points set aside can be tested by defining the statistic

$$F_A = \frac{1}{R} \frac{(N - p - 1)\tilde{s}_e^2 - (N - R - p - 1)\hat{s}_e^2}{\hat{s}_e^2} \tag{5.4.1}$$

Large values of F_A indicate that at least one of the $R = [0.5 + 0.2N]$ data points set aside is anomalous in comparison with the remaining $N - R$ data points. The null distribution of F_A can be approximated by a central F distribution with $(R, N - R - p - 1)$ degrees of freedom.

For Example 5.3.1,

$$F_A = \frac{1}{4} \frac{178.84 - 20.41}{1.57}$$

$$= 25.23 \quad (\text{df} = 4.13)$$

the 1% significance level being 5.21, the indication is that at least one of the data points x = 42 (group 1), x = 37 (group 3), x = 28 (group 4), and x = 15 (group 21) is seriously anomalous with the remaining 17 data points.

EXAMPLE 5.4.1 Draper and Smith (1966, p. 101) give the following values of x = suds height and the explanatory variable u = grams of product, u being error-free:

	Data point									
	1	2	3	4	5	6	7	8	9	10
x	24.4	32.1	37.1	40.4	43.3	51.4	61.9	66.1	77.2	79.2
u	3.5	4.0	4.5	5.0	5.5	6.0	6.5	7.0	7.5	8.0

Assume the linear model $x = \mu + \phi u + e$. Here $n = 10$, $\Sigma u_i = 57.5$, $\Sigma x_i = 513.1$, $\Sigma u_i^2 = 351.25$, $\Sigma u_i x_i = 3206.95$; $\tilde{\phi} = 12.442$ and $\tilde{\mu} = -20.234$. The fitted line is

$$x = -20.234 + 12.442u$$

and the estimated residuals $\tilde{e}_i$ are, in order,

1.086 2.564 1.343 -1.578 -4.899 -3.020 1.258 -0.763 4.116 -0.105

The mean error sum of squares is

$$\tilde{s}_e^2 = \sum_{i=1}^{10} \frac{\tilde{e}_i^2}{8} = 8.036$$

The data points that correspond to $R = [0.5 + 0.2(10)] = 2$ largest $\tilde{e}_i^2$ are groups 5 and 9. Setting aside these two points, we obtain ($K_i = x_i$ since there is only one observation in each group)

$$\hat{\mu} = -17.461 \quad \text{and} \quad \hat{\phi} = 11.962$$

The fitted line is

$$x = -17.461 + 11.962u$$

and the mean error sum of squares is

$$\hat{s}_e^2 = \sum_{i=1}^{8} \frac{\hat{e}_i^2}{6} = 3.215$$

The statistic

$$F_A = \frac{1}{2}\frac{64.288 - 19.29}{3.215}$$

$$= 7.00 \qquad (df = 2, 6)$$

The 1% significance level being 10.9, we conclude that neither of the data points 5 and 9 is anomalous compared with the remaining eight points.

The MML procedure above can easily be carried out even with nonlinear models, since it is essentially a least-squares procedure based on reduced data.

Comment: Data anomaly in regression analysis is subjective to the assumed model. For the data of Examples 5.2.2 and 5.3.1, for instance, if the models were assumed to be nonlinear, some other data points (or none at all) might turn out to be anomalous. However, if the experimentation is restricted to a small range of the explanatory variable, the linear model might be accurate, at least as a close approximation, and in that situation the foregoing procedures have a useful role to play. If the range of the explanatory variable u is not small, one may have to work with nonlinear models. The MML method above can readily be extended to nonlinear models, but the Huber method (described in the next section) will become computationally rather unmanageable.

We have assumed the explanatory variable to be error-free. The existence of error in u raises many problems which have not been completely resolved yet; we refer the reader to Atkinson (1982) and the references cited by him; see also Carroll and Gallo (1982).

5.5 HUBER'S ROBUST REGRESSION PROCEDURE

Consider the linear model

$$x_i = \phi_0 + \phi_1 u_{1i} + \cdots + \phi_p u_{pi} + e_i \quad (i = 1, 2, \ldots, n) \tag{5.5.1}$$

where e_i are assumed to be IID and have a symmetric distribution with location zero and scale σ. We assume that the explanatory variables u_{ji} are error-free, which is usually the case with regression problems. Huber's (1964) M estimators of ϕ_0, ϕ_j, j = 1, 2, ..., p, and σ^2 are obtained by solving the following p + 2 simultaneous equations (Huber, 1977, p. 36 or 1981, p. 176; Hogg, 1979; Ringland, 1983):

$$\sum_{i=1}^{n} \psi\left(\frac{e_i}{\sigma}\right) u_{ji} = 0 \quad (j = 0, 1, 2, \ldots, p, \; u_{0i} = 1) \tag{5.5.2}$$

and

$$\sum_{i=1}^{n} \psi^2\left(\frac{e_i}{\sigma}\right) = (n - p - 1) \int_{-\infty}^{\infty} \psi^2(x) f(x)\, dx \tag{5.5.3}$$

where $e_i = x_i - \phi_0 - \Sigma_{j=1}^{p} \phi_j u_{ji}$ and $f(x) = (2\pi)^{-\frac{1}{2}} \exp(-x^2/2)$.

Equation (5.5.3) has been referred to as Huber's proposal 2 for estimating σ^2. Alternatively, one may use (5.5.2) to estimate ϕ_0 and $\underset{\sim}{\phi}' = (\phi_1, \phi_2, \ldots, \phi_p)$ but use mad/0.6745 (mad = med $|\hat{e}_i - \text{med}(\hat{e}_i)|$) to estimate σ (Birch and Myers, 1982), the latter method being referred to as Huber's proposal 1 [Huber himself preferred proposal 2 in regression models; see Huber (1977, p. 36)]. From (5.5.2) and (5.5.3), if one defines the weight function $w_i = \psi(e_i/\sigma)/(e_i/\sigma)$, then the following iterative formulas provide the estimates of $\underset{\sim}{\theta}' = (\phi_0, \underset{\sim}{\phi}')$ and σ^2:

$$\hat{\underset{\sim}{\theta}} = (U'W^{-1}U)^{-1}U'W\underset{\sim}{X} \tag{5.5.4}$$

and

$$\hat{\sigma}^2 = \frac{\Sigma_{i=1}^{n} w_i^2 e_i^2}{(n - p - 1) \int_{-\infty}^{\infty} \psi^2(x) f(x)\, dx} \tag{5.5.5}$$

where $U = (\underset{\sim}{u}_0, \underset{\sim}{u}_1, \ldots, \underset{\sim}{u}_p)$ with $\underset{\sim}{u}_j' = (u_{j1}, \ldots, u_{jn})$, $W = \text{diag}(w_0, w_1, \ldots, w_n)$ and $\underset{\sim}{X}' = (x_1, x_2, \ldots, x_n)$ (w_i are functions of $\hat{\underset{\sim}{\theta}}$ and $\hat{\sigma}^2$ and e_i functions of $\hat{\underset{\sim}{\theta}}$).

It is clear that the M estimators above have to be computed iteratively. Hampel (1971) and Andrews (1974) study some of the ρ functions ($\rho = \psi^{-1}$) and discuss convergence properties of the iterations involved. The choice of the starting values for ϕ_i is usually very crucial for the iterations to converge reasonably quickly.

Andrews (1974), for example, considered the ρ function

$$\rho(z) = \begin{cases} 1 + c \cos \dfrac{z}{c} & |z| \le c\pi \\ 0 & |z| > c\pi \quad (c = 1.5) \end{cases} \tag{5.5.6}$$

and using a complex computational procedure obtained the M estimates from the data of Example 5.3.1. His procedure also discarded the four points (1, 3, 4, and 21) and resulted in the linear fit

$$x = -37.2 + 0.82u_1 + 0.52u_2 - 0.07u_3$$

with mean error sum of squares

$$\sum_{i=1}^{17} \frac{\hat{e}_{iM}^2}{13} = 1.58$$

It is interesting to note that the MML estimates (which are in this situation essentially the least-squares estimates calculated from a reduced set of data) and the M estimates based on the ρ function (5.5.6) are almost identical for the data of Example 5.3.1. The computation of the M estimates is, however, too involved (see Andrews, 1974) and that renders a detailed comparison between the two methods very difficult if not impossible.

Some of the authoritative works on least-squares estimators and their computations are Marquardt (1963), Golub and Reinsch (1970), Wilkinson (1970), Forsythe (1972), and Ruppert and Caroll (1980).

5.6 ROBUST ANALYSIS OF VARIANCES VIA MMLEs

As in Sections 5.2 and 5.3, one may develop robust procedures through the MMLE method for fixed-effects analysis of variance models (Tiku, 1973, 1981a). For illustration, consider the fixed-effects two-way classification model

$$x_{ij\ell} = \mu + g_i + \tau_j + \delta_{ij} + e_{ij\ell} \tag{5.6.1}$$

($\ell = 1, 2, \ldots, n_{ij}$; $i = 1, 2, \ldots, k$; $j = 1, 2, \ldots, c$), where $e_{ij\ell}$ are assumed to be IID variables with $E(e_{ij\ell}) = 0$ and $\operatorname{var}(e_{ij\ell}) = \sigma^2$. Let ($a_{ij} = r_{ij} + 1$, $b_{ij} = n_{ij} - r_{ij}$)

$$x_{(ij,a_{ij})},\ x_{(ij,a_{ij}+1)},\ \ldots,\ x_{(ij,n_{ij}-b_{ij})} \tag{5.6.2}$$

be the type II symmetrically censored sample available in the (i, j)th cell ($i = 1, 2, \ldots, k$ and $j = 1, 2, \ldots, c$) and let A_{ij}, B_{ij}, C_{ij}, m_{ij}, and K_{ij} be the A, B, C, m, and K (defined in Chapter 2) values, respectively, calculated from the observations in the (i, j)th cell. Assuming normality for the type II censored samples (5.6.2), and assuming without any loss of generality the constraints

$$\sum_i m_{i.} g_i = \sum_j m_{.j}\tau_j = \sum_i m_{ij}\delta_{ij} = \sum_j m_{ij}\delta_{ij} = 0$$

$$m_{i.} = \sum_{j=1}^{c} m_{ij} \qquad m_{.j} = \sum_{j=1}^{k} m_{ij} \qquad m_{..} = \sum_{i=1}^{k}\sum_{j=1}^{c} m_{ij}$$

the MMLEs of μ, g_i, τ_j, δ_{ij}, and σ are obtained as

$$\hat{\mu} = K_{..} \qquad \hat{g}_i = \bar{K}_{i.} - \bar{K}_{..} \qquad \hat{\tau}_j = \bar{K}_{.j} - \bar{K}_{..} \qquad \hat{\delta}_{ij} = K_{ij} - \bar{K}_{i.} - \bar{K}_{.j} + \bar{K}_{..}$$

and

$$\hat{\sigma} = \frac{B + (B^2 + 4AC)^{\frac{1}{2}}}{2[A(A - kc)]^{\frac{1}{2}}} \tag{5.6.3}$$

Here

$$\bar{K}_{i.} = \frac{\Sigma_j m_{ij} K_{ij}}{m_{i.}} \qquad \bar{K}_{.j} = \frac{\Sigma_i m_{ij} K_{ij}}{m_{.j}} \qquad \bar{K}_{..} = \frac{\Sigma_i \Sigma_j m_{ij} K_{ij}}{m_{..}}$$

$$A = \sum_i \sum_j A_{ij} \qquad B = \sum_i \sum_j B_{ij} \qquad C = \sum_i \sum_j C_{ij}$$

Like the MMLEs given in Section 5.2, the MMLEs $\hat{g}_i$, $\hat{\tau}_j$, and $\hat{\delta}_{ij}$ are unbiased estimators for symmetric distributions. For skew distributions, they are unbiased if $n_i = n$, $i = 1, 2, \ldots, k$, and are almost unbiased if $n_i \neq n$. The approximate standard errors of the MMLEs $\hat{g}_i$, $\hat{\tau}_j$, and $\hat{\delta}_{ij}$ are given, respectively, by

$$\pm\hat{\sigma}\left(\frac{1}{m_{i.}} - \frac{1}{m_{..}}\right)^{\frac{1}{2}} \qquad \pm\hat{\sigma}\left(\frac{1}{m_{.j}} - \frac{1}{m_{..}}\right)^{\frac{1}{2}} \qquad \pm\hat{\sigma}\left(\frac{1}{m_{ij}} - \frac{1}{m_{i.}} - \frac{1}{m_{.j}} + \frac{1}{m_{..}}\right)^{\frac{1}{2}}$$

For symmetric distributions, $\hat{\mu}$ is unbiased and its standard error is given by $\pm\hat{\sigma}/\sqrt{m_{..}}$.

For normal type II censored samples, as in Chapter 2, one can again show that the likelihood L is closely approximated by L*,

$$L^* = \sigma^{-A} \exp\left[-\frac{B}{\sigma} - \frac{C}{2\sigma^2} - \frac{1}{2\sigma^2}\sum_i \sum_j m_{ij}(K_{ij} - \mu - g_i - \tau_j - \delta_{ij})^2\right] H(\underset{\sim}{x}) \quad (5.6.4)$$

where $H(\underset{\sim}{x})$ is free of the parameters. Using (5.6.4), one can develop MML procedures (analogous to the classical ANOVA procedures) for testing hypotheses about g_i, τ_j, and δ_{ij}. These procedures have type I error robustness and are considerably more powerful overall than the classical ANOVA procedures (Tiku, 1981a; Tan and Tabatabai, 1984). Suppose, for example, that we want to test the null hypotheses H_{01}: $g_i = 0$ for all i, H_{02}: $\tau_j = 0$ for all j, and H_{03}: $\delta_{ij} = 0$ for all i and j. Then, one considers the component sums of squares as follows:

Rows:	$S_{row} = \sum_i m_{i.}(\bar{K}_{i.} - \bar{K}_{..})^2$	df = k - 1
Columns:	$S_{col} = \sum_j m_{.j}(\bar{K}_{.j} - \bar{K}_{..})^2$	df = c - 1
Interaction:	$S_I = \sum\sum m_{ij}(K_{ij} - \bar{K}_{i.} - \bar{K}_{.j} + \bar{K}_{..})^2$	df = (k - 1)(c - 1)
Error:	$S_E = (A - kc)\hat{\sigma}^2$	df = A - kc

The robust tests for H_{01}, H_{02}, and H_{03} are then given, respectively, by the statistics

$$F_1 = \frac{(A - kc)S_{row}}{(k - 1)S_E} \quad F_2 = \frac{(A - kc)S_{col}}{(c - 1)S_E} \quad F_3 = \frac{(A - kc)S_I}{(k - 1)(c - 1)S_E} \tag{5.6.5}$$

Thus one rejects H_{01} at level α^* if $F_1 \geq F_{\alpha^*}(k - 1, A - kc)$, rejects H_{02} at level α^* if $F_2 \geq F_{\alpha^*}(c - 1, A - kc)$, and rejects H_{03} at level α^* if $F_3 \geq F_{\alpha^*}((k - 1)(c - 1), A - kc)$.

To test that two treatment effects, say g_1 and g_2, are equal, one calculates

$$t_c = \frac{\hat{g}_1 - \hat{g}_2}{\hat{\sigma}(1/m_{1.} + 1/m_{2.})^{\frac{1}{2}}} \tag{5.6.6}$$

and rejects $H_0: g_1 = g_2$ against $H_1: g_1 > g_2$ at the $\alpha^*\%$ significance level if the calculated value of t_c is greater than the $\alpha^*\%$ significance level of Student's t distribution with $A - kc$ degrees of freedom.

To illustrate how the foregoing procedure works, we consider the following data from Box and Cox (1964, p. 220), which give the survival times x (in units of 10 hr) of 48 animals exposed to three different poisons and four different treatments. The experiment was set in a 3×4 factorial design with four observations per cell:

	Treatment			
Poison	A	B	C	D
I	0.31	0.82	0.43	0.45
	0.45	1.10	0.45	0.71
	0.46	0.88	0.63	0.66
	0.43	0.72	0.76	0.62
II	0.36	0.92	0.44	0.56
	0.29	0.61	0.35	1.02
	0.40	0.49	0.31	0.71
	0.23	1.24	0.40	0.38
III	0.22	0.30	0.23	0.30
	0.21	0.37	0.25	0.36
	0.18	0.38	0.24	0.31
	0.23	0.29	0.22	0.33

Several authors have discussed transformations to reduce interaction in these data, particularly a transformation of x to 1/x (see Brown, 1975, for example). Usually, it is difficult to interpret transformations, but here 1/x has a natural appeal and is the rate of dying.

An ordinary ANOVA gives the following component sums of squares:

	Original data				Transformed data			
Source	SS	df	MSS	F	SS	df	MSS	F
Poison	1.033	2	0.517	23.3**	34.88	2	17.44	72.7**
Treatment	0.921	3	0.307	13.8**	20.41	3	6.80	28.3**
Interaction	0.250	6	0.042	1.9	1.57	6	0.26	1.1
Error	0.801	36	0.022	—	8.64	36	0.24	—
	Scale estimate = 0.149				Scale estimate = 0.490			

**These values are highly significant.

The original data, unlike the transformed data, are known to have interaction (Schrader and McKean, 1977, p. 889) but the ordinary ANOVA fails to

Table 5.6.1 Values of K_{ij}

	A	B	C	D	$\bar{K}_{i.}$
			Original data		
I	0.440	0.850	0.540	0.640	0.6175
II	0.325	0.765	0.375	0.635	0.5250
III	0.215	0.335	0.235	0.320	0.2762
$\bar{K}_{.j}$	0.3267	0.6500	0.3833	0.5317	$\bar{K}_{..}$ = 0.4729
			Transformed data		
I	2.2739	1.1779	1.9048	1.5640	1.7302
II	3.1130	1.3632	2.6785	1.5970	2.1879
III	4.6536	3.0180	4.2573	3.1280	3.7642
$\bar{K}_{.j}$	3.3468	1.8530	2.9468	2.0964	$\bar{K}_{..}$ = 2.5608

reveal it, possibly because of nonnormality of the data. The transformation of x to 1/x is known to reduce interaction and here the ANOVA gives an F value of 1.1 for the interaction, which is not significant.

Let us now use the analogous procedure based on the MML estimators given above. Here

n_{ij} are all equal and equal to $n = 4$

r_{ij} are all equal and equal to $r = 1$ $(q_{ij} = q = r/n = 0.25)$

α_{ij} are all equal and equal to $\alpha = 0.7834$

β_{ij} are all equal and equal to $\beta = 0.8049$

$k = 3 \quad c = 4$

The values of K_{ij} for the original data and the transformed data are given in Table 5.6.1. From these values, we obtain the following component sums of squares:

ANOVA based on the MML estimators

	Original data				Transformed data			
Source	SS	df	MSS	F	SS	df	MSS	F
Poison	0.90	2	0.45	62.5**	32.882	2	16.44	151.8**
Treatment	0.696	3	0.232	32.2**	16.066	3	5.355	49.4**
Interaction	0.153	6	0.0255	3.54*	1.190	6	0.198	1.83
Error	0.0867	12	0.072	—	1.299	12	0.1083	—
	Scale estimate $\hat{\sigma} = 0.085$				Scale estimate $\hat{\sigma} = 0.329$			

**These values are highly significant.
*This value is significant at the 5% level.

The analysis above agrees with the findings of most authors that the main effects are highly significant in both the original and the transformed data, but the interaction is present in the original data but not in the transformed data (see Brown, 1975; Schrader and McKean, 1977, for example).

The estimated efficiency of the ANOVA based on the MML estimators to the ordinary ANOVA based on the least-squares estimators is 100(0.149/0.085) = 175.3% for the original data and 100(0.490/0.329) = 148.9% for the transformed data.

Schrader and McKean (1977) analyzed the data above by adopting the Huber method and a nonparametric R method. The Huber method involves a minimization of the dispersion of residuals of the form

$$\sum_{i=1}^{N} \rho(x_i - u_i'b)$$

where

$$\rho(z) = \begin{cases} z^2 & |z| < c \\ 2c|z| - c^2 & |z| \geq c \end{cases} \tag{5.6.7}$$

The nonparametric R method involves minimization of the dispersion of residuals of the form

$$\sum_{i=1}^{N} a[R(x_i - u_i'b)](x_i - u_i'b) \tag{5.6.8}$$

Table 5.6.2 Results of Schrader and McKean Analyses

	Original data				Transformed data			
Source	Dispersion	df	Mean dispersion	F	Dispersion	df	Mean dispersion	F
				M-method				
Poison	0.272	2	0.1358	46.0**	7.25	2	3.625	54.3**
Treatment	0.193	3	0.0642	21.8**	4.33	3	1.443	21.6**
Interaction	0.050	6	0.0083	2.8*	0.42	6	0.069	1.0
Error	—	36	0.0030	—	—	36	0.067	—
	Scale estimate = 0.097				Scale estimate = 0.455			
				R-method				
Poison	3.601	2	1.800	38.8**	24.74	2	12.37	53.8**
Treatment	2.902	3	0.967	20.9**	16.57	3	5.52	23.9**
Interaction	0.852	6	0.142	3.1*	1.90	6	0.31	1.4
Error	—	36	0.046	—	—	36	0.23	—
	Scale estimate $\hat{\sigma}$ = 0.093				Scale estimate $\hat{\sigma}$ = 0.461			

**These values are highly significant.
*This value is significant at the 5% level.

where $a(\cdot)$ is some score function and $R(\cdot)$ denotes rank;

$$a(i) = h\left(\frac{i}{N+1}\right) \quad \text{and} \quad h(z) = \sqrt{12}\left(z - \frac{1}{2}\right) \tag{5.6.9}$$

The score function (5.6.9) generates the familiar Wilcoxon and Mann-Whitney nonparametric tests; Jaeckel (1972) discusses regularity conditions for the R method. The results of Schrader and McKean analyses are given in Table 5.6.2. Note that the error degrees of freedom for the M method are at most 36.

The estimates for the efficiencies of the ANOVA based on the M and R methods relative to the ordinary ANOVA for the original and the transformed data are, respectively,

$$100\left(\frac{0.149}{0.097}\right) = 153.6\% \text{ and } 100\left(\frac{0.490}{0.455}\right) = 107.7\% \text{ for the M method}$$

and

$$100\left(\frac{0.149}{0.093}\right) = 160.2\% \text{ and } 100\left(\frac{0.490}{0.461}\right) = 106.5\% \text{ for the R method}$$

It is clear that for the Box-Cox data above, the ANOVA based on the MML estimators and the M and R methods give almost identical results, but the MML ANOVA is computationally straightforward. The M and R methods, particularly the M method, are computationally involved.

If some or all the cells in the experimental design have fewer than four observations each, one calculates the residuals [say that the model is given by (5.6.1)]

$$\tilde{e}_{ij\ell} = x_{ij\ell} - \tilde{\mu} - \tilde{g}_i - \tilde{\tau}_j - \tilde{\delta}_{ij}$$

and sets aside those observations $x_{ij\ell}$ which correspond to $R = [0.5 + 0.2N]$ (N is the total number of observations in the experiment), largest $\tilde{e}^2_{ij\ell}$; $\tilde{\mu} = \bar{x}_{..}$, $\tilde{g}_i = \bar{x}_{i.} - \bar{x}_{..}$, $\tilde{\tau}_j = \bar{x}_{.j} - \bar{x}_{..}$, and $\tilde{\delta}_{ij} = \bar{x}_{ij} - \bar{x}_{i.} - \bar{x}_{.j} + \bar{x}_{..}$, are the least-squares estimators. A few additional observations might have to be censored to achieve symmetric censoring as explained in Section 5.2, for ease of computations. The MML estimates K_{ij} are then calculated and the ANOVA proceeds as above. Note that if one ends up with only one observation in the (i, j)th cell, then $K_{ij} = x_{ij}$.

There is a vast literature discussing various aspects, particularly the asymptotic efficiencies, of M, R, and L estimators of location; see, for example, Bickel (1973, 1975, 1976), Chernoff and Gastwirth (1967), Hampel (1974), Hall (1980), Hodges and Lehmann (1963), Huber (1964, 1968), Hogg (1967, 1974), Holland and Welsch (1977), Jaeckel (1971a, b, 1972), Jureckova (1971), Joiner and Hall (1983), Relles and Rogers (1977), Reider (1978, 1979), Sacks and Ylvisaker (1972, 1978), Shorack (1972, 1974), Stein (1956), Stigler (1977), Stone (1975), Takeuchi (1971), Tiku (1980a, 1982a), Tiku and Singh (1982), Tukey (1958, 1960, 1977), and Yohai and Maronna (1979).

5.7 ROBUST TWO-WAY CLASSIFICATION

Let $x_{11}, x_{12}, \ldots, x_{1n_1}$ and $x_{21}, x_{22}, \ldots, x_{2n_2}$ be two independent random samples from populations π_1 and π_2, respectively. Assume that π_1 and π_2 are of the form $(1/\sigma)f((x - \mu)/\sigma)$ $(i = 1, 2)$ and have finite means μ_1 and μ_2 and a common finite variance σ^2. A new independent observation x_0 becomes available. It comes from π_1 or π_2 but it is not known which one. The problem is to classify x_0 in π_1 or π_2 in such a way that the errors of misclassification are minimized.

Errors of Misclassification

Let e_1 be the probability that x_0 is classified in π_2 when in fact it has come from π_1, and e_2 be the probability that x is classified in π_1 when in fact it has come from π_2; e_1 and e_2 are called errors of misclassification.

A procedure V_1 is said to be superior to another procedure V_2 if for a fixed e_1 (e_2), V_1 has larger $1 - e_2$ $(1 - e_1)$ values. Note that $1 - e_2$ and $1 - e_1$ are the probabilities of correct classification; $1 - e_2$ is the probability of classifying x_0 in π_2 when in fact it has come from π_2 and similarly for $1 - e_1$.

If π_1 and π_2 are normal $N(\mu_1, \sigma)$ and $N(\mu_2, \sigma)$, respectively, then the likelihood functions are given by

$$\frac{1}{\sqrt{2\pi}\sigma} \exp\left[-\frac{(x_0 - \mu_1)^2}{2\sigma^2}\right] \qquad x_0 \in \pi_1$$

and

$$\frac{1}{\sqrt{2\pi}\sigma} \exp\left[-\frac{(x_0 - \mu_2)^2}{2\sigma^2}\right] \qquad x_0 \in \pi_2$$

The likelihood ratio is, therefore,

$$\Lambda = \exp\left[(\mu_1 - \mu_2)\frac{x_0 - (\mu_1 + \mu_2)/2}{\sigma^2}\right]$$

That is, the log likelihood-ratio is

$$U = \left(x_0 - \frac{\mu_1 + \mu_2}{2}\right)\frac{\mu_1 - \mu_2}{\sigma^2} \tag{5.7.1}$$

(5.7.1) is called the Fisher linear discriminant function.

The classification procedure based on the log likelihood-ratio U is to classify x_0 in π_1 or π_2 according as

$$U \gtrless C_0 \tag{5.7.2}$$

The constant C_0 is so chosen that one of the errors of misclassification, say e_1, assumes a preassigned level (usually 0.05 or 0.01).

Since μ_1, μ_2, and σ^2 are not known, it is natural to replace them by their respective classical estimates

$$\bar{x}_1 = \sum_{i=1}^{n_1} \frac{x_{1i}}{n_1} \qquad \bar{x}_2 = \sum_{i=1}^{n_2} \frac{x_{2i}}{n_2} \qquad s^2 = \frac{(n_1 - 1)s_1^2 + (n_2 - 1)s_2^2}{n_1 + n_2 - 2}$$

where

$$s_1^2 = \sum_{i=1}^{n_1} \frac{(x_{1i} - \bar{x}_1)^2}{n_1 - 1} \quad \text{and} \quad s_2^2 = \sum_{i=1}^{n_2} \frac{(x_{2i} - \bar{x}_2)^2}{n_2 - 1}$$

This leads to the classical statistic (Anderson, 1951)

$$V = \left(x_0 - \frac{\bar{x}_1 + \bar{x}_2}{2}\right)\frac{\bar{x}_1 - \bar{x}_2}{s^2} \tag{5.7.3}$$

Under the assumption of normality, therefore, the classical procedure is to classify x_0 in π_1 or π_2 according as (Anderson, 1951)

$$V \gtrless C \tag{5.7.4}$$

where the constant C is so chosen that the error of misclassification e_1 assumes a preassigned level, say 0.05.

Although $\bar{x}_1$ and $\bar{x}_2$ are independent normals $N(\mu_1, \sigma/\sqrt{n_1})$ and $N(\mu_2, \sigma/\sqrt{n_2})$, respectively, and $(n_1 + n_2 - 2)s^2/\sigma^2$ is an independent chi-square, the distribution of V does not assume a manageable form. It is therefore difficult to evaluate the constant C for a given e_1 (or e_1 for a given C) and $1 - e_2$. In the asymptotic situation (n_1 and n_2 both tend to infinity), however, $\bar{x}_1$ and $\bar{x}_2$ converge to their expected values $E(\bar{x}_1) = \mu_1$ and $E(\bar{x}_2) = \mu_2$ and s converges to its expected value $E(s) = \sigma$, in which case the distribution of V is the same as that of the Fisher linear discriminant function U given by (5.7.1).

The discriminant function U is a linear function of the normal variate x_0, and therefore its distribution is normal with mean

$$E(U|x_0 \in \pi_1) = \frac{\delta}{2} \quad \text{and} \quad E(U|x_0 \in \pi_2) = -\frac{\delta}{2} \tag{5.7.5}$$

and variance $\text{var}(U) = \delta$, where $\delta = (\mu_1 - \mu_2)^2/\sigma^2$ is the standardized distance between the two populations π_1 and π_2.

Under the assumption of normality, the asymptotic value of C is given by

$$C = \frac{\delta}{2} - 1.64\sqrt{\delta} \qquad (e_1 = 0.05) \tag{5.7.6}$$

The difficulty is that δ is not known and, therefore, C cannot be determined. In practice, however, C is replaced by an estimate

$$\tilde{C} = \frac{\tilde{\delta}}{2} - 1.64\sqrt{\tilde{\delta}} \qquad (e_1 \simeq 0.05) \tag{5.7.7}$$

To obtain a suitable estimate of δ, we realize that if

$$t = \frac{\bar{x}_1 - \bar{x}_2}{s(1/n_1 + 1/n_2)^{\frac{1}{2}}}$$

then t^2 has a noncentral F distribution with $(1, n_1 + n_2 - 2)$ degrees of freedom and noncentrality parameter

$$\frac{(\mu_1 - \mu_2)^2}{\sigma^2(1/n_1 + 1/n_2)}$$

for example, the distribution of t^2 is noncentral F with $\nu_1 = 1$ and $\nu_2 = n_1 + n_2 - 2$, and noncentrality parameter $\lambda^2 = \delta/[(1/n_1) + (1/n_2)]$; this distribution is usually denoted by $F(\nu_1, \nu_2, \lambda^2)$.

For the noncentral F distribution $F(\nu_1, \nu_2, \lambda^2)$,

$$E(F) = \frac{\nu_2}{\nu_2 - 2}(\nu_1 + \lambda^2)$$

Thus

$$E(t^2) = \frac{\nu}{\nu - 2}\left(1 + \frac{\delta}{1/n_1 + 1/n_2}\right) \qquad \nu = n_1 + n_2 - 2 \tag{5.7.8}$$

An unbiased estimator of δ is therefore given by

$$\tilde{\delta} - \left(\frac{1}{n_1} + \frac{1}{n_2}\right) \tag{5.7.9}$$

where

$$\tilde{\delta} = \frac{n_1 + n_2 - 4}{n_1 + n_2 - 2}\,\frac{(\bar{x}_1 - \bar{x}_2)^2}{s^2} \tag{5.7.10}$$

Since (5.7.9) can take negative values, which are not permissible, $\tilde{\delta}$ itself may be regarded as a suitable estimator of δ, as suggested by Lachenbruch and Mickey (1968).

Thus the classical procedure is to classify x_0 in π_1 or π_2 according as

$$V \gtrless \tilde{C} \tag{5.7.11}$$

where $\tilde{C}$ is given by (5.7.7) with $\tilde{\delta}$ equal to the right-hand side of (5.7.10). Under the assumption of normality, this classification procedure has $e_1 \simeq 0.05$ (approximately) for large n_1 and n_2.

As said earlier, the underlying distribution will hardly ever be exactly normal (in fact, the underlying distribution will hardly ever be known exactly) and that raises the question whether the classification procedure above is robust to departures from normality. We will see later that it is not.

As for the two-sample regression and ANOVA procedures discussed earlier, we set the following goals for a robust classification procedure:

1. Under the assumption of normality, its $1 - e_2$ value is almost the same as that of the classical procedure (5.7.11), for a common prespecified level for e_1 (usually 0.05 or 0.01).

2. Its e_1 values are fairly stable from distribution to distribution, at any rate its e_1 value for any plausible alternative to the normal distribution is never substantially greater than a prespecified level.
3. For plausible alternatives to normality, its $1 - e_2$ values are substantially larger overall than those of the classical procedure (5.7.11).

Such a robust procedure is accomplished by replacing the sample means $\bar{x}_1$ and $\bar{x}_2$ and the pooled sample standard deviation s by the corresponding MML10 estimators $\hat{\mu}_1$, $\hat{\mu}_2$, and $\hat{\sigma}$ (given in Section 4.3). Thus a robust procedure is to classify x_0 in π_1 or π_2 according as

$$V_R \gtrless \hat{C} \tag{5.7.12}$$

where

$$V = \left(x_0 - \frac{\hat{\mu}_1 + \hat{\mu}_2}{2}\right) \frac{\hat{\mu}_1 - \hat{\mu}_2}{\hat{\sigma}^2} \tag{5.7.13}$$

and

$$\hat{C} = \frac{\hat{\delta}}{2} - 1.64\sqrt{\hat{\delta}} \tag{5.7.14}$$

$$\hat{\delta} = \frac{A_1 + A_2 - 4}{A_1 + A_2 - 2} \frac{(\hat{\mu}_1 - \hat{\mu}_2)^2}{\hat{\sigma}^2} \tag{5.7.15}$$

THEOREM 5.7.1 Under the assumption of normality, the distribution of V_R is exactly the same as that of V, for large A_1 and A_2.

Proof: This follows immediately from Lemma 2.7.2.

THEOREM 5.7.2 Under the assumption of normality, the V and V_R procedures have the same $1 - e_2$ for a common e_1, for large A_1 and A_2 (hence large n_1 and n_2).

Proof: For the classical procedure

$$e_1 = P(V < \tilde{C} | x_0 \in \pi_1) \quad \text{and} \quad 1 - e_2 = P(V < \tilde{C} | x_0 \in \pi_2) \tag{5.7.16}$$

and similarly for the robust procedure

$$e_1 = P(V_R < \hat{C} | x_0 \in \pi_1) \quad \text{and} \quad 1 - e_2 = P(V_R < \hat{C} | x_0 \in \pi_2) \tag{5.7.17}$$

Since for large A_1 and A_2, the distributions of V and V_R are identical, one need only show that $\tilde{C}$ and $\hat{C}$ are identical. As A_1 and A_2 become large, $\tilde{C}$ and $\hat{C}$ converge to their expected values in probability, where for large A_1 and A_2,

$$E(\tilde{C}) = \frac{1}{2}\delta + \frac{1}{2}\left(\frac{1}{n_1} + \frac{1}{n_2}\right) - 1.64\left[\delta + \left(\frac{1}{n_1} + \frac{1}{n_2}\right)\right]^{\frac{1}{2}} \tag{5.7.18}$$

and

$$E(\hat{C}) = \frac{1}{2}\delta + \frac{1}{2}\left(\frac{1}{m_1} + \frac{1}{m_2}\right) - 1.64\left[\delta + \left(\frac{1}{m_1} + \frac{1}{m_2}\right)\right]^{\frac{1}{2}} \tag{5.7.19}$$

For $n_1 = n_2 = 50$ and $\delta = 2$, for example,

$$E(\tilde{C}) = 1 + 0.02 - 1.64(2 + 0.04)^{\frac{1}{2}} = -1.3224$$

and since $m_1 = m_2 = n - 2r + 2r\beta = 50 - 10 + 8.313 = 48.313$ (Table B.1),

$$E(\hat{C}) = 1 + 0.0207 - 1.64(2 + 0.0414)^{\frac{1}{2}} = -1.3225$$

Since (5.7.18) and (5.7.19) are equal for large samples, the result follows.

THEOREM 5.7.3 For symmetric long-tailed distributions (kurtosis $\beta_2^* = \mu_4/\mu_2^2 > 3$), the V_R procedure has higher $1 - e_2$ values than the V procedure, asymptotically (A_1 and A_2 both tend to infinity).

Proof: For the V and V_R procedures

$$1 - e_2 = P(V < \tilde{C} \mid x_0 \in \pi_2) \quad \text{and} \quad 1 - e_2 = P(V_R < \hat{C} \mid x_0 \in \pi_2)$$

respectively. Asymptotically, $\bar{x}_1$ and $\hat{\mu}_1$ are equal to their expected values $E(\bar{x}_1) = E(\hat{\mu}_1) = \mu_1$, $\bar{x}_2$ and $\hat{\mu}_2$ are equal to their expected values $E(\bar{x}_2) = E(\hat{\mu}_2) = \mu_2$, s is equal to its expected value $E(s) = \sigma$, and $\hat{\sigma}$ is equal to its expected value $E(\hat{\sigma}) = h\sigma$ ($h < 1$ for symmetric long-tailed distributions). Asymptotically, therefore, E(V) and var(V) are given by (5.7.5), but

$$E(V_R \mid x_0 \in \pi_1) = \frac{\delta}{2h^2} \quad \text{and} \quad \operatorname{var}(V_R) = \frac{\delta}{h^4} \tag{5.7.20}$$

Moreover, $\tilde{C}$ and $\hat{C}$ are equal to their expected values

$$\frac{\delta}{2} - 1.64\sqrt{\delta} \quad \text{and} \quad \frac{\delta}{2h^2} - 1.64\sqrt{\frac{\delta}{h^4}} \tag{5.7.21}$$

respectively.

Define

$$Z_1 = \frac{V - E(V)}{\sqrt{\operatorname{var}(V)}} \quad \text{and} \quad Z_2 = \frac{V_R - E(V_R)}{\sqrt{\operatorname{var}(V_R)}} \tag{5.7.22}$$

Since Z_1 and Z_2 are asymptotically linear in x_0, they have identical distributions. Now, asymptotically,

$$P(V < \tilde{C} \mid x_0 \in \pi_2) = P(Z_1 < \sqrt{\delta} - 1.64) \quad \text{and} \quad P(V_R < \hat{C} \mid x_0 \in \pi_2) = P(Z_2 < \sqrt{\delta} - 1.64h) \tag{5.7.23}$$

Since $0 < h < 1$, the result follows.

Remark: If the underlying distributions are normal, then $h = 1$ (for large samples), in which case the V and V_R procedures have the same $1 - e_2$ values.

COROLLARY 5.7.1 For symmetric long-tailed distributions, the V procedure has actually smaller e_1 values than the V_R procedure, asymptotically.

Proof: Asymptotically, the e_1 values of the V and V_R procedures are given by $(x_0 \in \pi_1)$

$$P(V < \tilde{C} | x_0 \in \pi_1) = P(Z_1 < -1.64) \quad \text{and} \quad P(V_R < \hat{C} | x_0 \in \pi_1) = P(Z_2 < -1.64h) \tag{5.7.24}$$

respectively. Since for symmetric long-tailed distributions, $0 < h < 1$, the result follows.

As we mentioned earlier, the distributions of V and V_R do not assume manageable forms for small sample sizes n_1 and n_2, even under the assumption of normality (let alone nonnormality). Balakrishnan et al. (1985), however, report the simulated values of e_1 and $1 - e_2$ for numerous nonnormal distributions for sample sizes $n_1 = n_2 = 20$. These values are reproduced in Table 5.7.1 for the following distributions:

$$\text{(1) normal, (2) logistic; outlier models: (3) } (n-1)N(0,1) \text{ and } 1N(0,10), \text{ (4) } (n-2)N(0,1) \text{ and } 2N(0,4)\text{; mixture model: (5) } 0.90N(0,1) + 0.10N(0,4) \tag{5.7.25}$$

These distributions represent a wide class of symmetric distributions.

It is clear that the e_1 values of the V_R procedure are more stable (robust) from distribution to distribution than the classical V procedure. Equation (5.7.24) explains the nonrobust character of the classical V procedure. Since

Table 5.7.1 Simulated Values of e_1 (Presumed Level $e_1 = 0.05$) for the V- and V_R-Procedures; $n_1 = n_2 = 20$, $d = \mu_1 - \mu_2$

		Distribution				
d		(1)	(2)	(3)	(4)	(5)
0.5	V	0.057	0.055	0.011	0.020	0.037
	V_R	0.056	0.067	0.058	0.049	0.065
1.5	V	0.056	0.052	0.011	0.018	0.039
	V_R	0.055	0.060	0.047	0.042	0.064
3.0	V	0.051	0.046	0.005	0.017	0.039
	V_R	0.042	0.052	0.036	0.037	0.060

Table 5.7.2 Empirical Values of $1 - e_2$ (Presumed Level $e_1 = 0.05$); $n_1 = n_2 = 20$

		Distribution				
d		(1)	(2)	(3)	(4)	(5)
0.5	V	0.145	0.138	0.021	0.047	0.092
	V_R	0.156	0.169	0.122	0.109	0.164
1.5	V	0.453	0.448	0.107	0.198	0.252
	V_R	0.458	0.484	0.406	0.381	0.395
3.0	V	0.918	0.916	0.378	0.647	0.653
	V_R	0.915	0.924	0.894	0.859	0.794

the cutoff point -1.64 is constant, the probability $P(Z_1 < -1.64)$ assumes different values for different distributions of Z_1. On the other hand, the cutoff point -1.64h for the robust V_R procedure assumes different values for different distributions of Z_2, and this has the effect of maintaining the value of the probability $P(Z_2 < -1.64h)$ close to a specified level. For this reason, h might be called a "tuning" constant. Note that this tuning constant adjusts itself automatically with the distribution and no attempt whatsoever is needed on the part of a user to evaluate h. It is clear from the $1 - e_2$ values shown in Table 5.7.2 that the V_R procedure is clearly superior to the classical V procedure.

Unspecified Errors

The V and V_R procedures described above are designed to maximize $1 - e_2$ while e_1 is fixed at a prespecified level. In certain practical situations, however, both e_1 and e_2 are allowed to float, but one would like to adopt a classification procedure which has smallest possible error rates $(e_1 + e_2)/2$, irrespective of what values e_1 and e_2 individually assume. The only restriction is that if π_1 and π_2 are identical, then e_1 is equal to e_2. It is clear from the likelihood function that if $\mu_1 = \mu_2$ (e.g., the populations π_1 and π_2 are identical), the log-likelihood ratio is zero. This leads to the following classical and robust procedures (see also Cochran, 1968):

If

$$V^* \begin{cases} > 0 & \text{classify } x_0 \text{ in } \pi_1 \\ < 0 & \text{classify } x_0 \text{ in } \pi_2 \\ = 0 & \text{classify } x_0 \text{ arbitrarily in } \pi_1 \text{ or } \pi_2 \end{cases} \tag{5.7.26}$$

if

$$V_R^* \begin{cases} > 0 & \text{classify } x_0 \text{ in } \pi_1 \\ < 0 & \text{classify } x_0 \text{ in } \pi_2 \\ = 0 & \text{classify } x_0 \text{ arbitrarily in } \pi_1 \text{ or } \pi_2 \end{cases} \tag{5.7.27}$$

$$V^* = \left(x_0 - \frac{\bar{x}_1 + \bar{x}_2}{2}\right)(\bar{x}_1 - \bar{x}_2) \quad \text{and} \quad V_R^* = \left(x_0 - \frac{\hat{\mu}_1 + \hat{\mu}_2}{2}\right)(\hat{\mu}_1 - \hat{\mu}_2) \tag{5.7.28}$$

EXAMPLE 5.7.1 Consider the following data (Balakrishnan et al., 1985), which represent the values of $10(x - 2.0)$, x being the pollution levels (measurements of lead) in water samples from two lakes:

Sample from π_1: -1.48 1.25 -0.51 0.46 0.60 -4.27 0.63 -0.14
-0.38 1.28 0.93 0.51 1.11 -0.17 -0.79 -1.02
-0.91 0.10 0.41 1.11

Sample from π_2: 1.32 1.81 -0.54 2.68 2.27 2.70 0.78 -4.62
1.88 0.86 2.86 0.47 -0.42 0.16 0.69 0.78
1.72 1.57 2.14 1.62

The observation to be classified is 0.60 (this has in fact come from π_1). Here

$\bar{x}_1 = -0.063 \quad \bar{x}_2 = 1.04 \quad V^* = -0.12$

$\hat{\mu}_1 = 0.11 \quad \hat{\mu}_2 = 1.25 \quad V_R^* = 0.09$

The robust V_R^* procedure correctly classifies the observation 0.60 in π_1 but not the classical V^* procedure; the ineffectiveness of V^* might be due to the presence of outliers in the two samples (Chapter 7; see also Special Appendix C).

EXAMPLE 5.7.2 Consider the following data, which are similar to the data above but the water samples come from two lakes (Balakrishnan et al., 1985) at a different location:

Sample from π_1: 0.63 -0.90 -0.44 2.73 -0.21 0.86 -1.24 0.37
1.63 0.73 -0.88 -1.34 -0.13 -0.34 1.46 -1.00
0.09 0.11 1.10 -1.72

Sample from π_2: 1.16 2.41 1.43 -0.60 1.80 1.03 1.49 1.30 0.49
-0.54 -0.37 2.22 0.14 1.74 0.46 1.95 0.29 0.43
3.50 1.42

The observation to be classified is -1.23 (this has in fact come from π_1). Here

$\bar{x}_1 = 0.076 \quad \bar{x}_2 = 1.09 \quad V^* = 1.83$

$\hat{\mu}_1 = 0.031 \quad \hat{\mu}_2 = 1.04 \quad V_R^* = 1.78$

Both the V^* and V_R^* procedures correctly classify the observation -1.23 in π_1.

The distributions of the data above can be examined by using the methods given in Chapter 6 and shown to be close to normal, and that might explain the closeness of the V^* and V_R^* values above (see Exercise 6.3).

Balakrishnan et al. (1985) report the simulated values of the V^* and V_R^* procedures for numerous symmetric (both short- and long-tailed) and skew distributions and it is clear from their values that the V_R^* procedure has smaller error rates than the V^* procedure. For $n_1 = n_2 = 20$, for example, the error rates $(e_1 + e_2)/2$ are as follows:

	d = 3.0			d = 2.5		
	Normal	Logistic	(n - 2)N(0, 1) and 2N(0, 4)	$\chi_6^2/\sqrt{12}$	Exponential	d = 0.5, uniform
V^*	0.055	0.071	0.060	0.096	0.062	0.233
V_R^*	0.054	0.070	0.057	0.089	0.061	0.233

Although for finite sample sizes n_1 and n_2, the V_R^* procedure has smaller error rates than the V^* procedure, asymptotically (n_1 and n_2 both tend to infinity) the two procedures have exactly the same error rates. This follows from the following theorem.

THEOREM 5.7.4 Asymptotically, the V^* and V_R^* procedures have exactly the same error rates for symmetric distributions.

Proof: Asymptotically, $\bar{x}_1$ and $\hat{\mu}_1$ are equal to their expected values $E(\bar{x}_1) = E(\hat{\mu}_1) = \mu_1$ and $\bar{x}_2$ and $\hat{\mu}_2$ are equal to their expected values $E(\bar{x}_2) = E(\hat{\mu}_2) = \mu_2$. Both V^* and V_R^* are then identical with the statistic

$$Z = \left(x_0 - \frac{\mu_1 + \mu_2}{2}\right)(\mu_1 - \mu_2) \tag{5.7.29}$$

and the result follows.

Remark: For skew populations, V^* is asymptotically the same as (5.7.29) because $E(\bar{x}_1) = \mu_1$ and $E(\bar{x}_2) = \mu_2$. Since $E(\hat{\mu}_1 - \hat{\mu}_2)$ is almost equal to $\mu_1 - \mu_2$ as explained in Chapter 4, and

$$E\left(\frac{\hat{\mu}_1 + \hat{\mu}_2}{2}\right) = \frac{\mu_1 + \mu_2}{2} + \text{bias}$$

V_R^* is a little different from (5.7.29); however, this bias is very small (see Table 2.7.4) and therefore the V* and V_R^* procedures have asymptotically the same error rates (almost).

Distribution-Free Procedures

A number of nonparametric or distribution-free classification procedures are available and the ones that are simpler to understand are based on density estimates. These work out as follows. Let $\hat{f}(x)$ and $\hat{g}(x)$ be consistent estimators of the density functions of π_1 and π_2, respectively, based on the two samples $x_{i1}, x_{i2}, \ldots, x_{in_i}$ $(i = 1, 2)$. To classify a new independent observation x_0, calculate the ratio $\hat{f}(x_0)/\hat{g}(x_0)$.

If

$$\hat{f}(x_0)/\hat{g}(x_0) \begin{cases} > 1 & \text{classify } x_0 \text{ in } \pi_1 \\ < 1 & \text{classify } x_0 \text{ in } \pi_2 \\ = 1 & \text{classify } x_0 \text{ arbitrarily in } \pi_1 \text{ or } \pi_2 \end{cases} \qquad (5.7.30)$$

Any consistent density estimate can be used in (5.7.30), the Parzen-Cacoullos kernel estimator and the Loftsgaarden-Quesenberry (1965) estimator being the most popular. The Parzen-Cacoullos estimator, for example, is of the form (Parzen, 1962)

$$f_a'(x) = \frac{1}{ah(a)} \sum_{j=1}^{a} K\left(\frac{x - x_j}{h(a)}\right) \qquad (5.7.31)$$

where $h(a) = a^{-1/8}$ and $K(w) = \exp(-w^2/2)/\sqrt{2\pi}$.

A typical nonparametric classification procedure (based on Mann-Whitney statistic) works out as follows (Hudimoto, 1964). Let U_{n_1} be the number of x_{1i} $(i = 1, 2, \ldots, n_1)$ less than x_0, and let U_{n_2} be the number of x_{2i} $(i = 1, 2, \ldots, n_2)$ less than x_0. Define $\hat{p}_1 = U_{n_1}/n_1$ and $1 - \hat{p}_2 = U_{n_2}/n_2$. Classify x_0 in π_1 if and only if $\hat{p}_1 < \hat{p}_2$.

Several other classification procedures, similar to the distribution-free and nonparametric procedures above, and variations of them have been proposed by Anderson (1966), Das Gupta (1964), Fix and Hodges (1951), Gessaman (1970), Gupta and Kim (1978), and Stoller (1954). The distributions of all these procedures are not known, however. The computations of some of these procedures are very involved.

Balakrishnan et al. (1985) report the simulated values of the error rates $(e_1 + e_2)/2$ of all the procedures above for numerous symmetric and skew populations. It is clear from their values that all the distribution-free and nonparametric procedures have on the whole larger error rates than the

Table 5.7.3 Error Rates for Various Classification Procedures

Distribution	V_R^*	V*	F-H	L-Q	P	G-K	H	W	K	S
					d = 0.5					
Uniform	0.233	0.233	0.248	0.248	0.230	0.238	0.237	0.255	0.270	0.237
					d = 2.5					
$\chi_6^2/\sqrt{12}$	0.089	0.096	0.089	0.093	0.082	0.100	0.112	0.139	0.144	0.094
Exponential	0.061	0.062	0.069	0.065	0.066	0.058	0.084	0.101	0.107	0.069
					d = 3.0					
Normal	0.054	0.055	0.062	0.062	0.059	0.060	0.072	0.102	0.098	0.078
Logistic	0.070	0.071	0.072	0.069	0.071	0.069	0.076	0.089	0.099	0.078
(n − 2)N(0, 1) and 2N(0, 4)	0.057	0.060	0.057	0.064	0.059	0.065	0.070	0.099	0.092	0.083

robust V_R^* procedure. For $n_1 = n_2 = 20$, for example, Balakrishnan et al. (1985) report the values of the error rate $(e_1 + e_2)/2$ shown in Table 5.7.3; F-H, L-Q, P, G-K, H, W, K, and S denote the classification procedures due to Fix and Hodges, Loftsgaarden and Quesenberry, Parzen, Gupta and Kim, Hudimoto, Das Gupta (Wilcoxon W procedure), Das Gupta (Kolmogorov K procedure), and Stoller, respectively. The overall superiority of the robust V_R^* procedure is obvious.

5.8 EFFICIENCY OF THE MML ESTIMATOR IN SAMPLE SURVEY

Let π_N denote a finite population of N quantitative measurements X_1, X_2, ..., X_N. The mean

$$\bar{X} = \sum_{i=1}^{N} \frac{X_i}{N}$$

of π_N is unknown and one is interested in estimating $\bar{X}$ from a given random sample of size n (chosen without replacement) from π_N. Denote this simple random sample by $x_1, x_2, \ldots, x_n$. The classical estimator of $\bar{X}$ is the sample mean

$$\bar{x} = \sum_{i=1}^{n} \frac{x_i}{n} \tag{5.8.1}$$

Let $x_{(1)}, x_{(2)}, \ldots, x_{(n)}$ be the order statistics obtained by arranging $x_1, x_2, \ldots, x_n$ in ascending order of magnitude. As usual, let

$$\hat{\mu} = \frac{\Sigma_{i=r+1}^{n-r} x_{(i)} + r\beta(x_{(r+1)} + x_{(n-r)})}{m} \tag{5.8.2}$$

be the MML10 estimator; $r = [0.5 + 0.1n]$. What we will show here is that in the context of sample survey also, $\hat{\mu}$ is considerably more efficient (i.e., has considerably smaller mean square error) than $\bar{x}$ for long-tailed symmetric populations, although a little less efficient for the normal and short-tailed symmetric populations.

There are essentially two approaches in sample survey. One suggested approach is to consider the finite population as having been selected at random from a hypothetical population (referred to as a super-population); the observed simple random sample $x_1, x_2, \ldots, x_n$, being a random sample from π_N, is also a random sample from the super-population. A second approach does not make any reference to a super-population, in which case an empirical verification entails repeatedly sample the same objective finite population π_N. For a discussion of various aspects of these two approaches, we refer the reader to Cochran (1946) and Godambe (1955).

Super-Population Model

Regard $X_1, X_2, \ldots, X_N$ as a random sample from a super-population of the type $(1/\sigma)f((x - \mu)/\sigma)$; assume that f is symmetric with mean μ and variance σ^2. The observed simple random sample $x_1, x_2, \ldots, x_n$ being a random sample from π_N, is also a random sample from the super-population f. Let

$$\bar{x}_n = \sum_{i=1}^{n} \frac{x_i}{n}$$

denote the mean of this simple random sample and let

$$\bar{X}_{N-n} = \sum \frac{X_i}{N - n}$$

denote the mean of the remaining $N - n$ observations in π_N. If

$$\bar{X}_N = \sum_{i=1}^{N} \frac{X_i}{N}$$

denotes the mean of π_N, then clearly

$$\bar{X}_N = \frac{n}{N}\bar{x}_n + \left(1 - \frac{n}{N}\right)\bar{X}_{N-n} \tag{5.8.3}$$

Since $\bar{x}_n$ and $\bar{X}_{N-n}$ are means of independent random samples of size n and N - n, respectively, from the population f, they are unconditionally independent. Therefore,

$$E(\bar{x}_n - \bar{X}_N)^2 = \left(1 - \frac{n}{N}\right)^2 E[(\bar{x}_n - \mu) - (\bar{X}_{N-n} - \mu)]^2$$

$$= \left(1 - \frac{n}{N}\right)^2 \left(\frac{\sigma^2}{n} + \frac{\sigma^2}{N - n}\right) = \frac{\sigma^2}{n}\left(1 - \frac{n}{N}\right) \tag{5.8.4}$$

Writing $\hat{\mu}_n$ for $\hat{\mu}$ and realizing that $\hat{\mu}_n$ is an unbiased estimator of μ for symmetric populations,

$$E(\hat{\mu}_n - \bar{X}_N)^2 = E\left[(\hat{\mu}_n - \mu) - \frac{n}{N}(\bar{x}_n - \mu) - \left(1 - \frac{n}{N}\right)(\bar{X}_{N-n} - \mu)\right]^2$$

$$= V(\hat{\mu}_n) + \frac{n}{N}\frac{\sigma^2}{N} + \left(1 - \frac{n}{N}\right)\frac{\sigma^2}{N} - 2\frac{n}{N}\operatorname{cov}(\hat{\mu}_n, \bar{x}_n)$$

$$= V(\hat{\mu}_n) + \frac{\sigma^2}{N} - 2\frac{n}{N}\operatorname{cov}(\hat{\mu}_n, \bar{x}_n) \tag{5.8.5}$$

Another type of estimator that has been considered in the literature is

$$\mu_n^* = \frac{n}{N}\bar{x}_n + \left(1 - \frac{n}{N}\right)\hat{\mu}_n \tag{5.8.6}$$

For symmetric populations, μ_n^* is an unbiased estimator of μ. Estimators of this type have been proposed and studied by Fuller (1970), Rao (1975), and Tiku (1983b).

Now

$$E(\mu_n^* - \bar{X}_N)^2 = \left(1 - \frac{n}{N}\right)^2 E(\hat{\mu}_n - \bar{X}_{N-n})^2$$

$$= \left(1 - \frac{n}{N}\right)^2 \left[V(\hat{\mu}_n) + \left(1 - \frac{n}{N}\right)^{-1} \frac{\sigma^2}{N}\right] \tag{5.8.7}$$

From the equations above it immediately follows that

$$E(\hat{\mu}_n - \bar{X}_N)^2 \le E(\bar{x}_n - \bar{X}_N)^2$$

if

$$V(\hat{\mu}_n) \le \frac{\sigma^2}{n}\left(1 - 2\frac{n}{N}\right) + 2\frac{n}{N}\operatorname{cov}(\hat{\mu}_n, \bar{x}_n) \tag{5.8.8}$$

and

$$E(\hat{\mu}_n - \bar{X}_N)^2 \le E(\mu_n^* - \bar{X}_N)^2$$

if

$$V(\hat{\mu}_n) \le \left(2 - \frac{n}{N}\right)^{-1} \left[2 \operatorname{cov}(\hat{\mu}_n, \bar{x}_n) - \frac{\sigma^2}{N}\right] \tag{5.8.9}$$

Now

$$V(\hat{\mu}_n) = (\underset{\sim}{\ell}'\underset{\sim}{V}\underset{\sim}{\ell})\sigma^2$$

and

$$\operatorname{cov}(\hat{\mu}_n, \bar{x}_n) = (\underset{\sim}{\ell}'\underset{\sim}{V}\underset{\sim}{k})\sigma^2 \tag{5.8.10}$$

where $m\underset{\sim}{\ell}' = (0, \ldots, 0, 1 + r\beta, 1, 1, \ldots, 1, 1, 1 + r\beta, 0, \ldots, 0)$ and $n\underset{\sim}{k}' = (1, 1, \ldots, 1)$ are vectors with n elements and $\underset{\sim}{V}$ is the variance-covariance matrix of the standardized variates $Z_i = (x_{(i)} - \mu)/\sigma$, $i = 1, 2, \ldots, n$; note that $\underset{\sim}{V}$ is a symmetric matrix.

If f is normal $N(\mu, \sigma)$, then every row (column) of $\underset{\sim}{V}$ adds up to 1, that is,

$$\sum_{j=1}^{n} \operatorname{cov}(Z_i, Z_j) = 1$$

Consequently,

$$\operatorname{cov}(\hat{\mu}_n, \bar{x}_n) = \frac{\sigma^2}{n}$$

and since

$$V(\hat{\mu}_n) \simeq \frac{\sigma^2}{m} > \frac{\sigma^2}{n}$$

both the inequalities (5.8.8) and (5.8.9) are violated. For the normal population, therefore, $\hat{\mu}_n$ is less efficient than both $\bar{x}_n$ and μ_n^*. The loss in efficiency is slight, though, since

$$\left[\frac{E(\bar{x}_n - \bar{X}_N)^2}{E(\hat{\mu}_n - \bar{X}_N)^2}\right]^{\frac{1}{2}} \simeq \left(\frac{1 - n/N}{1.035 - n/N}\right)^{\frac{1}{2}} \tag{5.8.11}$$

$$= 0.981 \quad n/N = 0.1$$

$$= 0.979 \quad n/N = 0.2$$

$$= 0.976 \quad n/N = 0.3$$

and

$$\left[\frac{E(\mu_n^* - \bar{X}_N)^2}{E(\hat{\mu}_n - \bar{X}_N)^2}\right]^{\frac{1}{2}} \simeq \left\{\frac{(1 - n/N)[1.035(1 - n/N) + (n/N)]}{1.035(1 - n/N) + 0.035(n/N)}\right\}^{\frac{1}{2}} \qquad (5.8.12)$$

$$= 0.996 \quad n/N = 0.1$$

$$= 0.992 \quad n/N = 0.2$$

$$= 0.988 \quad n/N = 0.3$$

Note that (5.8.11) and (5.8.12) only depend on n/N, the ratio of the size of the observed sample to the size of π_N.

To get an idea about the efficiency of $\hat{\mu}_n$ relative to $\bar{x}_n$ and μ_n^*, suppose that the underlying distribution is one of the t family given by (2.3.1). We give below the exact values of the variance var = $V(\hat{\mu}_n)$ and the covariance cov = $\text{cov}(\hat{\mu}_n, \bar{x}_n)$ for samples of size 10 and 20, calculated from the values of the variances and covariances of order statistics of random samples from this family (Tiku and Kumra, 1981):

p:	2.0	2.5	3.5	4.5	5	∞
			n = 10			
var/σ^2	0.0653	0.0814	0.0921	0.0961	0.0972	0.1043
cov/σ^2	0.0699	0.0836	0.0917	0.0945	0.0953	0.1000
			n = 20			
var/σ^2	0.0314	0.0398	0.0455	0.0476	0.0482	0.0520
cov/σ^2	0.0340	0.0412	0.0456	0.0471	0.0475	0.0500

From these values, the bounds B_1 and B_2 given by the expressions on the right-hand sides of (5.8.8) and (5.8.9), respectively, can be calculated. For n = 20 and n/N = 0.1, for example, we have the following values:

p:	2.0	2.5	3.5	4.5	5	∞
B_1/σ^2	0.0468	0.0482	0.0491	0.0494	0.0495	0.0500
B_2/σ^2	0.0332	0.0408	0.0453	0.0469	0.0474	0.0500

Comparing $V(\hat{\mu}_n)$ with these bounds, it is clear that $\hat{\mu}_n$ is more efficient than $\bar{x}_n$ other than for the normal ($p = \infty$), in which case $\hat{\mu}_n$ is a little less efficient; $\hat{\mu}_n$ is a little less efficient than μ_n^* for near-normal populations but more efficient for extreme nonnormal populations.

If the underlying distribution is normal, then $V(\hat{\mu}_n) \simeq \sigma^2/m$ and $\operatorname{cov}(\hat{\mu}_n, \bar{x}_n) = \sigma^2/n$, in which case

$$E(\hat{\mu}_n - \bar{X}_N)^2 \simeq \frac{\sigma^2}{m}\left(1 - \frac{m}{N}\right)$$

$$\simeq \frac{\sigma^2}{m}\left(1 - \frac{n}{N}\right) \tag{5.8.13}$$

since m is only slightly smaller than n; in fact, m = 0.97n. Moreover, if $r/n = R/N$, ($r = [0.5 = 0.1n]$ and $R = [0.5 + 0.1N]$), then

$$E(\hat{\mu}_n - \bar{X}_N) \simeq \frac{\sigma^2}{m}\left(1 - \frac{n}{N}\right)$$

$$= \frac{\sigma^2}{m}\left(1 - \frac{a}{A}\right) \tag{5.8.14}$$

where $a = n - 2r$ and $A' = N - 2R$.

As usual, let $\hat{\sigma}_n = [b + (b^2 + 4ac)^{\frac{1}{2}}]/2[a(a-1)]^{\frac{1}{2}}$ be the MML10 estimator of σ:

$$b = r\alpha(x_{(n-r)} - x_{(r+1)}) \quad \text{and} \quad c = \sum_{i=r+1}^{n-r} x_{(i)}^2 + r\beta(x_{(r+1)}^2 + x_{(n-r)}^2) - m\hat{\mu}_n^2$$

Since $\hat{\mu}_n$ and $\hat{\sigma}_n$ are matching estimators as explained earlier, the mean-square error $E(\hat{\mu}_n - \bar{X}_N)^2$ can always be estimated by

$$\frac{\hat{\sigma}_n^2}{m}\left(1 - \frac{n}{N}\right) \tag{5.8.15}$$

We give below the values of the ratio

$$\text{ratio} = \frac{[E(\hat{\mu}_n - \bar{X}_N)^2]^{\frac{1}{2}}}{[E(\hat{\sigma}_n)/\sqrt{m}](1 - n/N)^{\frac{1}{2}}}$$

for $n = 20$ and $n/N = 0.1$, calculated from the exact values of the variances and covariances of $\hat{\mu}_n$ and $\bar{x}_n$ given above; note that for large n, $E(\hat{\sigma}_n) \simeq \{E(b) + [E(b^2) + 4aE(c)]^{\frac{1}{2}}\}/2[a(a-1)]^{\frac{1}{2}}$:

p:	2.0	2.5	3.5	4.5	5	∞
Ratio	1.09	1.05	1.04	1.03	1.02	1.01

It is clear that $(\hat{\sigma}_n/\sqrt{m})\sqrt{1 - n/N}$ provides a good estimate of the square root of the mean-square error of $\hat{\mu}_n$; see Tiku (1983b) for more values of this ratio.

EXAMPLE 5.8.1 The logarithm of the number of trees in orchards is assumed to be normal, at any rate close to a normal distribution. For a simple random sample of 10 orchards chosen from 75 orchards, Singh et al. (1982) give the following observed values of x; x = log of the number of trees:

1.7918 2.3026 2.7726 3.2581 3.5264 3.8067 3.9703 4.0943 4.2905 4.5747

Here

$\bar{x}_n$ = 3.439 with an estimate of the square root of the mean square error equal to

$$\frac{s}{\sqrt{10}}\sqrt{1 - \frac{10}{75}} = 0.266$$

To diminish the effect of possible nonnormality, we censor $r = [0.5 + 0.1(10)] = 1$ smallest and largest observations (i.e., 1.7918 and 4.5747) and obtain

$\hat{\mu}_n$ = 3.467 with an estimate of the square root of the mean square error equal to

$$\frac{\hat{\sigma}_n}{\sqrt{9.6626}}\sqrt{1 - \frac{10}{75}} = 0.282$$

The two approaches give almost identical results, due perhaps to the near-normality of the data.

Incidentally, $\bar{X}_N$ for the 75 orchards is known to be 3.344 (Singh et al., 1982).

EXAMPLE 5.8.2 Another simple random sample of 10 orchards resulted in the same observations as above except for the observation 4.5747, which got replaced by 5.6058; that is, the orchard with 97 trees got replaced by the orchard with 272 trees.

Here

$\bar{x}_n$ = 3.542 with an estimate of the square root of the mean square error equal to

$$\frac{s}{\sqrt{10}}\sqrt{1 - \frac{10}{75}} = 0.320$$

$\hat{\mu}_n = 3.467$ with an estimate of the square root of the mean square error equal to

$$\frac{\hat{\sigma}_n}{\sqrt{9.6626}} \sqrt{1 - \frac{10}{75}} = 0.282$$

As expected, the observation 5.6058, which is rather large compared to the other nine observations in the sample, inflates the estimate $\bar{x}_n$ and its mean-square error; $\hat{\mu}_n$ is not affected at all.

Finite Population Model

Under this model, the elements $X_1, X_2, \ldots, X_N$ of π_N are treated as fixed. Let $X_{(i)}$ $(i = 1, 2, \ldots, N)$ denote the elements of π_N arranged in ascending order of magnitude. Let $x_{(i)}$ $(i = 1, 2, \ldots, n)$ denote the order statistics of a simple random sample $x_1, x_2, \ldots, x_n$ (chosen without replacement) from π_N.

It is well known that the classical estimator $\bar{x}$ is an unbiased estimator of $\bar{X}$ and the variance of $\bar{x}$ is given by

$$V(\bar{x}) = \frac{S^2}{n}\left(1 - \frac{n}{N}\right) \tag{5.8.16}$$

where $S^2 = \Sigma_{i=1}^{N} (X_i - \bar{X})^2/(N - 1)$ is the population variance. An unbiased estimator of S^2 is the sample variance $s^2 = \Sigma_{i=1}^{n} (x_i - \bar{x})^2/(n - 1)$; therefore, the standard error of $\bar{x}$ is given by

$$\frac{s}{\sqrt{n}} \sqrt{1 - \frac{n}{N}} \tag{5.8.17}$$

To obtain the values of $E(\hat{\mu})$ and $V(\hat{\mu})$, and $E(\mu^*)$ and $V(\mu^*)$, we note that

$$E(\hat{\mu}) = \underset{\sim}{\ell}'\underset{\sim}{\mu}_{i:n} \qquad V(\hat{\mu}) = \underset{\sim}{\ell}'\underset{\sim}{V}\underset{\sim}{\ell}$$

and (5.8.18)

$$\text{cov}(\hat{\mu}, \bar{x}) = \underset{\sim}{\ell}'\underset{\sim}{V}\underset{\sim}{k}$$

where $\underset{\sim}{\mu}_{i:n}$ is the vector of expected values and $\underset{\sim}{V}$ is the variance-covariance matrix of order statistics $x_{(i)}$, $i = 1, 2, \ldots, n$; the vectors $\underset{\sim}{\ell}'$ and $\underset{\sim}{k}$ have already been defined [see expression (5.8.10)].

From a straightforward application of the following version of the hypergeometric distribution (Brownlee, 1965, Sec. 3.8; Wilks, 1962, pp. 243, 252)

$$p\{x = u\} = \frac{\binom{M-1}{u-1}\binom{N-M}{n-u}}{\binom{N}{n}} \qquad (u = 1, 2, \ldots, n)$$

it follows that the sampling distribution of $x_{(i)}$ is given by

$$p\{x_{(i)} = X_{(t)}\} = \frac{\binom{t-1}{i-1}\binom{N-t}{n-i}}{\binom{N}{n}} = p_{N,n,i}(t), \text{ say} \tag{5.8.19}$$

where $t = i, i + 1, \ldots, N - n + i$. The joint sampling distribution of $x_{(i)}$ and $x_{(j)}$ is given by $(i < j)$

$$p(x_{(i)} = X_{(t_1)}, x_{(j)} = X_{(t_2)}) = \frac{\binom{t_1-1}{i-1}\binom{t_2-t_1-1}{j-i-1}\binom{N-t_2}{n-j}}{\binom{N}{n}}$$

$$= p_{N,n,i,j}(t_1, t_2), \text{ say} \tag{5.8.20}$$

where $t_1 = i, i + 1, \ldots, N - n + i$ and $t_2 = j, j + 1, \ldots, N - n + j$. Thus

$$E(x_{(i)}) = \sum_{t=i}^{N-n+i} X_{(t)} p_{N,n,i}(t)$$

$$E(x_{(i)}^2) = \sum_{t=i}^{N-n+i} X_{(t)}^2 p_{N,n,i}(t) \tag{5.8.21}$$

and

$$E(x_{(i)}x_{(j)}) = \sum_{t_1=i}^{N-n+i} \sum_{t_2=j}^{N-n+j} X_{(t_1)} X_{(t_2)} p_{N,n,i,j}(t_1, t_2) \quad (i < j)$$

The expressions (5.8.21) can be computed from the given $X_{(i)}$-values of π_N. As a check on these computations, if $X_{(i)}$, $(i = 1, 2, \ldots, N)$, were to be replaced by the integers $1, 2, \ldots, N$, respectively, then

$$E\{x_{(i)}\} = \sum_{t=i}^{N-n+i} \frac{t\binom{t-1}{i-1}\binom{N-t}{n-i}}{\binom{N}{n}}$$

$$= \frac{i(N+1)}{(n+1)} \sum_{t=i}^{N-n+i} \frac{\binom{t}{i}\binom{N-t}{n-i}}{\binom{N+1}{n+1}}$$

$$= \frac{i(N+1)}{n+1}$$

and, similarly,

$$V\{x_{(i)}\} = E(x_{(i)}^2) - E^2(x_{(i)})$$
$$= \frac{i(N-n)(N+1)(n-i+1)}{(n+1)^2(n+2)} \qquad (5.8.22)$$

and $(i < j)$

$$\operatorname{cov}(x_{(i)}, x_{(j)}) = E(x_{(i)}x_{(j)}) - E(x_{(i)})E(x_{(j)})$$
$$= \frac{i(n-j+1)(N-n)(n+1)}{(n+1)^2(n+2)}$$

As a further check on these computations,

$$\underset{\sim}{k}'\underset{\sim}{\mu}_{i:n} = E(\bar{x}) = \bar{X} \quad \text{and} \quad \underset{\sim}{k}'\underset{\sim}{V}\underset{\sim}{k} = V(\bar{x}) = \frac{S^2}{n}\left(1 - \frac{n}{N}\right) \qquad (5.8.23)$$

To create a finite population π_N, we proceed as follows. Generate a random sample of size N from a population of the type $(1/\sigma)f((x - \mu)/\sigma)$, and regard the resulting X_i $(i = 1, 2, \ldots, N)$ as a fixed finite population π_N. The populations we consider here are those of Tiku (1983b) and Tiku and Bhasin (1982), that is,

(1) uniform; (2) normal; (3) logistic; (4) double exponential;

(5) Student's t_4; (6) mixture model: $0.90N(0,1) + 0.10N(0,4)$;

outlier model: (7) $(N - R)N(0,1)$ and $RN(0,4)$, $R = 0.1N$;

(8) $(N - R)N(0,1)$ and $RN(0,10)$, $R = 0.1N$; (9) Student's t_2;

(10) Cauchy

The mean and variance of the Cauchy distribution do not exist, but its location and scale parameters were taken to be equal to 0 and 1, respectively, that is,

$$f(x) = \frac{1}{\pi(1 + x^2)} \qquad -\infty < x < \infty$$

The generated values of X_i $(i = 1, 2, \ldots, N)$, when ordered in ascending order of magnitude and substituted in (5.8.21), give the expected values and variances and covariances of the order statistics $x_{(i)}$ $(i = 1, 2, \ldots, n)$ of a random sample of size n (chosen without replacement) from π_N. These expected values and variances and covariances can be substituted in (5.8.18) and $E(\hat{\mu})$ and $V(\hat{\mu})$, and $E(\mu^*)$ and $V(\mu^*)$, obtained. It is then a simple matter to calculate the mean-square errors of $\hat{\mu}$ and μ^* and the efficiencies of $\hat{\mu}$ and μ^* relative to $\bar{x}$,

$$RE(\hat{\mu}) = \left[\frac{\sqrt{MSE(\bar{x})}}{\sqrt{MSE(\hat{\mu})}}\right]100 \quad \text{and} \quad RE(\mu^*) = \left[\frac{\sqrt{MSE(\bar{x})}}{\sqrt{MSE(\mu^*)}}\right]100$$

Note that $MSE(\bar{x}) = V(\bar{x}) = (S^2/n)(1 - n/N)$ always.

Table 5.8.1 Relative Efficiencies of $\hat{\mu}$ and μ^*

π_N	$\bar{x}$	S	$E(\hat{\mu})$	$E(\mu^*)$	$cov(\hat{\mu},\bar{x})$	$V(\hat{\mu})$	$V(\mu^*)$	$RE(\hat{\mu})$	$RE(\mu^*)$
(1)	0.0192	1.000	0.0200	0.0200	0.0481	0.0518	0.0510	93.2	93.8
(2)	0.0522	1.000	0.0411	0.0422	0.0463	0.0489	0.0484	95.8	96.3
(3)	0.0312	0.933	0.0222	0.0231	0.0381	0.0389	0.0388	100.2	100.4
(4)	0.0353	0.904	0.0235	0.0247	0.0340	0.0341	0.0341	103.6	103.6
(5)	0.0680	1.019	0.0407	0.0435	0.0366	0.0354	0.0357	113.7	113.4
(6)	0.0034	1.008	0.0230	0.0210	0.0315	0.0274	0.0283	128.2	126.3
(7)	0.1031	0.985	0.0424	0.0485	0.0316	0.0284	0.0292	116.6	116.5
(8)	0.1165	0.978	0.0338	0.0421	0.0182	0.0129	0.0142	147.5	147.7
(9)	0.3729	7.075	0.0289	0.0633	0.2289	0.1291	0.1683	301.7	292.0
(10)	a	a	0.0838	0.1412	0.6206	0.3430	0.4315	∞	∞

[a]These values are very large.

Tiku (1983b) gives the exact values of $RE(\hat{\mu})$ and $RE(\mu^*)$ for several values of n and n/N, but we reproduce in Table 5.8.1 only the values for $n = 20$ and $n/N = 0.1$; for other values of n and n/N, the relative efficiencies $RE(\hat{\mu})$ and $RE(\mu^*)$ have similar magnitudes. It is clear that $\hat{\mu}$ and μ^* are both considerably more efficient overall than the classical estimator $\bar{x}$ (Tiku, 1983b; Tiku and Bhasin, 1982); Tiku (1983b) also illustrated that

$$[E(\hat{\mu} - \bar{X})^2]^{\frac{1}{2}} \simeq \frac{\hat{\sigma}}{\sqrt{m}}\sqrt{1 - \frac{n}{N}}$$

$$= \frac{\hat{\sigma}}{\sqrt{m}}\sqrt{1 - \frac{a}{A}} \qquad (\text{if } r/n = R/N)$$

($a = n - 2r$, $A = N - 2R$) provides a good estimate of the square root of the mean square error of $\hat{\mu}$ [e.g., $\sqrt{E(\hat{\mu} - \bar{X})^2}$]. It is not clear, however, how the mean-square error of μ^* can be estimated.

Confidence Interval

In sample survey, the sample size is usually large, in which case $\hat{\sigma}$ converges to its expected value $E(\hat{\sigma})$. Since $\sqrt{m}\,\hat{\mu}$ is a linear function of order statistics, its distribution is normal for large n with $E(\hat{\mu}) \simeq \bar{X}$ and standard deviation $\sqrt{V(\hat{\mu})} \simeq [E(\hat{\sigma})/\sqrt{m}]\sqrt{1 - n/N}$. Therefore, a 95% confidence interval for $\bar{X}$ is given by

$$\left\{\hat{\mu} - 1.96 \frac{\hat{\sigma}}{\sqrt{m}} \sqrt{1 - \frac{n}{N}},\ \hat{\mu} + 1.96 \frac{\hat{\sigma}}{\sqrt{m}} \sqrt{1 - \frac{n}{N}}\right\} \tag{5.8.24}$$

The expected length of this confidence interval is

$$3.92 \frac{E(\hat{\sigma})}{\sqrt{m}} \sqrt{1 - \frac{n}{N}}$$

The expected length of the corresponding confidence interval based on $\bar{x}$ and s is

$$3.92 \frac{S}{\sqrt{n}} \sqrt{1 - \frac{n}{N}}$$

since $E(s) = S$ always for large n.

The ratio of the two lengths above is

$$\frac{E(\hat{\sigma})}{S} \frac{\sqrt{n}}{\sqrt{m}} = 1.017 \frac{E(\hat{\sigma})}{S} \tag{5.8.25}$$

We give below the values of this ratio for the fixed finite populations π_N, created from distributions (1) to (9) as explained above; $n = 20$ and $n/N = 0.1$. The values of S are given in Table 5.8.1, and the values of $E(\hat{\sigma}) \simeq [E(b) + \sqrt{E(b^2) + 4aE(c)}]/2\sqrt{a(a-1)}$ were calculated from (5.8.21):

	Distribution								
	(1)	(2)	(3)	(4)	(5)	(6)	(7)	(8)	(9)
$E(\hat{\sigma})$	1.126	1.037	0.913	0.825	0.853	0.746	0.765	0.445	1.501
Ratio	1.145	1.055	0.995	0.928	0.851	0.753	0.790	0.463	0.216

It is clear that the confidence interval for $\bar{X}$ based on $\hat{\mu}$ and $\hat{\sigma}$ is considerably shorter overall than the one based on the classical estimators $\bar{x}$ and s.

Skew Populations

The estimator $\hat{\mu}$ (based on symmetrically censored samples) may also be used as an estimator of $\bar{X}$ if the population π_N has only a moderate amount of skewness. However, if π_N has more than moderate amount of skewness, use of the MML estimators based on samples with $r = [0.5 + 0.2n]$ number of observations censored only on one side of the sample (in the direction of the long tail) is recommended. Such an estimator might have smaller mean-square error than $\bar{x}$. Of course, $\bar{x}$ provides an unbiased estimator always.

EXAMPLE 5.8.3 From a population of 75 orchards, 10 orchards were chosen and the number of trees counted in each (Singh et al., 1982):

No. of trees: 60 272 16 73 53 26 10 45 34 6

The sample mean $\bar{x} = 595/10 = 59.5$; the population mean is known to be $\bar{X} = 43.81$ (Singh et al., 1982).

Since the logarithm of the number of trees is close to normal, the distribution of the number of trees is skew and has a long tail on the right-hand side. We, therefore, censor $r = [0.5 + 0.2(10)] = 2$ largest observations (i.e., 73 and 272). From the remaining eight observations, we obtain from (2.7.7) and (2.7.11) with $n = 10$, $r_1 = 0$, $r_2 = r = 2$, $\alpha_2 = \alpha = 0.7418$ and $\beta_2 = \beta = 0.7818$,

$$\hat{\mu} = K + D\hat{\sigma}$$

where ($m = n - r + r\beta = 9.5636$)

$$K = \frac{250 + 93.816}{9.5636} = 35.951 \qquad D = \frac{1.4836}{9.5636} = 0.1551$$

and

$$\hat{\sigma} = \frac{35.6798 + \sqrt{1273.0481 + 125640.1805}}{14.9666} = 26.188$$

Therefore,

$$\hat{\mu} = 35.951 + 4.062 = 40.013$$

This estimate is much closer to the population mean than the estimate $\bar{x} = 59.5$.

EXAMPLE 5.8.4 Consider another sample of 10 orchards from the population of 75 orchards above with the following number of trees:

No. of trees: 60 97 16 73 53 26 10 45 34 6

Here

$$\bar{x} = \frac{420}{10} = 42$$

This estimate is much closer to the population mean $\bar{X} = 43.81$ than the previous estimate $\bar{x} = 59.5$.

The MML estimate is $\hat{\mu} = 40.013$ in both cases and is reasonably close to the population mean and, unlike $\bar{x}$, is not affected by an observation which is apparently too large compared with the bulk of observations.

In each of the two samples above, let us censor only one largest observation. Then the common value of the MML estimator $\hat{\mu} = K + D\hat{\sigma}$ is

$$\hat{\mu} = 39.027 + 1.929 = 40.956$$

This estimate is close to the above MML estimate of 40.013. This again illustrates that censoring (deflating) all the "bad" observations is very crucial for obtaining robust estimators, and if in the process a few "good" observations are sacrificed, that does not necessarily do too much harm.

Most of the univariate procedures discussed so far can in a straightforward fashion be generalized to bivariate and multivariate distributions; see Tiku and Singh (1982) and Tiku and Balakrishnan (1984b).

5.9 CONCLUSIONS AND SUMMARY

In this chapter the MML estimation and hypothesis-testing procedures of Chapters 2 and 4 are extended to the important areas of regression analysis, analysis of variance (ANOVA), classification (of a new observation), and sample survey (estimation problem of the population mean). For the linear regression model $E(x_{ij}) = \mu + \phi u_i$, $i = 1, 2, \ldots, k$, $j = 1, 2, \ldots, n_i$ ($n_i \geq 1$), the MML10 estimator $\hat{\phi}$ is derived and its relative efficiency compared to the least-squares estimator is determined. The estimator $\hat{\phi}$ is shown to be considerably more efficient overall than the least-squares estimator. For testing $\phi = 0$, a variance-ratio test is developed in terms of the MML10 matching pair of estimators $(\hat{\phi}, \hat{\sigma})$ and this test is shown to be robust in its type I error to deviations from normality and remarkably powerful compared to the classical variance-ratio F test based on the least-squares estimators. These MML procedures are extended to a multiple regression model $E(x_{ij}) = \mu + \Sigma_{\ell=1}^{p} \phi_\ell u_{\ell i}$, $i = 1, 2, \ldots, k$, $j = 1, 2, \ldots, n_i$ ($n_i \geq 1$). A numerical example is presented to illustrate the performance of the MML method compared to the Huber method. The MML method is found to yield essentially the same results as the Huber method; a detailed comparison of the two methods has not been possible due to the tedious computations involved in the Huber method. The problem of estimation and hypothesis testing in the two-way-classification ANOVA model is tackled through the MML method. The MML estimators are obtained and the variance-ratio F tests are developed to test for main effects and interactions. A numerical example is presented to illustrate the computations of three robust methods: the MML method, the Huber method, and a nonparametric R-method. The MML method is computationally straightforward and is shown to yield essentially the same results as the other two methods. The problem of classification of a new observation in one of the two populations is considered and a discriminant function based on the MML estimators is defined. This discriminant function is shown to be superior to the Fisher-Anderson discriminant function in having smaller errors of misclassification. Finally, the problem of estimation of the population mean in the framework of sample survey is considered. The mean-square error of the MML10 estimator $\hat{\mu}$ is determined over numerous populations and shown to be considerably smaller overall than the mean-square error of the classical estimator $\bar{x}$, the sample mean. Numerous examples are presented to illustrate the applications.

EXERCISES

5.1 In an assay of a preparation of insulin (Brownlee, 1965, p. 394) of unknown potency, three doses were injected into rabbits, and the percentage fall in blood sugar after a certain period was observed; these data are given below. The plot of the variable against log-dose is usually assumed to be a straight line.

Log-dose u:	0.36	0.56	0.76
% fall x	33, 37, 40, 16, 21, 18, 25	41, 64, 34, 64, 48, 34	56, 62, 57, 72 73, 72, 81, 60

(a) Estimate the parameters in the model $E(x) = \mu + \phi u$, using the classical and the MML10 (e.g., censor $r = [0.5+0.1n_i]$ smallest and largest observations in each group) methods.
(b) Calculate the standard errors of the classical and the MML10 estimates of ϕ and comment on the relative efficiency of the latter.
(c) Test the null hypothesis $H_0: \phi = 0$ against the alternative hypotheses $H_1: \phi > 0$ using the classical as well as the MML estimates of ϕ and their standard errors. Are the two methods in agreement in rejecting H_0 at the 5% significance level?

5.2 For the data of Example 5.3.1, plot the observed x values against the corresponding estimated x values estimated from the fitted linear equation

$$x = -39.9 + 0.72u_1 + 1.29u_2 - 0.15u_3$$

Do you see any unusual behavior of the data points 1, 3, 4, and 21? Now, set aside these four data points and plot the remaining x values against the corresponding estimated values, estimated from the fitted equation

$$x = -37.6 + 0.80u_1 + 0.58u_2 - 0.07u_3$$

Do you see any unusual behavior in this plot?

5.3 In a parallel assay, the log-dose u of a drug is usually assumed to have a linear relationship with the potency x of the drug, for example

$$E(x_1) = \mu_1 + \phi_1 u_1 \quad \text{and} \quad E(x_2) = \mu_2 + \phi_2 u_2$$

Calculate the classical and the MML10 estimates of the slopes ϕ_1 and ϕ_2

from the following data and test the null hypothesis $H_0: \phi_1 = \phi_2$ at the 5% significance level.

Standard log-dose u_1:	0.36	0.56	0.76
Potency x_1	0.17, 0.21, 0.49, 0.54, 0.16	0.64, 0.48, 0.34, 0.63, 0.34	0.62, 0.72, 0.61, 0.91, 0.56

Unknown log-dose u_2:	0.36	0.56	0.76
Potency x_2	0.33, 0.37, 0.40, 0.16, 0.21, 0.25	0.41, 0.64, 0.34, 0.64, 0.48, 0.34	0.56, 0.62, 0.58, 0.72, 0.73, 0.81

5.4 The following data give the scores of 56 subjects assigned randomly to four drugs A, B, C, and D (Brownlee, 1965, p. 395):

A:	24	30	39	41	27	46	56	25	18	25	31	52	38	45
B:	18	32	33	35	19	28	41	37	33	39	38	41	38	36
C:	35	31	55	43	44	28	33	13	39	58	18	17	41	25
D:	28	40	34	27	47	30	24	28	21	28	39	26	46	42

(a) Use the classical and the MML10 ANOVA to test $H_0: \mu_A = \mu_B = \mu_C = \mu_D$, at the 5% significance level. Are the two procedures in agreement in rejecting H_0 ?

(b) Compute the classical and the MML 95% confidence intervals for the contrast $\mu_B - 2\mu_C + \mu_D$, compare the two, and comment.

5.5 Given below are the number of pepper standards for the selected villages (a simple random sample with units chosen without replacement) of a particular zone in Trivandrum (Singh et al., 1982):

Total number of villages	Number of villages selected	Number of pepper standards
441	11	41, 116, 19, 15, 144, 159, 212, 57, 28, 76, 119

Estimate the average number of pepper standards in the villages, using both the classical and the MML estimators of location. Calculate the standard errors of the estimates and comment.

6
Goodness-of-Fit Tests

6.1 INTRODUCTION

It is important to check the adequacy of models on which inferences are based. If, for example, one finds that a normal distribution fits the data very adequately, one might use the familiar classical procedures based on the sample mean $\bar{x}$ and the sample variance s^2. Thus there might be no need to switch over to robust procedures even if the latter are under normality only slightly less efficient than the classical procedures. Assume, then, that one wants to test the null hypothesis

H_0: the underlying distribution is $f_0(x; \underset{\sim}{\theta})$

The functional form f_0 is specified but the parameters $\underset{\sim}{\theta}$ are not known. For example, f_0 might be normal with unknown mean μ and unknown variance σ^2. A test of H_0, based on a random sample $x_1, x_2, \ldots, x_n$, is frequently referred to as a goodness-of-fit test. It is useful to distinguish between three types of tests. A test that is designed to test f_0 against a specified alternative $f_1(x; \underset{\sim}{\theta})$ is called a test of model discrimination. A test that is designed to test f_0 against a wide class of alternatives is called an omnibus test.

6.2 MODEL DISCRIMINATION

Given a random sample $x_1, x_2, \ldots, x_n$, one wants to test the null hypothesis

H_0: the underlying distribution is $f_0(x; \underset{\sim}{\theta})$ (f_0 specified)

against the alternative hypothesis

H_1: the underlying distribution is $f_1(x; \underset{\sim}{\theta})$ (f_1 specified)

$\underset{\sim}{\theta}$ is a vector of unknown parameters.

To test H_0 against H_1, one of the most popular tests is based on the relative maximum likelihood statistic (Cox, 1961, 1962)

$$\mathrm{RML} = \max_{(\underset{\sim}{\theta})} \prod_{i=1}^{n} f_1(x_i; \underset{\sim}{\theta}) \Big/ \max_{(\underset{\sim}{\theta})} \prod_{i=1}^{n} f_0(x_i; \theta) \tag{6.2.1}$$

Large values of RML lead to the rejection of H_0 in favor of H_1.

The main difficulty with the statistic RML is that its distribution under H_0 (or H_1) is usually intractable; even the asymptotic theory is difficult (Cox, 1961, 1962; Hogg et al., 1972; Dyer, 1973, 1974).

If both the distributions f_0 and f_1 are of the type $(1/\sigma)f((x - \mu)/\sigma)$, then the relative maximum likelihood statistic is

$$G = \max_{(\mu,\sigma)} \prod_{i=1}^{n} \frac{1}{\sigma} f_1\left(\frac{x_i - \mu}{\sigma}\right) \Big/ \max_{(\mu,\sigma)} \prod_{i=1}^{n} \frac{1}{\sigma} f_0\left(\frac{x_i - \mu}{\sigma}\right) \qquad (6.2.2)$$

Large values of G lead to the rejection of H_0 in favor of H_1. The statistic G is location and scale invariant and its distribution is known to be free of μ and σ (Dumonceaux et al., 1973). In most situations, the statistic G reduces to the ratio of the ML estimator of σ under H_0 to the ML estimator of σ under H_1.

EXAMPLE 6.2.1 Suppose that one wants to test the normal

$$H_0: \frac{1}{\sqrt{2\pi}\,\sigma} \exp\left[-\frac{(x - \mu)^2}{2\sigma^2}\right] \qquad -\infty \le x \le \infty$$

against the exponential

$$H_1: \frac{1}{\sigma} \exp\left[-\frac{(x - \mu)}{\sigma}\right] \qquad x \ge \mu$$

The maximum values of f_0 and f_1 are attained at the corresponding ML estimators of μ and σ. Under H_0, the ML estimators are given by

$$\hat{\hat{\mu}}_0 = \bar{x} = \sum_{i=1}^{n} \frac{x_i}{n} \quad \text{and} \quad \hat{\hat{\sigma}}_0 = \sum_{i=1}^{n} \left[\frac{(x_i - \bar{x})^2}{n}\right]^{\frac{1}{2}}$$

Under H_1, the ML estimators are given by (Kendall and Stuart, 1973)

$$\hat{\hat{\mu}}_1 = x_{(1)} \quad \text{and} \quad \hat{\hat{\sigma}}_1 = \sum_{i=1}^{n} \frac{x_i - x_{(1)}}{n}$$

$x_{(1)}$ being the smallest observation in the random sample $x_1, x_2, \ldots, x_n$. Replacing μ and σ in (6.2.2) by these ML estimators, we obtain

$$G = \left(\sqrt{2\pi e}\,\frac{\hat{\hat{\sigma}}_0}{\hat{\hat{\sigma}}_1}\right)^n$$

Hence the test may be performed with the statistic D given by

$$D = \frac{\hat{\hat{\sigma}}_0}{\hat{\hat{\sigma}}_1} \qquad (6.2.3)$$

Large values of D lead to the rejection of H_0 in favor of H_1.

It is difficult to work out the distribution of D, but Dumonceaux et al. (1973) have given the simulated values of its $100(1 - \alpha^*)\%$ points and power.

Consider the following set of 20 observations (Dumonceaux et al., 1973):

35.15	44.62	40.85	45.32	36.08	38.97	32.48	34.36	38.05
26.84	33.68	42.90	33.57	36.64	33.82	42.26	37.88	38.57
32.05	41.50							

Under H_0, the ML estimates are given by

$$\hat{\hat{\mu}}_0 = \sum_{i=1}^{20} \frac{x_i}{20} = 37.23 \quad \text{and} \quad \hat{\hat{\sigma}}_0 = \left[\sum_{i=1}^{20} \frac{(x_i - \bar{x})^2}{20}\right]^{\frac{1}{2}} = 4.6$$

Under H_1, the ML estimates are given by

$$\hat{\mu}_1 = x_{(1)} = 26.84 \quad \text{and} \quad \hat{\sigma}_1 = \sum_{i=1}^{20} \frac{x_i - x_{(1)}}{20} = 10.39$$

The statistic

$$D = \frac{4.6}{10.39} = 0.44$$

which is smaller than 0.67, the 90% point of D (Dumonceaux et al., 1973). We do not reject normality in favor of exponentiality, at the 10% significance level.

EXAMPLE 6.2.2 Suppose that one wants to test the normal

$$H_0: \frac{1}{\sqrt{2\pi}\,\sigma} \exp\left[-\frac{(x - \mu)^2}{2\sigma^2}\right] \qquad -\infty \leq x \leq \infty$$

against the double exponential

$$H_1: \frac{1}{2\sigma} \exp\left(-\frac{|x - \mu|}{\sigma}\right) \qquad -\infty \leq x \leq \infty$$

Under H_0,

$$\max \prod_{i=1}^{n} \frac{1}{\sigma} f\left(\frac{x_i - \mu}{\sigma}\right) = \left(\frac{1}{\sqrt{2\pi}\,\hat{\hat{\sigma}}_0}\right)^n \exp\left(-\frac{n}{2}\right)$$

Under H_1, the ML estimators of μ and σ are given by (Kendall and Stuart, 1973)

$$\hat{\mu}_1 = x_{med} = \text{med}(x_i) \quad \text{and} \quad \hat{\hat{\sigma}}_1 = \sum_{i=1}^{n} \frac{|x_i - x_{med}|}{n}$$

Under H_1, therefore,

$$\max \prod_{i=1}^{n} \frac{1}{\sigma} f\left(\frac{x_i - \mu}{\sigma}\right) = \left(\frac{1}{2\hat{\hat{\sigma}}_1}\right)^n \exp\left(-\frac{n}{2}\right)$$

Hence the test may be performed with the statistic

$$G = \frac{\hat{\hat{\sigma}}_0}{\hat{\hat{\sigma}}_1} = \frac{[n \Sigma_{i=1}^{n} (x_i - \bar{x})^2]^{\frac{1}{2}}}{\Sigma_{i=1}^{n} |x_i - \text{med}(x_i)|} \tag{6.2.4}$$

It is difficult to work out the distribution of G, but Dumonceaux et al. (1973) have given simulated values of its $100(1 - \alpha^*)\%$ points and power.

Although G is easy to compute, its distribution is in most situations intractable. The distribution and power properties of G have therefore been investigated primarily by simulation (see, e.g., Uthoff, 1970; Hogg et al., 1972; Dumonceaux et al., 1973; Dumonceaux and Antle, 1973; Bain and Engelhardt, 1980).

UMPI Test. For location and scale families, there exists a uniformly most powerful invariant (UMPI) test (Lehmann, 1959) for testing H_0 against H_1 (or H_1 against H_0). This test is based on the statistic (Hajek and Sidak, 1967)

$$\text{LRI} = \int_0^{\infty}\int_{-\infty}^{\infty} \left(\frac{1}{\sigma}\right)^n \prod_{i=1}^{n} f_1\left(\frac{x_i - \mu}{\sigma}\right) \frac{d\mu\, d\sigma}{\sigma} \Bigg/ \int_0^{\infty}\int_{-\infty}^{\infty} \left(\frac{1}{\sigma}\right)^n \prod_{i=1}^{n} f_0\left(\frac{x_i - \mu}{\sigma}\right) \frac{d\mu\, d\sigma}{\sigma} \tag{6.2.5}$$

Large values of LRI lead to the rejection of H_0 in favor of H_1. In some situations, but not in general, the statistics RML and LRI are equivalent. Often, the computation of LRI has to be carried out numerically and is therefore quite involved.

EXAMPLE 6.2.3 Consider testing the normal

$$H_0: \frac{1}{\sqrt{2\pi}\,\sigma} \exp\left[-\frac{(x - \mu)^2}{2\sigma^2}\right] \qquad -\infty \leq x \leq \infty$$

against the extreme-value distribution

$$H_1: \frac{1}{\sigma}\left[\exp\left(\left(\frac{x - \mu}{\sigma}\right) - \exp\left(\frac{x - \mu}{\sigma}\right)\right)\right] \qquad -\infty \leq x \leq \infty$$

Here the numerator of (6.2.5) is readily expressible as

$$\int_0^{\infty} \frac{\sigma^{-(n-1)}\Gamma(n) \exp\left(\Sigma_{i=1}^n x_i/\sigma\right)}{\left[\Sigma_{i=1}^n \exp(x_i/\sigma)\right]^n} \frac{d\sigma}{\sigma} \tag{6.2.6}$$

and the denominator is given by

$$\frac{\Gamma[(n-1)/2]}{2\sqrt{n}\ \pi^{(n-1)/2}\left[\Sigma_{i=1}^n (x_i - \bar{x})^2\right]^{(n-1)/2}}$$

Equation (6.2.6) has to be evaluated by numerical integration.

Dumonceaux et al. (1973) simulated values of the power of the G and LRI tests and found the G test quite satisfactory in most situations; that is, its power compares favorably with the LRI test, even for small n. For testing H_0: normal, against H_1: Cauchy, for example, the values of the power of these two tests are as follows, n = 10:

	Significance level		
	0.01	0.05	0.10
G	0.51	0.65	0.73
LRI	0.56	0.72	0.78

Comment: In most practical situations, it is difficult to specify the functional form f_1 exactly. One would therefore think that the following "directional" and "omnibus" tests would be more useful for testing a specified functional form f_0.

6.3 DIRECTIONAL TESTS

A procedure designed to test f_0 against certain types of alternatives is called a directional goodness-of-fit test. For example, one might like to test f_0 against positively skew distributions.

Directional Tests of Normality

Under H_0, let f_0 be the normal $N(\mu, \sigma)$ distribution; μ and σ are not known. To test this H_0 against skew distributions, a well-known test is given by the sample skewness, namely,

$$\sqrt{b_1} = \frac{(1/n)\,\Sigma_{i=1}^{n}(x_i - \bar{x})^3}{[(1/n)\,\Sigma_{i=1}^{n}(x_i - \bar{x})^2]^{3/2}} \qquad -\infty < \sqrt{b_1} < \infty \tag{6.3.1}$$

Large positive values of $\sqrt{b_1}$ are indicative of positively skew distributions, whereas large negative values are indicative of negatively skew distributions. Under normality, Pearson (1963) obtained the first four moments of $\sqrt{b_1}$ up to terms in n^{-3} and these are given by

$$E(\sqrt{b_1}) = 0$$

$$\sqrt{\mu_2(\sqrt{b_1})} = \sigma(\sqrt{b_1}) = \sqrt{\frac{6}{n}}\left(1 - \frac{3}{n} + \frac{6}{n^2} - \frac{15}{n^3} + \cdots\right) \tag{6.3.2}$$

and

$$\beta_2(\sqrt{b_1}) = \frac{\mu_4(\sqrt{b_1})}{\mu_2^2(\sqrt{b_1})} = 3 + \frac{36}{n} - \frac{864}{n^2} + \frac{12{,}096}{n^3} - \cdots$$

All the odd moments of $\sqrt{b_1}$ are zero.

To test normality against symmetric distributions, a well-known test is given by the sample kurtosis, namely,

$$b_2 = \frac{(1/n)\,\Sigma_{i=1}^{n}(x_i - \bar{x})^4}{[(1/n)\,\Sigma_{i=1}^{n}(x_i - \bar{x})^2]^2} \qquad (0 < b_2 < \infty) \tag{6.3.3}$$

Large values of b_2 are indicative of symmetric long-tailed distributions (i.e., distributions with kurtosis $\beta_2^* = \mu_4/\mu_2^2$ greater than 3), whereas small values are indicative of symmetric short-tailed distributions (β_2^* less than 3). Under normality, Pearson (1963) obtained the first four moments up to terms in n^{-3} and these are given by

$$E(b_2) = \frac{3(n-1)}{n+1}$$

$$\sqrt{\mu_2(b_2)} = \sigma(b_2) = \frac{24}{n}\left(1 - \frac{15}{2n} + \frac{271}{8n^2} - \frac{2319}{16n^3} + \cdots\right)$$

$$\beta_1(b_2) = \frac{\mu_3^2(b_2)}{\mu_2^3(b_2)} = \frac{216}{n}\left(1 - \frac{29}{n} + \frac{519}{n^2} - \frac{7637}{n^3} + \cdots\right) \tag{6.3.4}$$

and

$$\beta_2(b_2) = \frac{\mu_4(b_2)}{\mu_2^2(b_2)} = 3 + \frac{540}{n} - \frac{20{,}196}{n^2} + \frac{470{,}412}{n^3} - \cdots$$

Thus the null distribution of $\sqrt{b_1}$ is symmetric and leptokurtic; that is, $\beta_2(\sqrt{b_1})$ is greater than 3, which tends to normal fairly rapidly as n increases.

The null distribution of b_2, however, is very skew and cannot be approximated by a normal distribution even for n as large as 1000.

The percentage points of $\sqrt{b_1}$ and b_2 can be approximated by fitting Pearson curves to their first four moments, using Johnson et al. (1963) tables of the percentage points of Pearson curves. Various other approximations have been developed and discussed by D'Agostino (1970), D'Agostino and Tietjen (1971, 1973), D'Agostino and Pearson (1973), Bowman and Shenton (1973a,b), and Mulholland (1977), among others. In particular, D'Agostino and Pearson (1973) developed a Johnson S_U approximation to the null distribution of $\sqrt{b_1}$, namely, the distribution of

$$\delta \sinh^{-1} \frac{\sqrt{b_1}}{\lambda} \tag{6.3.5}$$

is regarded as normal N(0,1); values of δ and λ are available for $n \geq 8$ (D'Agostino and Pearson, 1973). Charts for determining the approximate percentage points of b_2 are available for n = 20(5)50, 100, 200 (D'Agostino and Pearson, 1973).

Another test of normality which is particularly powerful against symmetric short-tailed distributions (see Tiku, 1974a, for example) was given by David et al. (1954), namely,

$$u = \frac{\sqrt{n}\, w}{S} \qquad (0 < u < \infty) \tag{6.3.6}$$

where $w = x_{(n)} - x_{(1)}$ is the sample range and $S^2 = \Sigma_{i=1}^{n}(x_i - \bar{x})^2$. Large values of u are indicative of symmetric long-tailed distributions, whereas small values are indicative of symmetric short-tailed distributions. It is well known that u and S are independently distributed under normality (see David, 1981, p. 111, for example) and, therefore,

$$E(u^k) = \frac{E(w^k)n^{k/2}}{E(S^k)} \qquad (k = 1, 2, \ldots) \tag{6.3.7}$$

The approximate percentage points of the null distribution of u can be obtained by fitting Pearson curves to the first four moments of u and are given by Pearson and Hartley (1972, Table 29).

Directional Test for Any Distribution

Under H_0, let f_0 be any specified distribution of the type $(1/\sigma)f((x - \mu)/\sigma)$, μ and σ are unknown. To test f_0 against skew or symmetric alternatives, "directional" goodness-of-fit tests were developed by Tiku (1975b) as follows. Since the extreme observations in a random sample $x_1, x_2, \ldots, x_n$ are particularly sensitive to differences between the tails of the null and alternative distributions, censor r_1 smallest and r_2 largest observations (to deflate the influence of the tails) and consider the censored sample

$$x_{(r_1+1)},\ x_{(r_1+2)},\ \ldots,\ x_{(n-r_2)} \tag{6.3.8}$$

Let $\hat{\hat{\sigma}}_c$ be the ML estimator, or an estimator that is identical, at least asymptotically, to the ML estimator (MML estimator, for example), of σ calculated from the censored sample (6.3.8), and let $\hat{\hat{\sigma}}$ be the ML estimator calculated from the complete sample $x_1, x_2, \ldots, x_n$. Define the statistic

$$T = \frac{h\hat{\hat{\sigma}}_c}{\hat{\hat{\sigma}}} \qquad h = \frac{E(\hat{\hat{\sigma}})}{E(\hat{\hat{\sigma}}_c)} \qquad (0 < T < \infty) \tag{6.3.9}$$

T is location and scale invariant.

Since $\hat{\hat{\sigma}}$ is more sensitive to departures from the null distribution f_0 than $\hat{\hat{\sigma}}_0$ due to the presence of the dominating extreme observations in $\hat{\hat{\sigma}}$, $\hat{\hat{\sigma}}_c/\hat{\hat{\sigma}}$, and hence T, will be particularly sensitive to changes in the tails of the assumed distribution. As a directional goodness-of-fit test, the statistic T is used with the following choice of r_1 and r_2 to achieve high power (Tiku, 1975b):

1. To test f_0 against positively skew distributions, $r_1 = 0$ and $r_2 = [0.5 + 0.6n]$; $[g]$ denotes the integer value of g.
2. To test f_0 against negatively skew distributions, $r_1 = [0.5 + 0.6n]$ and $r_2 = 0$.
3. To test f_0 against symmetric distributions, $r_1 = r_2 = [0.5 + 0.3n]$.

THEOREM 6.3.1 Assuming that the mean and variance of f_0 exist, the asymptotic (n tends to infinity) null distribution of T is normal for fixed $q_1 = r_1/n$ and $q_2 = r_2/n$.

Proof: Since $\hat{\hat{\sigma}}$ is based on a larger number of observations than $\hat{\hat{\sigma}}_c$, it converges to its expected value $E(\hat{\hat{\sigma}})$ faster than $\hat{\hat{\sigma}}_c$ as n becomes large. As n tends to infinity, the null distribution of T is therefore the same as that of $\hat{\hat{\sigma}}_c$. The theorem then follows immediately from the well-known result that the large-sample distribution of an ML estimator is normal under some very general regularity conditions (Kendall and Stuart, 1973, Chap. 18).

LEMMA 6.3.1 For large n, the mean and variance of the null distribution of T are given by

$$E(T) \simeq 1$$

and (6.3.10)

$$V(T) \simeq \frac{V(\hat{\hat{\sigma}}_c)}{[E(\hat{\hat{\sigma}}_c)]^2} + \frac{V(\hat{\hat{\sigma}})}{[E(\hat{\hat{\sigma}})]^2} - 2\frac{\operatorname{cov}(\hat{\hat{\sigma}}_c, \hat{\hat{\sigma}})}{E(\hat{\hat{\sigma}}_c)E(\hat{\hat{\sigma}})}$$

Proof: This follows from the fact that if $\hat{\theta}_1$ and $\hat{\theta}_2$ are two estimators of θ based on a sample of size n, then for large n (Kendall and Stuart, 1969, p. 232)

$$E\left(\frac{\hat{\theta}_1}{\hat{\theta}_2}\right) \simeq \frac{E(\hat{\theta}_1)}{E(\hat{\theta}_2)}$$

and

$$V\left(\frac{\hat{\theta}_1}{\hat{\theta}_2}\right) \simeq \left[\frac{E(\hat{\theta}_1)}{E(\hat{\theta}_2)}\right]^2 \left\{\frac{V(\hat{\theta}_1)}{[E(\hat{\theta}_1)]^2} + \frac{V(\hat{\theta}_2)}{[E(\hat{\theta}_2)]^2} - 2\frac{\text{cov}(\hat{\theta}_1, \hat{\theta}_2)}{E(\hat{\theta}_1)E(\hat{\theta}_2)}\right\}$$

Testing Normality

Here the desired estimators $\hat{\hat{\sigma}}_c$ and $\hat{\hat{\sigma}}$ are given respectively by

$$\tilde{\sigma}_c = \frac{B + (B^2 + 4AC)^{\frac{1}{2}}}{2A} \quad \text{and} \quad \tilde{\sigma} = \left[\sum_{i=1}^{n} \frac{(x_i - \bar{x})^2}{n}\right]^{\frac{1}{2}} \tag{6.3.11}$$

$A = n - r_1 - r_2$ and B and C are given by (2.7.9). The coefficients α and β to calculate B and C are as follows (see also Table B.1):

q = r/n:	0.2	0.3	0.4	0.5	0.6
α	0.7418	0.7733	0.7914	0.7979	0.7902
β	0.7818	0.7355	0.6885	0.6377	0.5773

The statistic T is given by

$$T = \frac{(1 - 1/n)\tilde{\sigma}_c}{(1 - 1/A)\tilde{\sigma}} \qquad 0 < T < \infty \tag{6.3.12}$$

Small values of T lead to the rejection of normality; values of T close to zero are indicative of extreme long-tailed distributions, Cauchy distribution, for example.

For large n, the null distribution of T is normal with mean $E(T) \simeq 1$. To obtain the large-sample variance of T, we note that for large n

$$V(\tilde{\sigma}_c) \cong \frac{1}{-E(d^2 \log L^*/d\sigma^2)} = \frac{\sigma^2}{2(n - r_1 - r_2) - (r_2\alpha_2 t_2 - r_1\alpha_1 t_1)}$$

(L^* is the MML function given in Section 2.7) and since $\tilde{\sigma}$ is the estimator $\tilde{\sigma}_c$ with $r_1 = r_2 = 0$,

$$V(\tilde{\sigma}) \simeq \frac{\sigma^2}{2n}$$

t_1 and t_2 are determined by the equations

$$\int_{-\infty}^{t_1} f(t)\,dt = q_1 \quad \text{and} \quad \int_{-\infty}^{t_2} f(t)\,dt = 1 - q_2 \qquad (q_1 = r_1/n,\ q_2 = r_2/n)$$

$$f(t) = \frac{1}{\sqrt{2\pi}} \exp\left(-\frac{t^2}{2}\right) \tag{6.3.13}$$

In the context of testing goodness of fit, r_1 and r_2 are large and, consequently, $A = n - r_1 - r_2$ may be small, and therefore the formula $V(\tilde{\sigma}_c) \simeq \sigma^2/2A$ given in Chapter 2 may not provide good approximations. Instead, one uses the formula above for $V(\tilde{\sigma}_c)$ when $q_1 = r_1/n$ and $q_2 = r_2/n$ are large, say $q_1 \geq 0.3$ and $q_2 \geq 0.3$.

If r_1 and r_2 were both zero, $\tilde{\sigma}_c$ would be exactly equal to $\tilde{\sigma}$, in which case $\text{cov}(\tilde{\sigma}_c, \tilde{\sigma}) = V(\tilde{\sigma}) \simeq \sigma^2/2n$. An approximation to the variance of T is therefore given by

$$n\,V(T) \simeq \left(\frac{1 - 1/n}{1 - 1/A}\right)^2 \left[\frac{1}{2(1 - q_1 - q_2) - (q_2\alpha_2 t_2 - q_1\alpha_1 t_1)} - \frac{1}{2}\right]$$

Based on empirical values which when plotted against n (q_1 and q_2 fixed) produce almost a straight line, Tiku (1974b) obtained an improved approximation, namely,

$$n\,V(T) \simeq \left(\frac{1 - 1/n}{1 - 1/A}\right)^2 \left[\frac{1}{2(1 - q_1 - q_2) - (q_2\alpha_2 t_2 - q_1\alpha_1 t_1)} - \frac{1}{2} - 0.532(q_1 + q_2) + 1.1284(q_2\alpha_2 - q_1\alpha_1)\right] \tag{6.3.14}$$

The formula (6.3.14) gives remarkably accurate approximations. This is clear from the following values of V(T); the empirical values are based on 15,000 Monte Carlo runs for $n \leq 30$ and 10,000 runs for $n > 30$:

	20	24	30	40	50	60	80
			$r_1 = 0,\ r_2 = [0.5 + 0.6n]$				
Empirical	0.0660	0.0492	0.0404	0.0297	0.0238	0.0190	0.0143
Approx.	0.0640	0.0503	0.0404	0.0295	0.0232	0.0191	0.0142
			$r_1 = r_2 = [0.5 + 0.3n]$				
Empirical	0.0731	0.0559	0.0469	0.0340	0.0277	0.0221	0.0164
Approx.	0.0746	0.0566	0.0469	0.0342	0.0269	0.0222	0.0164

It may be noted that the null distribution of T with $r_1 = r$ and $r_2 = 0$ is exactly the same as that of T with $r_1 = 0$ and $r_2 = r$; this follows from symmetry of the normal distribution.

EXAMPLE 6.3.1 Consider the data of Example 5.8.1, namely,

1.7918 2.3026 2.7726 3.2581 3.5264 3.8067 3.9703 4.0943

4.2905 5.6058

One suspects that these data come, if not from a distribution close to normal, then from a positively skew distribution. We therefore use a directional test of normality.

We recommend that a goodness-of-fit test for normality (or for any other functional form f_0) be carried out at a high value of the significance level α^*, say $\alpha^* = 0.10$. This is prompted by the fact that the robust procedures described in Chapters 2 to 4 have almost the same efficiencies as the classical procedures for a normal distribution and distributions close to normal but have considerably higher efficiencies for more nonnormal distributions. Thus, when normality is wrongly rejected with a high probability α^* and consequently a robust procedure is adopted, this will result in only a slight loss in efficiency. On the other hand, a high value of α^* will yield high power $1 - \beta^*$ (probability of correctly rejecting the assumed normality) and this will lead to considerable gains in efficiencies through the adaptation of a robust procedure.

For the data above, the appropriate r_1 and r_2 values for the T statistic are $r_1 = 0$ and $r_2 = [0.5 + 0.6(10)] = 6$. From the four smallest observations, we then obtain

$$K = \frac{10.1251 + 6(0.5773)(3.2581)}{7.4638} = 2.8686$$

$$A = 4 \qquad B = 6(0.7902)(3.2581 - 2.8686) = 1.8467$$

$$C = 26.8150 + 6(0.5773)(10.6152) - 7.4638(8.2289) = 2.1650$$

$$\tilde{\sigma}_C = \frac{1.8467 + (3.4103 + 34.6400)^{\frac{1}{2}}}{8} = 1.0019$$

$$\tilde{\sigma} = \left[\frac{136.1015 - (35.4191)^2/10}{10}\right]^{\frac{1}{2}} = 1.0320$$

$$T = \frac{(1 - 1/10)(1.0019)}{(1 - 1/4)(1.0320)}$$

$$= 1.165$$

Since this value is not smaller than the 10% point 0.476 (see Table 6.3.1), it is reasonable to conclude that the underlying distribution is not positively skew but is perhaps normal or very close to it. This explains the closeness

Table 6.3.1 Lower 100α^*% Points for the T Test, Determined Empirically

		α^*				α^*		
n	nV(T)	0.01	0.05	0.10	nV(T)	0.01	0.05	0.10
		$r_1 = 0$, $r_2 = [0.5 + 0.6n]$				$r_1 = r_2 = [0.5 + 0.3n]$		
10	1.587	0.204	0.369	0.476	1.702	0.192	0.350	0.462
11	1.818	0.194	0.354	0.468	1.262	0.293	0.451	0.554
12	1.362	0.278	0.444	0.551	2.336	0.181	0.336	0.441
13	1.554	0.276	0.440	0.545	1.712	0.271	0.428	0.527
14	1.240	0.351	0.504	0.604	1.348	0.347	0.500	0.596
15	1.419	0.337	0.500	0.598	2.167	0.256	0.412	0.514
16	1.568	0.336	0.500	0.597	1.702	0.321	0.477	0.573
17	1.284	0.393	0.545	0.639	1.447	0.370	0.525	0.614
18	1.397	0.387	0.537	0.635	1.226	0.424	0.564	0.654
19	1.205	0.440	0.586	0.671	1.748	0.362	0.506	0.601
20	1.320	0.432	0.576	0.661	1.462	0.415	0.561	0.647

of the two estimates, the classical and the robust, and their standard errors in Example 5.8.1.

For the test of normality based on sample skewness,

$$\sqrt{b_1} = \frac{0.17343}{(1.0320)^3} = 0.158$$

Since this value is not larger than the 90% point 0.950 (D'Agostino and Tietjen, 1973, Table 2), the $\sqrt{b_1}$ test agrees with the T test.

EXAMPLE 6.3.2 Consider the data of Example 2.7.1, which we have arranged in ascending order of magnitude, namely,

-67 -48 6 8 14 16 23 24 28 29 41 49 56 60 72

Some of the smallest and largest observations are suspected of being outliers, although the bulk of data is supposed to come from a distribution very close to a normal.

Since the effect of outliers is to create long tails, we will test normality against long-tailed distributions, having long tails on both sides. For the directional T test, the appropriate values are $r_1 = r_2 = [0.5 + 0.3(15)] = 5$;

$q_1 = q_2 = 5/15$, $\alpha_1 = \alpha_2 = 0.7896$, and $\beta_1 = \beta_2 = 0.7223$. From the middle five observations

16 23 24 28 29

we obtain

$$K = \frac{120 + 3.5915(45)}{12.183} = 23.1156$$

$$A = 5 \quad B = 3.902(29 - 16) = 50.726$$

$$C = 2986 + 3.5915(1097) - 6509.7541 = 416.1214$$

$$\tilde{\sigma}_c = \frac{50.726 + (2573.1271 + 8322.428)^{\frac{1}{2}}}{10} = 15.511$$

$$\tilde{\sigma} = \sum_{i=1}^{15} \frac{(x_i - \bar{x})^2}{15} = 36.4645$$

$$T = \frac{(1 - 1/15)(15.511)}{(1 - 1/5)(36.4645)}$$

$$= 0.486$$

Since this value is smaller than the 10% point 0.514 (see Table 6.3.1), we conclude that the underlying distribution is long-tailed and that would perhaps account for the substantial differences in the sample mean and the MML10 estimate of μ and the standard errors of these estimates (see Example 2.7.1).

For $n > 20$, the percentage points of T are obtained from a normal approximation; for example, $(T - 1)/\sqrt{\text{var}(T)}$ is regarded as a standard normal variate; var (T) is given by (6.3.14).

The T test is remarkably powerful against skew and symmetric long-tailed distributions (Tiku, 1974b), but it is ineffective against symmetric short-tailed distributions. The u test is particularly powerful against distributions of the latter type. However, in the context of robustness discussed in Chapters 3 to 5, one has particularly to guard against skew and symmetric long-tailed distributions, and here the T test is very effective as a directional test of normality.

Testing Exponentiality

Since alternatives to the exponential distribution $E(\theta, \sigma)$ with density

$$f(x) = \frac{1}{\sigma} \exp\left(-\frac{x - \theta}{\sigma}\right) \qquad \theta \le x < \infty$$

are usually distributions which are positively skew, the directional T test of exponentiality is based on samples censored only on the right-hand side. Thus the T test of exponentiality is based on the statistic (Tiku et al., 1974)

$$T_E = \frac{\tilde{\sigma}_c}{\tilde{\sigma}} \quad (0 \leq T < \infty) \tag{6.3.15}$$

where

$$\tilde{\sigma}_c = \frac{\Sigma_{i=1}^{n-r} x_{(i)} + rx_{(n-r)} - nx_{(1)}}{n - r - 1}$$

and (6.3.16)

$$\tilde{\sigma} = \frac{\Sigma_{i=1}^{n} x_i - nx_{(1)}}{n - 1}$$

Small and large values of T_E lead to the rejection of exponentiality in favor of a positively skew distribution. Note that T_E is location and scale invariant.

To obtain the null distribution of T_E, we define sample spacings as follows (Karlin, 1966; Pyke, 1965):

Let $x_{(1)}, x_{(2)}, \ldots, x_{(n)}$ be the order statistics of a random sample of size n from a location and scale distribution $(1/\sigma)f(x - \mu)/\sigma)$. The sample spacings are defined as (Karlin, 1966; Pyke, 1965)

$$G_i = \frac{x_{(i+1)} - x_{(i)}}{\mu_{i+1:n} - \mu_{i:n}} \quad (i = 1, 2, \ldots, n - 1) \tag{6.3.17}$$

where $\mu_{r:n}$ is the expected value of the rth standardized order statistic $Z_r = [x_{(r)} - \mu]/\sigma$, $(r = 1, 2, \ldots, n)$.

For the exponential, $E(Z_r)$ is given by (B.15), so that the sample spacings G_i reduce to the exponential spacings D_i,

$$D_i = (n - i)(x_{(i+1)} - x_{(i)}) \quad (i = 1, 2, \ldots, n - 1) \tag{6.3.18}$$

THEOREM 6.3.2 Under exponentiality, the null distribution of $y = [(n - r - 1)/(n - 1)]T_E$ is the beta distribution beta(a,b) with density function

$$f(y) = \frac{1}{B(a,b)} y^{a-1}(1 - y)^{b-1} \quad 0 \leq y \leq 1$$

where $a = n - r - 1$ and $b = r$.

Proof: Now

$$y = \frac{\Sigma_{i=1}^{n-r} x_{(i)} + rx_{(n-r)} - nx_{(1)}}{\Sigma_{i=1}^{n} x_i - nx_{(1)}}$$

$$= \frac{\Sigma_{i=1}^{n-r-1} D_i}{\Sigma_{i=1}^{n-r-1} D_i + \Sigma_{i=n-r}^{n-1} D_i} \tag{6.3.19}$$

Noting that $2D_i/\sigma$ ($i = 1, 2, \ldots, n - 1$) are IID chi-square with 2 degrees of freedom (Appendix B), the distribution of y is the same as that of

$$z = \frac{\chi_1^2}{\chi_1^2 + \chi_2^2}$$

where χ_1^2 and χ_2^2 are two independent chi-square variates with $2(n - r - 1)$ and 2r degrees of freedom, respectively. It is easy to show that the distribution of z is the beta(a, b) and the result follows.

Remark: Note that

$$E(y) = \frac{a}{a + b} \quad \text{and} \quad V(y) = \frac{ab}{(a + b)^2(a + b + 1)}$$

and hence

$$E(T_E) = 1 \quad \text{and} \quad V(T_E) = \frac{r}{n(n - r - 1)} \tag{6.3.20}$$

under exponentiality.

Choice of r. The T_E test is quite powerful if r is chosen to be large; Tiku et al. (1974) chose $r = [0.5 + 0.5n]$ since for this value of r the beta(a, b) distribution above is also symmetric (almost) and, consequently, tends to a normal very quickly as n increases.

LEMMA 6.3.2 Under exponentiality, the asymptotic null distribution of T_E is normal with mean and variance given by (6.3.20).

Proof: This follows immediately from the fact that both a and b become large as n tends to infinity, in which case the above beta(a, b) distribution tends to a normal.

In fact, the normal distribution with mean and variance given by (6.3.20) provides accurate approximations to the percentage points of T_E. For $n = 20$, for example, we have the following values; here $r = [0.5 + 0.5(20)] = 10$:

	Lower			Upper		
α*:	0.01	0.05	0.10	0.01	0.05	0.10
Exact	0.478	0.615	0.694	1.30	1.39	1.54
Approx.	0.452	0.612	0.698	1.30	1.39	1.55

LEMMA 6.3.3 The asymptotic power function of T_E can be expressed as $P(Z \geq d)$, where Z is a standard normal variate and d is a constant.

Proof: The power function of T_E is given by

$$P(T_E \leq c_1 | H_1) + P(T_E \geq c_2 | H_1)$$

where H_1 represents any distribution. Now

$$\begin{aligned} P(T_E \geq h) &= P[(\tilde{\sigma}_c - h\tilde{\sigma}) \geq 0] \\ &= P\left\{\sum_{i=1}^{n} \ell_i x_{(i)} \geq 0\right\} \\ &= P(Z \geq d) \end{aligned} \tag{6.3.21}$$

where $\Sigma_{i=1}^n \ell_i x_{(i)}$ is a linear function of order statistics and $d = -\Sigma_{i=1}^n \ell_i E(x_{(i)})/\sqrt{V[\Sigma_{i=1}^n \ell_i x_{(i)}]}$ is a constant. Since the asymptotic distribution of a linear function of order statistics is asymptotically normal under some very general regularity conditions (Theorem 4.2.1), the result follows.

Tiku et al. (1974) and Dyer and Harbin (1981) have shown that the T_E test has good power properties.

<u>The Case θ Known.</u> In some practical situations, θ is known, say $\theta = 0$, and one wants to test the exponential $E(0, \sigma)$ with density

$$f(x) = \frac{1}{\sigma} \exp\left(-\frac{x}{\sigma}\right) \qquad 0 \leq x < \infty$$

Here the T test is based on the statistic

$$T_{E_0} = \frac{\tilde{\sigma}_{c_0}}{\tilde{\sigma}_0} \tag{6.3.22}$$

where

$$\tilde{\sigma}_{c_0} = \frac{\Sigma_{i=1}^{n-r} x_{(i)} + r x_{(n-r)}}{n - r} \quad \text{and} \quad \tilde{\sigma}_0 = \sum_{i=1}^{n} \frac{x_i}{n}$$

and $r = [0.5 + 0.5n]$. The null distribution of $[(n - r)/n]T_{E_0}$ is the beta(a,b) distribution with $a = n - r$ and $b = r$ (Tiku et al., 1974). This follows from the fact that

$$\sum_{i=1}^{n-r} x_{(i)} + r x_{(n-r)} = \sum_{i=1}^{n-r} y_i \quad \text{and} \quad \sum_{i=1}^{n} x_i = \sum_{i=1}^{n-r} y_i + \sum_{i=n-r+1}^{n} y_i$$

where

$$y_i = (n - i + 1)(x_{(i)} - x_{(i-1)}) \quad x_{(0)} = 0 \tag{6.3.23}$$

are the exponential sample spacings, and $2y_i/\sigma$ ($i = 1, 2, \ldots, n$) are IID chi-square with 2 degrees of freedom.

It is easy to show that

$$E(T_{E_0}) = 1 \quad \text{and} \quad V(T_{E_0}) = \frac{r}{(n - r)(n + 1)} \tag{6.3.24}$$

For $n \geq 15$, the percentage points of T_{E_0} are approximated by the corresponding percentage points of the normal distribution with mean and variance given by (6.3.24).

EXAMPLE 6.3.3 Consider the following 15 observations, which represent the time to breakdown of a type of electrical insulating fluid subject to a constant-voltage stress (Nelson, 1970):

1.97 0.59 2.58 1.69 2.71 25.50 0.35 0.99 3.99 3.67 2.07 0.96

5.35 2.90 13.77

Suppose one wants to test that the data come from the exponential $E(0,\sigma)$, against alternatives of positively skew distributions. For the T_{E_0} test, $r = [0.5 + 0.5(15)] = 8$; from the first $n - r = 7$ smallest observations

0.35 0.59 0.96 0.99 1.69 1.97 2.07

we obtain

$$\sum_{i=1}^{n-r} x_{(i)} + r x_{(n-r)} = 8.62 + 16.56 = 25.18 \qquad \sum_{i=1}^{n} x_i = 69.09$$

Therefore,

$$T_{E_0} = \frac{25.18/7}{69.09/15} = 0.781$$

$$E(T_{E_0}) = 1 \quad \text{and} \quad V(T_{E_0}) = \frac{8}{7(16)} = 0.0714$$

The standard variate

$$Z = \frac{0.781 - 1}{\sqrt{0.0714}} = -0.82$$

lies between -1.64 and 1.64 and we do not, therefore, reject exponentiality at 10% significance level.

Numerous other directional tests of the exponential $E(0,\sigma)$ are available; see Epstein (1960a), Cox and Lewis (1966, Chap. 6), Wang and Chang (1977), and Lee et al. (1980). Many of these tests are designed to test $E(0,\sigma)$ against monotone increasing (IFR) or monotone decreasing (DFR) hazard functions (failure rates) $f(z)/[1 - F(z)]$; for the exponential, the failure rate is constant. One of the most popular tests is based on the so-called Gini statistic

$$G = \frac{\Sigma_{i=1}^{n-1} iy_{i+1}}{(n-1)\,\Sigma_{i=1}^{n} y_i} \quad (0 \leq G \leq 1) \tag{6.3.25}$$

where y_i are the exponential sample spacings given by (6.3.23). Values of G close to zero are indicative of DFR and values close to 1 are indicative of IFR. Gail and Gastwirth (1978) studied the distribution and power properties of G. They showed that the null distribution of

$$\sqrt{12(n-1)}\,(G - 0.5)$$

is approximately normal $N(0,1)$ for large n (effectively $n \geq 20$). Gail and Gastwirth (1978) also showed that the G test has good power properties against distributions with monotone hazard functions.

For the data of Example 6.3.3, the spacings $y_i = (n - i + 1)(x_{(i)} - x_{(i-1)})$ are given by

5.25 3.36 4.81 0.36 7.7 2.8 0.9 4.08 0.91 1.14 3.85 1.28

4.08 16.84 11.73

Hence

$$G = \frac{595.04}{14(71.01)} = 0.60$$

The lower and upper 5% points of G being 0.127 and 0.627 (Gail and Gastwirth, 1978), we do not reject exponentiality at the 10% significance level.

Testing Uniformity

To test the uniform distribution $U(\theta_1, \theta_2)$ with density

$$f(x) = \frac{1}{\theta} \quad (\theta = \theta_2 - \theta_1) \quad (\theta_1 < x < \theta_2)$$

θ_1 and θ_2 unknown, against positively skew distributions, the directional T test is based on the statistic

$$T_U = \frac{\tilde{\theta}_c}{\tilde{\theta}} \quad (0 < T_U < \infty) \tag{6.3.26}$$

where $\tilde{\theta}_c = (x_{(n-r)} - x_{(1)})/(n - r - 1)$ and $\tilde{\theta} = (x_{(n)} - x_{(1)})/(n - 1)$. Small values of T_U lead to the rejection of uniformity in favor of a positively skew distribution.

THEOREM 6.3.3 Under uniformity, the null distribution of $y = [(n - r - 1)/(n - 1)]T_U$ is the beta distribution with $a = n - r - 1$ and $b = r$.

Proof: Since $\mu_{r:n} = r/(n + 1)$, the uniform sample spacings are given by [equation (6.3.17)]

$$u_i = (n + 1)(x_{(i+1)} - x_{(i)}) \quad (i = 1, 2, \ldots, n - 1)$$

Now

$$y = \frac{x_{(n-r)} - x_{(1)}}{x_{(n)} - x_{(1)}} = \frac{\Sigma_{i=1}^{n-r-1} u_i}{\Sigma_{i=1}^{n-1} u_i}$$

It can readily be shown that, like $D_i/\Sigma_{i=1}^{n-1} D_i$, $u_i/\Sigma_{i=1}^{n-1} u_i$ $(i = 1, 2, \ldots, n - 1)$ are jointly distributed as $n - 1$ spacings generated by the order statistics of a random sample of size $n - 2$ from the uniform (0, 1) (see Karlin, 1966, for example). But then y is distributed as the $(n - r - 1)$th order statistic of a random sample of size $n - 2$ from the uniform (0, 1) and this distribution is the beta(a, b) with $a = n - r - 1$ and $b = r$.

The T_U test above is carried out with $r = [0.5 + 0.5n]$, which gives it the same distributional properties as the T_E test; that is, the null distribution of T_U is symmetric (almost) and converges to normal very quickly. Besides, the T_U test has high power.

To test the uniform (θ_1, θ_2) against symmetric alternatives, the directional T test is based on the statistic

$$T_U = \frac{(n - 1)(x_{(n-r)} - x_{(r+1)})}{(n - 2r - 1)(x_{(n)} - x_{(1)})} \quad (0 < T_U < \infty) \tag{6.3.27}$$

Small values of T_U lead to the rejection of uniformity in favor of a symmetric distribution.

THEOREM 6.3.4 Under uniformity, the null distribution of $y = [(n - 2r - 1)/(n - 1)]T_U$ is the beta(a,b) with $a = n - 2r - 1$ and $b = 2r$.

Proof: Realizing that

$$y = \frac{\Sigma_{i=r+1}^{n-r+1} u_i}{\Sigma_{i=1}^{n-1} u_i}$$

the proof follows exactly on the same lines as that of Theorem 6.3.3.

The T_U test (6.3.27) is done with $r = [0.5 + 0.3n]$, which gives it high power against symmetric alternatives.

The asymptotic power functions of the T_U tests above have expressions exactly similar to (6.3.21).

Suppose that θ_1 and θ_2 are known, say $\theta_1 = 0$ and $\theta_2 = 1$, and one wants to test the uniform (0, 1) against positively skew distributions. The directional T statistic (6.3.26) reduces to

$$T_{U_0} = \frac{n + 1}{n - r} x_{(n-r)} \qquad r = [0.5 + 0.5n] \tag{6.3.28}$$

Small values of T_{U_0} lead to the rejection of uniformity. The null distribution of $[(n - r)/(n + 1)]T_{U_0}$ is the beta (a,b) distribution with $a = n - r$ and $b = r + 1$.

The T_{U_0} test, although amazingly simple, is very effective as a test of uniformity. For example, we have the following values of the power of the tests based on T_{U_0} and the Kolmogorov-Smirnov statistic D which is known to be a very powerful test of uniformity and will be introduced in Section 6.4.

Power against $k(1 - x)^{k-1}$; significance level 10%

k	Skewness	Kurtosis		n = 10	n = 20	n = 40
1.5	0.12	2.05	T_{U_0}	0.30	0.46	0.67
			D	0.23	0.38	0.60
2	0.32	2.40	T_{U_0}	0.53	0.78	0.96
			D	0.53	0.78	0.97

Testing Log-Normality

Let the assumed distribution be log-normal (Kotz, 1973) $LN(\mu, \sigma, \tau)$ with density

$$\frac{1}{\sqrt{2\pi}\,\sigma(x - \tau)} \exp\left\{-\frac{[\log(x - \tau) + \mu]^2}{2\sigma^2}\right\} \quad (\tau < x < \infty)$$

The location and scale parameters μ and σ are not known, but the threshold parameter τ is assumed to be known. In most practical situations τ is zero. Since $y = -\log(x - \tau)$ is the normal $N(\mu, \sigma)$, the T test is based on the statistic T given by (6.3.12); here $\tilde{\sigma}$ is calculated from the complete normal sample $y_i = -\log(x_i - \tau)$ $(i = 1, 2, \ldots, n)$, and realizing that alternatives to the log-normal are usually distributions which are positively skew, $\tilde{\sigma}_c$ is calculated from the one-sided censored sample (Tiku, 1975b)

$$y_{(1)}, y_{(2)}, \ldots, y_{(n-r)} \qquad r = [0.5 + 0.6n]$$

This T test has remarkably high power for testing the log-normal distribution against positively skew distributions (Tiku, 1975b).

Another interesting feature of the directional T test is that it admits a very straightforward generalization to multisample situations. Consider, for example, the problem of testing the hypothesis that k independent random samples

$$x_{i1}, x_{i2}, \ldots, x_{in_i} \qquad (i = 1, 2, \ldots, k) \tag{6.3.29}$$

come from the populations $(1/\sigma_i)f_0((x - \mu_i)/\sigma_i)$, respectively; the functional form f_0 is specified but the location and scale parameters μ_i and σ_i are not known. The generalized T test is based on the statistic (Tiku, 1974b)

$$T^* = \frac{1}{k}\sum_{i=1}^{k} T_i \tag{6.3.30}$$

where T_i is the statistic T calculated from the ith sample $x_{i1}, x_{i2}, \ldots, x_{in_i}$. Note that T^* is location and scale invariant.

THEOREM 6.3.5 The asymptotic (n_i tend to infinity) null distribution of T^* is normal with mean $E(T^*) = 1$ and variance $V(T^*) = (1/k^2)\,\Sigma_{i=1}^{k}\,V(T_i)$.

Proof: This follows immediately from the fact that the asymptotic distribution of T_i is normal and T_i $(i = 1, 2, \ldots, k)$ are independent.

6.4 OMNIBUS TESTS

A procedure designed to test a specified distribution f_0 against a wide class of alternatives is called an omnibus goodness-of-fit test. For example, one might like to test the normal $N(\mu, \sigma)$ against both symmetric and skew alternatives. Various omnibus tests have been developed and we describe them now.

EDF Statistics

Given a random sample $x_1, x_2, \ldots, x_n$, one wants to test the null hypothesis

H_0: the underlying distribution is $f_0(x, \underset{\sim}{\theta})$

The functional form f_0 is specified and, in the first place, suppose that the parameters $\underset{\sim}{\theta}$ are known. Let the EDF (empirical distribution function) be defined as

$$u_i = \int_{-\infty}^{x_i} f_0(x; \underset{\sim}{\theta})\, dx \qquad (i = 1, 2, \ldots, n) \tag{6.4.1}$$

It is easy to show that u_i $(i = 1, 2, \ldots, n)$ are IID uniform (0, 1). The problem of testing H_0 then reduces to testing the uniformity of the IID random variates u_i $(i = 1, 2, \ldots, n)$.

Let $u_{(1)}, u_{(2)}, \ldots, u_{(n)}$ be the order statistics of the random sample $u_1, u_2, \ldots, u_n$. Since the distribution of $u_{(r)}$ is the beta $(r, n - r + 1)$, the expected value of $u_{(r)}$ is given by

$$E(u_{(r)}) = \frac{r}{n+1} \qquad (r = 1, 2, \ldots, n)$$

Thus a suitable function of the distances $u_{(r)} - r/(n + 1)$, $r = 1, 2, \ldots, n$, can be used as a test of uniformity.

A very important "omnibus" test is given by the Kolmogorov-Smirnov statistic

$$D = \max(D^+, D^-) \tag{6.4.2}$$

where

$$D^+ = \max_{1 \le i \le n} \left(\frac{i}{n} - u_{(i)}\right) \quad \text{and} \quad D^- = \max_{1 \le i \le n} \left(u_{(i)} - \frac{i-1}{n}\right)$$

Large values of D lead to the rejection of H_0; D^+ and D^- can, in fact, be used as directional tests (Pearson and Hartley, 1972).

Other well-known omnibus goodness-of-fit tests (based on the EDF) are given by the following statistics: the Cramér-von Mises statistic,

$$W^2 = \sum_{i=1}^{n} \left(u_{(i)} - \frac{2i-1}{2n}\right)^2 + \frac{1}{12n} \tag{6.4.3}$$

the Kuiper statistic,

$$V = D^{+} + D^{-} \tag{6.4.4}$$

the Watson statistic,

$$U^2 = W^2 - n(\bar{u} - 0.5)^2 \qquad \bar{u} = \sum_{i=1}^{n} \frac{u_i}{n} \tag{6.4.5}$$

and the Anderson-Darling statistic,

$$A = -\frac{1}{n}\left\{\sum_{i=1}^{n} (2i-1)\left[\log u_{(i)} + \log(1 - u_{(n-i+1)})\right]\right\} - n \tag{6.4.6}$$

Large values of these statistics lead to the rejection of H_0.

The null distributions of D^+, D^-, and D are known for all n. The distributions of all other EDF statistics are very difficult to work out (see Durbin, 1973, for example), but their asymptotic distributions are known (see Anderson and Darling, 1952, for example). Approximations to their null distributions are available (Marshall, 1958; Pearson and Stephens, 1962; Stephens, 1963, 1964; Tiku, 1965a). A simple table (Pearson and Hartley, 1972, Table 54) allows easy determination of the percentage points of the null distributions of these EDF statistics.

All the EDF statistics above are, however, of very limited use because of the requirement that the unknown parameters $\underset{\sim}{\theta}$ be known. A natural thing to do, as suggested by Lilliefors (1967) and Srinivasan (1970), is to replace $\underset{\sim}{\theta}$ by a suitable estimate, say the maximum likelihood estimate $\hat{\underset{\sim}{\theta}}$, calculated from the given sample $x_1, x_2, \ldots, x_n$, and then calculate the EDF statistics above with u_i replaced by $\hat{u}_i$, where

$$\hat{u}_i = \int_{-\infty}^{x_i} f(x;\hat{\theta})\,dx \qquad (i = 1, 2, \ldots, n)$$

The resulting statistics are usually called modified EDF statistics and denoted by $\hat{D}$, $\hat{W}^2$, and so on. However, the difficulty is that $\hat{u}_i$ $(i = 1, 2, \ldots, n)$ are no more uniformly distributed, and therefore the distribution theory becomes very complex. The null distributions of the modified EDF statistics also depend on the assumed distribution f_0. Although methods exist for determining the asymptotic distributions of these modified statistics (Durbin, 1973; Durbin et al., 1975; Stephens, 1976), it is generally very difficult to work out their null distributions for finite sample sizes. Durbin (1975) and Margolin and Maurer (1976) have given the percentage points of some of the modified EDF statistics for finite sample sizes in some special cases, when f_0 is exponential, for example. Using Monte Carlo methods, Stephens (1974) constructed tables, which are very convenient to use, of the approximate percentage points of the null distributions of $\hat{D}$, $\hat{V}$, $\hat{W}^2$, $\hat{U}^2$, and $\hat{A}$, when f_0 is the normal $N(\mu,\sigma)$ or the exponential $E(0,\sigma)$; these tables are given by Pearson and Hartley (1972, Table 54).

Shapiro-Wilk Statistics

Let $\mu_{r:n} = E(x_{(r)} - \mu)/\sigma$ ($r = 1, 2, \ldots, n$), where $x_{(r)}$ is the rth order statistic of a random sample of size n from the assumed distribution $(1/\sigma)f_0((x - \mu)/\sigma)$; μ and σ are not known. Since

$$E(x_{(r)}) = \mu + \sigma\mu_{r:n}$$

the regression of $x_{(r)}$ on $\mu_{r:n}$ is linear, provided of course that f_0 is the true underlying distribution. Shapiro-Wilk goodness-of-fit test is essentially based on the BLUE estimator of the slope σ described in Chapter 2.

Testing Normality. Shapiro-Wilk test of normality is based on the statistic

$$W = \frac{(\Sigma_{i=1}^{n} a_{i:n} x_{(i)})^2}{\Sigma_{i=1}^{n} (x_i - \bar{x})^2} \qquad \left(\sum_{i=1}^{n} a_{i:n}^2 = 1\right) \tag{6.4.7}$$

Small values of W are indicative of nonnormality. The statistic W is location and scale invariant.

The coefficients $a_{i:n}$ are expressions in terms of the expected values $\mu_{i:n}$ and variances and covariances $\sigma_{i,j:n}$ of normal order statistics. Tables of $a_{i:n}$ originally computed by Shapiro and Wilk (1965), are given by Pearson and Hartley (1972, Table 15); note that $a_{n-i+1:n} = -a_{i:n}$ because of the symmetry of the normal distribution.

The distribution of W is intractable, and therefore the percentage points of the null distribution of W have been obtained primarily by Monte Carlo methods. Tables of these percentage points are given by Pearson and Hartley (1972, Table 16).

Since the computation of the coefficients $a_{i:n}$ is formidable, especially for large n, several modifications of W have been proposed; Mardia (1980) discusses these modifications. D'Agostino (1971b), for example, proposed the modified statistic

$$W^* = \frac{\Sigma_{i=1}^{n} [i - (n + 1)/2] x_{(i)}}{n \, \Sigma_{i=1}^{n} (x_i - \bar{x})^2} \tag{6.4.8}$$

Small and large values of W* are indicative of nonnormality. The numerator in (6.4.8) is essentially Downton's (1966b) linear estimate of scale.

The asymptotic distribution of W* is normal with mean $E(W^*) \simeq 0.282$ and variance $V(W^*) \simeq (0.03)^2/n$; D'Agostino (1971b) gives approximate percentage points of the null distribution of W* for samples of size $n \leq 1000$.

Testing Exponentiality. For testing the exponential $E(\theta, \sigma)$, Shapiro-Wilk statistic is given by

$$W = \left(1 - \frac{1}{n}\right)^{-1} \frac{(\bar{x} - x_{(1)})^2}{\Sigma_{i=1}^{n} (x_i - \bar{x})^2} \tag{6.4.9}$$

Small and large values of W_E lead to the rejection of exponentiality. The distribution of W_E is not known, but Shapiro and Wilk (1972) have provided simulated percentage points of the null distribution of W_E.

In principle, the Shapiro-Wilk method can be adopted for testing any assumed distribution $(1/\sigma)f_0((x - \mu)/\sigma)$, but the computations are generally too involved. Besides, the distribution theory is difficult.

Csörgö-Seshadri-Yalovsky Statistics

Csörgö et al. (1973, 1975) used data transformations to give random variables with completely known distributions. For testing the exponential $E(\theta, \sigma)$, for example, Seshadri et al. (1969) defined the transforms

$$V_{(i)} = \frac{\Sigma_{j=1}^{i} D_j}{\Sigma_{j=1}^{n-1} D_j} \quad (i = 1, 2, \ldots, n - 2) \tag{6.4.10}$$

where D_j, $j = 1, 2, \ldots, n - 1$, are the exponential sample spacings. Under exponentiality, $V_{(i)}$ are jointly distributed as the order statistics of a random sample of size $n - 2$ from the uniform $(0, 1)$ distribution. Thus the exponentiality of the random sample $x_1, x_2, \ldots, x_n$ can be tested by using any one of the EDF statistics based on $V_{(i)}$, $i = 1, 2, \ldots, n - 2$, or by using the simpler statistic

$$\chi^2 = -2 \sum_{i=1}^{n-2} \log V_{(i)} \tag{6.4.11}$$

The null (e.g., x_i are IID exponential $E(\theta, \sigma)$) distribution of χ^2 is chi-square with $2(n - 2)$ degrees of freedom.

Filliben and Smith-Bain Statistics

Let $F(z) = \int_{-\infty}^{z} f(z)\, dz$, $z = (x - \mu)/\sigma$, denote the cumulative distribution function of $(1/\sigma)f_0((x - \mu)/\sigma)$. To test the assumed functional form $(1/\sigma)f_0((x - \mu)/\sigma)$, μ and σ unknown, Filliben (1975) and Smith and Bain (1976) defined the correlation statistic

$$R = 1 - \hat{\rho}^2 \quad (0 < R < 1) \tag{6.4.12}$$

where $\hat{\rho}$ is the ordinary product moment correlation coefficient between the order statistics $x_{(i)}$ $(i = 1, 2, \ldots, n)$ of a random sample of size n, and the corresponding k_i values determined by the equations

$$F_0(k_i) = \frac{i}{n+1} \quad (i = 1, 2, \ldots, n) \tag{6.4.13}$$

Large values of R lead to the rejection of the assumed distribution f_0. Note that R is location and scale invariant.

Although the computation of R is straightforward, the distribution theory is difficult. Therefore, the percentage points of the null distribution of R have to be obtained by Monte Carlo methods.

Tiku Statistics

Tiku (1980b) developed an omnibus test of exponentiality (and other distributional forms) by combining the directional T_E tests as follows. Noting that T_E does, in fact, provide a test statistic for each value of $r = 1, 2, \ldots, n-2$, not merely for $r = [0.5 + 0.5n]$, a more powerful test of exponentiality may be achieved by combining these T_E's, that is, combining the estimators $\tilde{\sigma}_c$'s [equation (6.3.16)]. A linear combination with weights inversely proportional to the variances $V(\tilde{\sigma}_c) = \sigma^2/(n - r - 1)$ yields the statistic

$$Z = \frac{2 \Sigma_{i=1}^{n-2} (n - 1 - i)D_i}{(n - 2) \Sigma_{i=1}^{n-1} D_i} \tag{6.4.14}$$

D_i being the exponential spacings given by (6.3.18). Small and large values of Z lead to the rejection of the exponential $E(\theta, \sigma)$.

THEOREM 6.4.1 Under exponentiality, the distribution of $u = Z/2$ is the same as the distribution of the mean of $n - 2$ IID uniform (0, 1) variates and is given by

$$p(u) = \frac{(n-2)^{n-2}}{(n-3)!} \sum_{r=0}^{i} (-1)^r \binom{n-2}{r} \left(u - \frac{r}{n-2}\right)^{n-3} \quad \text{for } \frac{i}{n-2} \le u \le \frac{i+1}{n-2}$$

$$(i = 0, 1, \ldots, n-3) \tag{6.4.15}$$

Proof: It is easy to show that

$$\sum_{i=1}^{n-2} (n - 1 - i)D_i = \sum_{r=1}^{n-2} \sum_{i=1}^{n-r-1} D_i$$

and, therefore,

$$\frac{1}{2}(n - 2)Z = \sum_{r=1}^{n-2} \sum_{i=1}^{n-r-1} u_i \qquad u_i = \frac{D_i}{\Sigma_{i=1}^{n-1} D_i}$$

Since D_i are IID exponential $E(0, \sigma)$, it is easy to show that u_i ($i = 1, 2, \ldots, n - 1$) are jointly distributed as $n - 1$ spacings generated by $n - 2$ order statistics of a random sample from the uniform (0, 1) (see Seshadri et al., 1969, for example). But then $\Sigma_{i=1}^{n-r-1} u_i$ is the (n - r - 1)th order statistic in a random sample of size n - 2 from the uniform (0, 1), and $\Sigma_{r=1}^{n-2} (\Sigma_{i=1}^{n-r-1} u_i)$ is the sum of the n - 2 order statistics of a random sample of size n - 2. Since the sum $\Sigma_{i=1}^{N} x_{(i)}$ of all the order statistics of a random sample $x_1, x_2, \ldots, x_N$ is exactly equal to $\Sigma_{i=1}^{N} x_i$, the result follows immediately.

The distribution of the mean of n - 2 IID uniform (0, 1) variates has been worked out by Kendall and Stuart (1969, p. 258) and is given by (6.4.15).

COROLLARY 6.4.1 Under exponentiality, the cumulative distribution function of the null distribution of $u = Z/2$ is given by

$$P(u \geq u_0) = \frac{(n-2)^{n-2}}{(n-2)!} \sum_{r=0}^{i} (-1)^r \binom{n-2}{r} \left(u_0 - \frac{r}{n-2}\right)^{n-2} \quad \text{for } \frac{i}{n-2} \leq u_0 \leq \frac{i+1}{n-2}$$

$$(i = 0, 1, \ldots, n-3) \quad (6.4.16)$$

Since the mean and variance of the uniform (0, 1) are 1/2 and 1/12, respectively, it immediately follows that

$$E(Z) = 1 \quad \text{and} \quad V(Z) = \frac{1}{3(n-2)} \quad (6.4.17)$$

The Case θ Known. In certain practical situations θ is known, say $\theta = 0$, in which case the statistic Z_0, obtained by combining T_{E_0}'s, is given by

$$Z_0 = \frac{2 \Sigma_{i=1}^{n-1} (n-i) D_{i0}}{(n-1) \Sigma_{i=1}^{n} D_{i0}} \quad (6.4.18)$$

where $D_{i0} = (n + 1 - i)(x_{(i)} - x_{(i-1)})$ $(i = 1, 2, \ldots, n)$ are the exponential spacings; $x_{(0)} = 0$.

LEMMA 6.4.1 Under exponentiality, the null distribution of $Z_0/2$ is the same as the distribution of the mean of n - 1 IID uniform (0, 1) variates and is given by (6.4.15) with n replaced by n + 1.

Proof: This follows exactly on the same lines as above. It immediately follows that

$$E(Z_0) = 1 \quad \text{and} \quad V(Z_0) = \frac{1}{3(n-1)} \quad (6.4.19)$$

It is well known that the distribution of the mean of IID uniform variates tends to a normal distribution very rapidly. For large n (effectively $n \geq 5$), therefore, the null distributions of Z and Z_0 are normal with means and variances given by (6.4.17) and (6.4.19), respectively.

To have an idea about the accuracy of these normal approximations, we give below the exact and approximate upper percentage points of Z_0:

Upper % Points	Sample Size, n: 5 Exact	5 Approx.	7 Exact	7 Approx.	11 Exact	11 Approx.	21 Exact	21 Approx.
10	1.376	1.370	1.306	1.302	1.236	1.234	1.166	1.165
5	1.477	1.475	1.389	1.388	1.301	1.300	1.213	1.212
1	1.650	1.672	1.536	1.548	1.419	1.425	1.299	1.300

It is easy to show that the asymptotic power function of Z has exactly the same form as that of T_E given by (6.3.21).

Tiku and Tamhankar (1980), Tiku (1980b), and Balakrishnan (1983) determined the values of the power of the Z and T_E tests of exponentiality against numerous alternatives and compared them with the corresponding values of the tests based on the Shapiro-Wilk statistic W_E and Kolmogorov-Smirnov and Cramér-von Mises EDF statistics (see also Seshadri et al., 1969; Srinivasan, 1970; Shapiro and Wilk, 1965, 1972; Durbin, 1975; and Csörgö et al., 1975). It is clear from these computations that the Z test is at least as powerful as the T_E test and is on the whole somewhat more powerful than the Shapiro-Wilk, Kolmogorov-Smirnov, and Cramér-von Mises tests (see specifically Tiku, 1980b, p. 269; Balakrishnan, 1983). Consider, for example, the problem of testing $E(\theta, \sigma)$ against the 10 alternatives considered by Csörgö et al. (1975), namely, the chi-squares with df = 1, 3, 4, 6, 8; the log-normal ($\sigma = 1$) (LN); the half-normal (HN); the Weibull ($\frac{1}{2}$) (W_1); the Weibull (2) (W_2); and the beta (2, 1) (B); for sample size n = 20 and significance level $\alpha^* = 0.10$. The values of the power of the Z test, the test based on χ^2 statistic given by (6.4.11), and the Shapiro-Wilk W_E test, are given below, reproduced from Tiku (1980b):

	χ_1^2	χ_3^2	χ_4^2	χ_6^2	χ_8^2	LN	HN	W_1	W_2	B	Sum
Z	0.59	0.20	0.34	0.54	0.66	0.23	0.30	0.91	0.78	0.39	4.94
χ^2	0.65	0.21	0.36	0.55	0.65	0.16	0.28	0.95	0.71	0.23	4.79
W_E	0.39	0.20	0.33	0.52	0.64	0.27	0.34	0.83	0.79	0.43	4.74

It is clear that the Z test is more powerful overall.

EXAMPLE 6.4.1 Consider the data of Example 6.3.3. Here the assumption is that these 15 random observations come from the exponential $E(0,\sigma)$. The spacings D_{i0} $(i = 1, 2, \ldots, 15)$ are given by

5.25 3.36 4.81 0.36 7.7 2.8 0.9 4.08 0.91 1.14 3.85 1.28

4.08 16.84 11.73

and, therefore,

$$Z_0 = \frac{2(372.22)}{14(69.09)} = 0.770$$

based on the normal distribution with mean 1 and variance $1/3(14) = 0.02381$, the lower and upper 5% points are given by 0.747 and 1.253, respectively. Since the calculated value 0.770 lies between these percentage points, we do not reject the exponential $E(0,\sigma)$ at 10% significance level.

Omnibus Z Test for Any Distribution

To test an assumed distribution of the type $(1/\sigma)f_0((x - \mu)/\sigma)$, f_0 specified and μ and σ unknown, Tiku (1980b) defined the statistic

$$Z^* = \frac{2\,\Sigma_{i=1}^{n-2}(n - 1 - i)G_i}{(n - 2)\,\Sigma_{i=1}^{n-1} G_i} \tag{6.4.20}$$

where $G_i = (x_{(i+1)} - x_{(i)})/(\mu_{i+1:n} - \mu_{i:n})$ $(i = 1, 2, \ldots, n - 1)$ are the sample spacings (6.3.17). The statistic Z^* is in fact the statistic Z with the exponential spacings D_i being replaced by the spacings G_i of a random sample of size n from the distribution f_0. Small and large values of Z^* lead to the rejection of f_0. Note that Z^* is location and scale invariant.

If we write

$$w_1 = 2\sum_{i=1}^{n-2}(n - 1 - i)G_i \quad \text{and} \quad w_2 = (n - 2)\sum_{i=1}^{n-1} G_i$$

then

$$E(w_1) = 2\sum_{i=1}^{n-2}(n-1-i)\sigma = (n-1)(n-2)\sigma \quad \text{and} \quad E(w_2) = (n-1)(n-2)\sigma$$

For large n, therefore (Kendall and Stuart, 1969, p. 232),

$$E(Z^*) = E\left(\frac{w_1}{w_2}\right) \simeq \frac{E(w_1)}{E(w_2)} = 1 \tag{6.4.21}$$

and [equation (6.3.10)]

$$V(Z^*) \simeq \frac{V(w_1) + V(w_2) - 2\,\mathrm{cov}\,(w_1, w_2)}{[(n-1)(n-2)\sigma]^2} \tag{6.4.22}$$

where

$$V(w_1) = 4(\ell_1' \underset{\sim}{V} \ell_1)\sigma^2 \qquad V(w_2) = (\ell_1' \underset{\sim}{V} \ell_2)\sigma^2$$

and (6.4.23)

$$\mathrm{cov}\,(w_1, w_2) = 2(\ell_1' \underset{\sim}{V} \ell_2)\sigma^2$$

$\ell_1' = (n-2, n-3, \ldots, 1, 0)$ and $\ell_2' = (n-1, n-1, \ldots, n-1)$ are vectors each having $n-1$ elements and $\underset{\sim}{V}$ is the variance-covariance matrix of the variates G_i/σ $(i = 1, 2, \ldots, n-1)$.

If the exact values of $\mu_{r:n}$ and $\sigma_{r,s:n}$ are not available, then, in place of these exact values, their approximate values which are obtained from (B.27)-(B.29) are used in (6.4.20)-(6.4.23).

The null distribution of Z* is approximated by a normal distribution with mean 1 and variance given by (6.4.22). Tiku (1980b) showed that this normal approximation gives accurate values for the percentage points as well as the probability integral of Z*. He determined these approximate values and compared them with the corresponding simulated values and found them to be close. We reproduce these values here in Table 6.4.1 for the following distributions:

Normal: $\dfrac{1}{\sqrt{2\pi}\sigma} e^{-(x-\mu)^2/2\sigma^2}$ $\quad -\infty < x < \infty$ (μ and σ unknown)

Gamma (p): $\dfrac{1}{\Gamma(p)\sigma^p} e^{-(x-\theta)/\sigma}(x-\theta)^{p-1}$ $\quad \theta \le x < \infty$ (θ and σ unknown)

Logistic: $\dfrac{1}{\sigma}\dfrac{e^{-(x-\mu)/\sigma}}{(1+e^{-(x-\mu)/\sigma})^2}$ $\quad -\infty < x < \infty$ (μ and σ unknown)

Thus, to determine the percentage points of Z* from the normal approximation

$$\frac{Z^* - 1}{\sqrt{\mathrm{var}\,(Z^*)}} \approx N(0, 1)$$

Table 6.4.1 Values of the Probability $P(Z^* \leq C \mid f_0)$

	Normal			Gamma(2)			Logistic		
n	C	Sim.	Approx.	C	Sim.	Approx.	C	Sim.	Approx.
10	0.65	0.025	0.026	0.64	0.023	0.031	0.63	0.025	0.031
	0.71	0.052	0.054	0.70	0.050	0.059	0.69	0.046	0.058
	1.29	0.949	0.946	1.30	0.947	0.941	1.31	0.946	0.942
	1.35	0.977	0.974	1.36	0.975	0.969	1.37	0.977	0.969
20	0.77	0.023	0.024	0.76	0.024	0.025	0.75	0.024	0.026
	0.81	0.049	0.051	0.80	0.047	0.053	0.79	0.047	0.051
	1.19	0.949	0.949	1.20	0.946	0.947	1.21	0.956	0.949
	1.23	0.977	0.976	1.24	0.975	0.974	1.25	0.979	0.974
50	0.86	0.023	0.022	0.86	0.027	0.026	0.85	0.026	0.026
	0.88	0.043	0.041	0.88	0.050	0.048	0.87	0.047	0.046
	1.12	0.956	0.959	1.12	0.951	0.952	1.13	0.951	0.974
	1.14	0.977	0.978	1.14	0.974	0.974	1.15	0.972	0.974

one only needs the values of the variance var (Z^*). When f_0 is normal $N(\mu,\sigma)$, the values of var (Z^*) are given in Table 6.4.2; these values were determined empirically and found to agree closely with the values obtained from (6.4.22). Also given in this table are the values of $E_r = \mu_{r+1:n} - \mu_{r:n}$ needed to calculate the Z^* statistic for testing that f_0 is normal $N(\mu,\sigma)$, μ and σ unknown.

The asymptotic power function of Z^* is exactly of the same form as that of T_E, given by (6.3.21), and will not therefore be discussed any further.

For the uniform (θ_1, θ_2), $\mu_{r:n} = r/(n+1)$, in which case the statistic Z^* is given by

$$Z^* = \frac{2 \sum_{i=1}^{n-2} (n-1-i)(x_{(i+1)} - x_{(i)})}{(n-2)(x_{(n)} - x_{(1)})} \qquad (6.4.24)$$

The null distribution of $Z^*/2$ is the same as the distribution of the mean of $n - 2$ IID uniform (0, 1) variates; the proof of this result follows exactly on the same lines as Theorems 6.3.3 and 6.4.1.

Table 6.4.2 Values of n var(Z*) and n var(U*), and $E_r = \mu_{r+1:n} - \mu_{r:n}$, for Testing that f_0 is Normal $N(\mu, \sigma)$; μ and σ Unknown[a]

E_r, r = 1, 2, ..., [n/2]; $E_{n-r} = E_r$

	n: 5	6	7	8	9	10	11	12	13
n var(Z*)	0.476	0.414	0.373	0.347	0.333	0.324	0.318	0.303	0.295
n var(U*)	0.925	0.794	0.587	0.561	0.490	0.482	0.432	0.427	0.392
r: 1	0.6679	0.6254	0.5948	0.5714	0.5527	0.5374	0.5245	0.5135	0.5039
2	0.4950	0.4402	0.4047	0.3794	0.3603	0.3453	0.3331	0.3229	0.3142
3		0.4031	0.3527	0.3203	0.2974	0.2803	0.2669	0.2560	0.2470
4				0.3050	0.2745	0.2531	0.2371	0.2246	0.2145
5						0.2453	0.2249	0.2097	0.1978
6								0.2052	0.1905

	n: 14	15	16	17	18	19	20	21	22
n var(Z*)	0.290	0.292	0.282	0.277	0.274	0.277	0.270	0.266	0.268
n var(U*)	0.387	0.370	0.363	0.364	0.342	0.325	0.327	0.326	0.330
r: 1	0.4955	0.4880	0.4812	0.4752	0.4696	0.4645	0.4599	0.4556	0.4515
2	0.3068	0.3002	0.2945	0.2893	0.2847	0.2805	0.2766	0.2732	0.2699
3	0.2394	0.2328	0.2271	0.2221	0.2176	0.2136	0.2100	0.2067	0.2036
4	0.2062	0.1992	0.1932	0.1879	0.1833	0.1792	0.1756	0.1723	0.1693
5	0.1883	0.1804	0.1738	0.1681	0.1632	0.1589	0.1551	0.1517	0.1486
6	0.1791	0.1700	0.1625	0.1561	0.1507	0.1461	0.1420	0.1383	0.1351
7	0.1763	0.1653	0.1565	0.1492	0.1431	0.1379	0.1334	0.1294	0.1260
8			0.1546	0.1460	0.1389	0.1330	0.1280	0.1236	0.1198
9					0.1376	0.1307	0.1250	0.1200	0.1158
10							0.1240	0.1184	0.1136
11									0.1128

(continued)

Table 6.4.2 (continued)

	n: 23	24	25	26	27	28	29	30
n var(Z*)	0.266	0.258	0.255	0.258	0.252	0.256	0.251	0.256
n var(U*)	0.323	0.319	0.316	0.308	0.306	0.315	0.308	0.304
r: 1	0.4478	0.4443	0.4410	0.4379	0.4350	0.4323	0.4296	0.4272
2	0.2669	0.2641	0.2616	0.2591	0.2568	0.2547	0.2527	0.2508
3	0.2009	0.1983	0.1960	0.1938	0.1917	0.1898	0.1880	0.1863
4	0.1666	0.1641	0.1618	0.1596	0.1577	0.1558	0.1541	0.1525
5	0.1458	0.1433	0.1410	0.1388	0.1368	0.1350	0.1333	0.1317
6	0.1322	0.1296	0.1272	0.1250	0.1229	0.1211	0.1193	0.1177
7	0.1229	0.1201	0.1176	0.1153	0.1132	0.1112	0.1094	0.1078
8	0.1164	0.1134	0.1108	0.1083	0.1061	0.1041	0.1022	0.1005
9	0.1121	0.1088	0.1059	0.1033	0.1010	0.0988	0.0968	0.0950
10	0.1094	0.1058	0.1026	0.0998	0.0972	0.0949	0.0928	0.0909
11	0.1081	0.1041	0.1005	0.0974	0.0946	0.0921	0.0899	0.0879
12		0.1035	0.0995	0.0961	0.0930	0.0903	0.0878	0.0856
13				0.0956	0.0922	0.0892	0.0865	0.0841
14						0.0888	0.0859	0.0832
15								0.0830

[a]For $n \geq 20$, the variance of Z* is closely approximated by $\mathrm{var}(Z^*) \simeq 0.72/3(n - 2)$ and the variance of U* is close approximated by $\mathrm{var}(U^*) \simeq 0.79/3(n - 4)$.

Censored Samples

The statistic Z* readily generalizes to censored samples. To test the assumed distribution $(1/\sigma)f_0((x - \mu)/\sigma)$ from a type II censored sample of size $n - r - s$,

$$x_{(r+1)},\ x_{(r+2)},\ \ldots,\ x_{(n-s)} \tag{6.4.25}$$

the statistic is

$$Z_c^* = \frac{2\,\Sigma_{i=r+1}^{n-s-2}\,(n - s - 1 - i)G_i}{(n - r - s - 2)\,\Sigma_{i=r+1}^{n-s-1}\,G_i} \tag{6.4.26}$$

The distribution theory of Z_c^* is exactly similar to that of Z*. For example, if f_0 is the exponential $E(\theta, \sigma)$, or the uniform (θ_1, θ_2), the null distribution of $Z_c^*/2$ is the same as the distribution of the mean of n - r - s - 2 IID uniform (0, 1) variates (Tiku, 1980b).

Tiku (1980b) investigated the power properties of the statistic Z^* (and Z_c^*) for testing the exponential, gamma, uniform, logistic, and normal distributions and compared their power values with the corresponding values of other prominent statistics mentioned earlier. From these comparisons, he arrived at the following conclusions.

The statistic Z^* (reduces to Z for the exponential) has good power properties for (1) testing the exponential and gamma (skew) distributions against skew and symmetric alternatives from complete or censored samples; (2) testing the uniform, normal, and logistic (symmetric) distributions against skew alternatives from complete or censored samples; and (3) testing the uniform, normal, and logistic distributions against symmetric alternatives from asymmetrically censored samples [i.e., $r \neq s$ in (6.4.25)]. The weakness of Z^* lies in testing symmetric distributions against symmetric, particularly short-tailed, alternatives from complete or symmetrically censored samples [i.e., $r = s$ in (6.4.25)]. A statistic similar to Z^* which is particularly powerful for testing symmetric distributions against symmetric alternatives from complete or symmetrically censored samples will be given later.

Of course, the EDF, the Shapiro-Wilk, and the correlation-type statistics can also be generalized to censored samples. Their distribution theory is however very difficult; see Aho et al. (1983), Barr and Davidson (1973), Pettit (1976, 1977), Shapiro and Brain (1981), and Smith and Bain (1976). For some interesting applications of goodness-of-fit tests based on censored samples, the reader is referred to Nelson (1982).

For testing the normal $N(\mu, \sigma)$ from complete samples, W is more powerful overall than R and Z^* (Tiku, 1980b). In the context of robustness, however, the differences between the sample mean $\bar{x}$ and the MML estimator $\hat{\mu}$ and their standard errors are expected to be appreciable only when the underlying distribution has long tail(s) and here the statistic Z^* is particularly useful in explaining these differences. Moreover, the statistic Z^* can also be readily generalized to multisample situations as will be shown later. Besides, the null distribution of Z^* is closely approximated by a normal. Since $E(Z^*) \simeq 1$, one only needs the values of the variance $\text{var}(Z^*)$ to work out this normal approximation; the values of $\text{var}(Z^*)$ are given in Table 6.4.2.

EXAMPLE 6.4.2 Consider the data of Example 6.3.1; one wants to test whether these 10 observations come from a normal $N(\mu, \sigma)$ distribution; μ and σ are not known. The values of $E_r = \mu_{r+1:n} - \mu_{r:n}$, r = 1, 2, ..., 9, needed to calculate normal spacings are given in Table 6.4.2. Here the normal spacings G_i, i = 1, 2, ..., 9, are given in order:

0.9505 1.3611 1.7321 1.0601 1.1422 0.6464 0.4424 0.5682 2.4475

Consequently,

$$Z^* = \frac{2(40.7858)}{8(10.3505)} = 0.985$$

The lower and upper 5% points of the null distribution of Z* (Table 6.4.2) being $1 - 1.64\sqrt{0.0324} = 0.69$ and $1 + 1.64\sqrt{0.0324} = 1.29$, there is no evidence against the normality of this data at the 10% significance level.

We have already commented on the effectiveness of the Z* test for testing the normal, logistic, exponential, and uniform distributions. We now illustrate the effectiveness of the Z* test for testing the important Weibull distribution.

Testing the Weibull

Consider now the problem of testing the Weibull distribution with density $W(\alpha, \beta)$ given by

$$f(x) = \frac{\alpha}{\beta} e^{-(x/\beta)^{\alpha}} \left(\frac{x}{\beta}\right)^{\alpha-1} \qquad 0 < x < \infty \quad (\alpha > 0,\ \beta > 0) \tag{6.4.27}$$

α and β are not known. Since $W(\alpha, \beta)$ is not of the form $(1/\sigma)f((x - \mu)/\sigma)$, the Z* test cannot be employed. However, the distribution of $y = \log x$ is the extreme-value distribution $EV(\delta, \eta)$ with density

$$f(x) = \frac{1}{\eta} \exp\left(-e^{-(y-\delta)/\eta}\right) e^{-(y-\delta)/\eta} \qquad -\infty < y < \infty \quad (\delta = \log \beta,\ \eta = 1/\alpha)) \tag{6.4.28}$$

Since this distribution is of the form $(1/\sigma)f((x - \mu)/\sigma)$, the Z* test can now be employed. The expected values of order statistics of random samples from the EV(0, 1) distribution are needed to calculate the sample spacings G_i. These expected values are available for samples of size $n \le 25$ (Mann et al., 1973). For $n > 25$, their approximate values can of course be obtained from (B.27).

To test the $EV(\delta, \eta)$ distribution, the modified EDF statistics can of course be used, with δ and η being replaced by their ML estimates. These estimates are the solutions of the ML equations

$$\frac{\partial \log L}{\partial \delta} = 0 \quad \text{and} \quad \frac{\partial \log L}{\partial \eta} = 0 \tag{6.4.29}$$

that is,

$$\hat{\eta} = \left(\frac{1}{n}\right) \sum_{i=1}^{n} y_i - \frac{\Sigma_{i=1}^{n} y_i \exp(-y_i/\hat{\eta})}{\Sigma_{i=1}^{n} \exp(-y_i/\hat{\eta})}$$

and (6.4.30)

Table 6.4.3 Empirical Values of the Power for the 5% Significance Level

Alternative	n	$\hat{A}^2$	R	S	Z*
Normal	10	0.084	0.087	0.175	0.133
	25	0.286	0.119	0.384	0.392
	40	0.462	0.168	0.540	0.617
Logistic	10	0.131	0.127	0.203	0.181
	25	0.376	0.207	0.427	0.466
	40	0.587	0.278	0.591	0.682
Double exponential	10	0.201	0.199	0.252	0.244
	25	0.568	0.346	0.485	0.548
	40	0.774	0.456	0.671	0.748
Cauchy	10	0.545	0.608	0.399	0.548
	25	0.919	1.000	0.555	0.761
	40	0.984	1.000	0.638	0.824
χ_1^2	10	0.095	0.068	0.031	0.063
	25	0.094	0.091	0.018	0.100
	40	0.113	0.113	0.014	0.142
χ_2^2	10	0.048	0.044	0.078	0.057
	25	0.058	0.030	0.102	0.079
	40	0.078	0.026	0.126	0.113

$$\hat{\delta} = -\hat{\eta} \log\left[\frac{\Sigma_{i=1}^{n} \exp(-y_i/\hat{\eta})}{n}\right]$$

The first equation is solved iteratively, and then $\hat{\delta}$ can be found; the computations are formidable, however (Thomas et al., 1969). The approximate percentage points of the modified EDF statistics are available (Stephens, 1977).

Two other prominent goodness-of-fit statistics that have been used for testing the $EV(\delta, \eta)$ distribution are the Smith-Bain correlation statistic R given by (6.4.12) and the Mann-Scheuer-Fertig statistic

$$S = \frac{\Sigma_{i=[n/2]+1}^{n-1} G_i}{\Sigma_{i=1}^{[n/2]} G_i} \tag{6.4.31}$$

where G_i ($i = 1, 2, \ldots, n - 1$) are the sample spacings mentioned above.

Littell et al. (1979) have given the empirical values of the power of the tests based on the modified EDF statistics $\hat{D}$, $\hat{W}^2$, and $\hat{A}^2$, and the statistics R and S, for testing the $EV(\delta, \eta)$ against six alternatives, namely the normal, logistic, double exponential, Cauchy, χ_1^2, and χ_4^2. We reproduce these values in Table 6.4.3; since $\hat{A}^2$ is generally more powerful than $\hat{D}$ and $\hat{W}^2$, we only reproduce the values for $\hat{A}^2$ out of the EDF statistics. The corresponding values of the power of the Z* test (Tiku and Singh, 1981b) are also given. It is clear that the Z* test is more powerful overall than its prominent competitors. For example, out of the 18 power comparisons (Table 6.4.3), Z* leads the statistic $\hat{A}^2$ 13 times.

Symmetric Versus Symmetric

As we mentioned earlier, the statistic Z* has good power properties other than for testing a symmetric distribution against symmetric alternatives from complete or symmetrically censored samples. For this latter situation, Tiku (1981b) developed a statistic U* by combining T_U's [given by (6.3.27)], following exactly the lines that led to the statistic Z*. To test an assumed symmetric distribution $(1/\sigma)f_0((x - \mu)/\sigma)$ from a complete sample $x_1, x_2, \ldots, x_n$, the statistic U* is given by

$$U^* = \frac{(n-1)\,\Sigma_{j=1}^{[n/2]-1}\,\Sigma_{i=j+1}^{n-j-1} G_i}{([n/2] - 1)(n - [n/2] - 1)\,\Sigma_{i=1}^{n-1} G_i} \tag{6.4.32}$$

where as usual G_i are the sample spacings; [g] denotes the integer part of g. Small and large values of U* lead to the rejection of f_0.

If f_0 is the uniform $U(\theta_1, \theta_2)$ distribution, the exact first four moments of U* can be obtained (Tiku, 1981b) and are given by the following values:

n	Mean	Variance	$\beta_1^* = \mu_3^2/\mu_2^3$	$\beta_2^* = \mu_4/\mu_2^2$
4	1	0.5000	0.3200	2.400
6	1	0.1458	0.0578	2.795
8	1	0.0802	0.0146	2.857
10	1	0.0547	0.0048	2.883

In fact, Tiku (1981b) showed that for large n the variance $\text{var}(U^*) = 1/3(n-4)$ and that the null distribution of U^* is normal with mean $E(U^*) = 1$.

If f_0 is any other symmetric distribution, then the null distribution of U^* is approximated by a normal with mean 1 and variance

$$V(U^*) \simeq \frac{V(w_1) + V(w_2) - 2\,\text{cov}(w_1, w_2)}{\{([n/2] - 1)(n - [n/2] - 1)\sigma\}^2} \qquad (6.4.33)$$

w_1 is the numerator and w_2 is the denominator in (6.4.32). Since w_1 and w_2 are both linear functions of order statistics, their variances and the covariance are given by equations exactly similar to (6.4.23).

To have an idea about the accuracy of this normal approximation, we give in Table 6.4.4 the values, reproduced from Tiku (1981b), of the 95% points of the null distribution of U^*, for the uniform, normal, logistic, and Student's t with 4 df distributions.

For testing a symmetric distribution against symmetric alternatives from a type II symmetrically censored sample of size $n - 2r$, namely

$$x_{(r+1)}, x_{(r+2)}, \ldots, x_{(n-r)}$$

the U^* statistic generalizes to (Tiku, 1981b)

Table 6.4.4 Approximate and Empirical 95% Points of U^*

	Uniform		Normal		Logistic		Student's t	
n	Approx.	Four-moment[a]	Approx.	Emp.	Approx.	Emp.	Approx.	Emp.
10	1.388	1.389	1.359	1.365	1.363	1.370	1.378	1.390
16	1.274	1.275	1.248	1.248	1.258	1.255	1.270	1.270
20	1.237	1.237	1.215	1.216	1.226	1.230	1.238	1.239

[a]These values are based on four-moment Pearson curves; the corresponding approximate values are based on a normal distribution with mean 1 and variance $1/3(n - 4)$.

$$U_c^* = \frac{(m-1)\,\Sigma_{j=1}^{[m/2]-1}\;\Sigma_{i=r+j+1}^{n-r-j-1} U_i}{([m/2]-1)(n-[n/2]-r-1)\,\Sigma_{i=r+1}^{n-r-1} U_i} \tag{6.4.34}$$

$m = n - 2r$. The distribution theory of U_c^* is exactly similar to that of U^* (Tiku, 1981b). The asymptotic power function of U^* (and U_c^*) has exactly the same form as (6.3.21).

The power properties of various tests commonly used for testing symmetric distributions against symmetric alternatives were investigated by Tiku (1981b), who showed that the U^* test is remarkably powerful.

The Z^* and U^* tests beautifully complement each other and are available for complete as well as censored samples. Besides, their null distributions are approximately normal. The only real effort involved is in calculating their approximate variances from equations of the form (6.4.22). For complete samples of size $n = 5(1)30$, the approximate values of the variances $\operatorname{var}(Z^*)$ and $\operatorname{var}(U^*)$, when f_0 is normal $N(\mu, \sigma)$, are given in Table 6.4.2.

EXAMPLE 6.4.3 Suppose that we want to test the normal $N(\mu,\sigma)$ against symmetric alternatives from the data of Example 6.3.1. Under normality, the sample spacings G_i are worked out in Example 6.4.2. Here

$$\sum_{j=1}^{[n/2]-1}\sum_{i=j+1}^{n-j-1} G_i = G_2 + 2G_3 + 3G_4 + 4G_5 + 3G_6 + 2G_7 + G_8$$
$$= 15.9666$$

Therefore,

$$U^* = \frac{9(15.9666)}{4(4)(10.3505)} = 0.868$$

The respective lower and upper 5% points of the null distribution of U^* being $1 - 1.64\sqrt{0.0482} = 0.640$ and $1 + 1.64\sqrt{0.0482} = 1.360$ (Table 6.4.2), we do not reject normality at 10% significance level.

The Z^* and U^* tests can readily be generalized to multisample situations as follows.

Multisample Situation

To test that k independent random samples

$$x_{i1}, x_{i2}, \ldots, x_{in_i} \qquad (i = 1, 2, \ldots, k)$$

come from the populations $(1/\sigma_i)f_0((x - \mu_i)/\sigma_i)$, the generalized statistic is

$$Z^{**} = \frac{1}{k}\sum_{i=1}^{k} Z_i^* \tag{6.4.35}$$

where Z_i^* is the Z* statistic calculated from the ith sample. Small and large values of Z** indicate that at least one of the samples does not come from the assumed distribution f_0.

Let $N = \Sigma_{i=1}^k n_i$. If the assumed distributions are exponentials $E(\theta_i, \sigma_i)$, or uniforms $U(\theta_{1i}, \theta_{2i})$, then the null distribution of Z**/2 is the same as the mean of $\Sigma_{i=1}^k (n_i - 2) = N - 2k$ IID uniform (0, 1) variates; hence, E(Z**) = 1 and V(Z**) = 1/3(N - 2k). This follows immediately from Theorems 6.3.2 and 6.3.3.

The null distribution of Z** for large n_i (i = 1, 2, ..., k) is approximately normal with mean E(Z**) ≅ 1 and variance $V(Z^{**}) = (1/k)^2 \Sigma_{i=1}^k V(Z_i)$; $V(Z_i)$ is given by (6.4.22).

Dyer and Harbin (1981) and Balakrishnan (1983) investigated the power properties of the multisample Z and T tests, the multisample Shapiro-Wilk test (which like Z and T is location and scale invariant), and the Kolmogorov-Smirnov tests when parameters are estimated (Durbin, 1975), for testing the exponential $E(\theta_i, \sigma_i)$ distributions (i = 1, 2, ..., k); they found that both Z** and T*, particularly Z**, are considerably more powerful overall than these prominent competitors. Note that the Shapiro-Wilk multisample test (and similarly, the Kolmogorov-Smirnov multisample test) is executed by calculating the values of W_i from each sample and defining (Dyer and Harbin, 1981)

$$\chi^2 = -2 \sum_{i=1}^{k} \log P_i$$

where $P_i = P(W \leq W_i)$; the null distribution of χ^2 is chi-square with 2k degrees of freedom. The values of P_i, however, have to be obtained from the available tables of the percentage points by using tedious interpolation formulas and the computations are formidable (see Dyer and Harbin, 1981).

While extreme values of Z** (or any similar multisample statistic) indicate that at least one of the samples does not come from the corresponding assumed distribution, values of Z** which are not extreme do not necessarily imply that all the samples come from the corresponding distributions. It is therefore always advisable to follow it up by testing each sample individually.

EXAMPLE 6.4.4 Consider the data of Exercise 2.4. Here one would like to test normality against distributions which have long tails on both sides. The appropriate statistic is the generalized T [e.g., T* given by (6.3.30)]; the estimates $\hat{\sigma}_{c_i}$ needed for calculating $T_i = h_i \hat{\sigma}_{c_i} / \hat{\sigma}_i$ are based on samples with $r = r_1 = r_2 = [0.5 + 0.3(8)] = 2$ smallest and largest observations censored. Since $n_i = n = 8$ and $h_i = h = (1 - 1/8)/(1 - 1/4)$, i = 1, 2, 3, 4, the values of T_i are given by

0.233 1.418 0.571 1.451

for laboratory A, B, C, and D, respectively. Consequently,

$$T^* = \frac{0.233 + 1.418 + 0.571 + 1.451}{4}$$

$$= 0.918$$

This value is not significantly different from its expected (under normality) value $E(T^*) \simeq 1$, since its approximate 10% significance level is $1 - 1.64\sqrt{V(T^*)} = 1 - 1.64\sqrt{0.2128/4} = 0.77$. [Note that for a fixed $q = r/n$, $nV(T)$ is almost constant for different values of n and hence the approximate variance of T_i for $n = 8$ is given by $1.702/8$, $nV(T)$ being equal to 1.702 for $n = 16$; see Table 6.3.1.]

Although the value of T^* is not significantly different from 1, it is clear that the first T value, 0.233, is significantly smaller than the expected (under normality) value $E(T) \simeq 1$ (in fact, the 10% significance level of T is $1 - 1.64\sqrt{0.2128}$), which is indicative of a long-tailed distribution.

Consequently, if censoring were done in the first sample only, the MML approach would lead to the estimate

$$44.783 + 41.675 - 40.375 - 37.250 = 8.833$$

with standard error

$$3.285\left(\frac{1}{7.656} + \frac{3}{8}\right) = 2.336$$

which is a slight improvement over the MML10 estimate given by Exercise 2.4.

The procedure above readily extends to censored samples. For example, to test that k independent type II censored samples

$$x_{(i,r_i+1)}, x_{(i,r_i+2)}, \ldots, x_{(i,n_i-s_i)}$$

come from the populations $(1/\sigma_i)f((x - \mu_i)/\sigma_i)$ $(i = 1, 2, \ldots, k)$, the generalized statistic is given by

$$Z_c^{**} = \frac{1}{k}\sum_{i=1}^{k} Z_{c_i}^* \tag{6.4.36}$$

where $Z_{c_i}^*$ is the statistic Z_c^* [equation (6.4.26)] calculated from the ith censored sample; the distribution theory of Z_c^{**} is exactly similar to that of Z^{**}.

EXAMPLE 6.4.5 Proschan (1963) observed the successive failures (in hours) of the air-conditioning system of each member of three fleets of Boeing 720 jet airplanes. Each fleet consisted of 23 airplanes, but only the first 15 successive failures were recorded in each fleet due to time limitations. The first 15 ordered observations for each fleet are given below:

I:	1 4 11 16 18 24 31 39 46 51 54 63 68 77 80
II:	10 14 20 23 24 25 26 29 44 49 56 59 60 61 62
III:	14 15 27 29 32 33 34 41 54 57 59 61 66 67 102

One wants to test exponentiality of these failure times.

Here the exponential sample spacings D_i (i = 1, 2, ..., 14) are given by

I:	42 91 60 22 60 63 64 49 30 15 36 15 18 3
II:	56 78 36 11 10 9 24 105 30 35 12 3 2 1
III:	14 156 24 33 10 9 56 91 18 10 12 15 2 35

Consequently, the values of $Z_{c_i}^*$ [equation (6.4.26)] are

1.265 1.294 0.765

for the three censored samples, respectively. Therefore,

$$Z_c^* = \frac{1.265 + 1.294 + 0.765}{3}$$
$$= 1.108$$

The null distribution of $Z_c^{**}/2$ is exactly the same as the distribution of the mean of 3(13) = 39 IID uniform (0, 1) variates and, therefore,

$$E(Z_c^{**}) = 1 \quad \text{and} \quad V(Z_c^{**}) = \frac{1}{3(39)} = \frac{1}{117}$$

The lower and upper 5% points of the null distribution of Z_c^{**} being $1 - 1.64\sqrt{1/117} = 0.845$ and $1 + 1.64\sqrt{1/117} = 1.152$, respectively, the assumption of "overall" exponentiality does not seem to be unreasonable at the 10% significance level. However, if one tests exponentiality of the three samples individually, the first two values of $Z_{c_i}^*$ (e.g., 1.265 and 1.294) are just significant at the 10% significance level.

The literature on goodness-of-fit testing is huge and some of the interesting works, not already mentioned, are the following: Atkinson (1969, 1970, 1973), Barnett (1975, 1976), Bowman and Shenton (1975), Chernoff and Lieberman (1954, 1956), D'Agostino and Rosman (1974), Durbin (1961), Durbin et al. (1975), Dyer (1974), Gear (1947), Hogg (1972), Lewis (1965), Locke and Spurrier (1976, 1977), Montfort (1970), Pettitt (1976, 1977), Pettitt and Stephens (1976), Prescott (1976a,b), Puri and Rao (1976), Sarkadi (1975), Schafer et al. (1972), Shapiro and Francia (1972), Shapiro et al. (1968), Shenton and Bowman (1977), Spiegelhalter (1977), Uthoff (1973), Watson (1961), and Weisberg and Bingham (1975).

6.5 CONCLUSIONS AND SUMMARY

It is important to ascertain whether the assumption of a particular distribution is valid. This is usually done through a formal goodness-of-fit test where a significant value of the test statistic indicates that the underlying distribution may in fact be much different from the one assumed. Three types of goodness-of-fit tests are developed in this chapter: (1) for testing a specified f_0 against a specified alternative f_1, (2) for testing a specified f_0 against certain types of alternatives f_1, and (3) for testing a specified f_0 against a wide range of alternatives f_1. The three types are called tests of model discrimination, directional tests, and omnibus tests, respectively. For model discrimination, a test based on the likelihood-ratio statistic is presented and its distributional and power properties are discussed. A number of directional tests for testing an assumed distribution, in particular a normal distribution, are developed. The directional tests of normality based on the sample skewness and kurtosis are given. A directional test due to Tiku (1975b) for testing any assumed distribution of the type $(1/\sigma)f_0((x-\mu)/\sigma)$ is presented and its distributional and power properties are discussed. Several omnibus goodness-of-fit procedures for testing an assumed distribution $(1/\sigma)f_0((x-\mu)/\sigma)$ are presented and their distributional and power properties are discussed in detail. An omnibus test based on sample spacings is shown to be very versatile and well suited for testing any distribution from complete as well as censored samples. Numerical examples are given to illustrate the applications of goodness-of-fit tests. It is also illustrated how goodness-of-fit tests can be used to explain the differences in results obtained from certain classical and robust procedures.

EXERCISES

6.1 For the data of Example 5.2.2,

69 175 272 335 490 415 72 265 492 180

perform tests of normality based on the omnibus Shapiro-Wilk statistic W and the omnibus Z* statistic and show that they are in agreement in rejecting normality at the 10% significance level.

6.2 For the data of Example 5.7.1 from the population π_1,

-1.48 1.25 -0.51 0.46 0.60 -4.27 0.63 -0.14 -0.38 1.28

0.93 0.51 1.11 -0.17 -0.79 -1.02 -0.91 0.10 0.41 1.11

use the directional $\sqrt{b_1}$ test and the directional T test to test normality against alternatives of negatively skew distributions. Are these tests in agreement in rejecting normality at the 10% significance level?

6.3 For the data of Example 5.7.2, perform a multisample Z** test of normality and show that the assumption of normality is not unreasonable.

7
Tests for Outliers

7.1 INTRODUCTION

An outlier is that observation in a sample which is widely different from the bulk of observations in that sample. In the data of Example 5.8.3,

60 272 16 73 53 26 10 45 34 6

for instance, the observation 272 appears to be an outlier. We have seen in Chapter 2 that in the presence of outliers the variance of the sample mean gets inflated, rendering it inefficient as an estimator of the population mean (location parameter). It is therefore important to check whether a given sample contains outliers or not. If outliers are present, it would be advantageous to use robust procedures since they are more efficient than the classical procedures, as we have seen in Chapters 2 to 5. This brings the problem of outlier detection in focus.

A book-length account of the subject of outliers has been given by Barnett and Lewis (1978), while David (1981) has discussed some of the recent works (see also Hawkins, 1980). Our aim is therefore limited to discussing some of the more recent developments in the area of outlier detection. Several models for outliers have been proposed; see Barnett (1978), who gives a full discussion of these models. The most popular model perhaps is the one proposed by Dixon (1950) as follows.

For a single outlier to the right of the sample, this model stipulates that in a random sample of size n, n - 1 observations

$$x_1, \ldots, x_{i-1}, x_{i+1}, \ldots, x_n \text{ come from a normal } N(\mu, \sigma) \qquad (\mu \text{ and } \sigma \text{ unknown}) \qquad (7.1.1)$$

and the ith observation

$$x_i \text{ comes from } N(\mu + \lambda\sigma, \sigma) \qquad (i \text{ unknown})$$

λ is unknown and greater than zero under H_1, and equal to zero under H_0. This is called a location-shift model. A scale-shift outlier model stipulates that

x_i comes from $N(\mu, \lambda\sigma)$ $\lambda > 0$ (i unknown)

Of course, these models can readily be generalized to multiple outliers to the right or left of the sample.

Another location-shift outlier model that has received attention was introduced by Tiku (1975c). For a single outlier to the right this model stipulates that (see also Hawkins, 1977)

$x_{(i)}$ $(i = 1, 2, \ldots, n-1)$ are the first $n - 1$ order statistics of a random sample of size n from f_0; f_0 is any specified distribution of the type $(1/\sigma)f((x - \mu)/\sigma)$ (μ and σ unknown) (7.1.2)

and

$x_{(n)}$ is the largest order statistic of this random sample plus $\delta\sigma$

δ is unknown and greater than zero under H_1, and equal to zero under H_0. Barnett (1978) calls this model the label-slippage model. Of course, this model can readily be generalized to multiple outliers to the right or left. Scale-shift outlier models can similarly be defined (Tiku, 1975c, 1977; Mann, 1982).

7.2 TESTING OUTLIERS

For testing a single outlier to the right of a sample, the most popular statistic perhaps is that of Pearson and Chandra Sekar (1936), namely,

$$D = \frac{x_{(n)} - \bar{x}}{s} \qquad (7.2.1)$$

where $\bar{x}$ and s^2 are the usual sample mean and sample variance, respectively, and $x_{(n)}$ is the largest order statistic in the random sample $x_1, x_2, \ldots, x_n$. Large values of D lead to the conclusion that $x_{(n)}$ is an outlier. For reasons given in Chapter 6, we suggest that a test for outliers be performed at a high significance level, say $\alpha^* = 0.10$.

If we write

$$D_1 = \frac{\Sigma_{i=1}^{n-1} (x_{(i)} - \bar{x}_n)^2}{\Sigma_{i=1}^{n} (x_i - \bar{x})^2} \qquad (7.2.2)$$

$\bar{x}_n = \Sigma_{i=1}^{n-1} x_{(i)}/(n - 1)$, then

$$D_1 = 1 - \frac{nD^2}{(n-1)^2} \qquad (7.2.3)$$

The statistics D and D_1 are therefore equivalent.

The generalization of D_1 for testing r outliers to the right is due to Grubbs (1950, 1969) for $r = 2$ and Tietjen and Moore (1972) for general r with $r \leq [n/2]$. The Tietjen and Moore statistic is given by

$$L_r = \frac{\Sigma_{i=1}^{n-r} (x_{(i)} - \bar{x}_{n-r})^2}{\Sigma_{i=1}^{n} (x_i - \bar{x})^2} \tag{7.2.4}$$

$\bar{x}_{n-r} = \Sigma_{i=1}^{n-r} x_{(i)}/(n - r)$. Small values of L_r lead to the conclusion that the r largest order statistics $x_{(i)}$ $(i = n - r + 1, \ldots, n)$ are outliers. Assuming normality, Grubbs (1950, 1969) has worked out the null distribution of L_1 and L_2 and Tietjen and Moore (1972) have given the simulated percentage points of the null distribution of L_r for $r = 1(1)10$.

For testing r outliers to the left of the sample, the numerator in L_r is replaced by

$$\sum_{i=r+1}^{n} (x_{(i)} - \bar{x}_{n-r})^2 \qquad \bar{x}_{n-r} = \frac{\Sigma_{i=r+1}^{n} x_{(i)}}{n - r}$$

and under normality the null distribution of the resulting statistic L_r^* is exactly the same as that of L_r; this follows from symmetry.

EXAMPLE 7.2.1 Consider the data of Example 5.8.1. The largest observation 4.5747 appears to be an outlier. Here

$$n = 10 \qquad x_{(10)} = 4.5747 \qquad \bar{x} = 3.439 \qquad s = 0.9038$$

$$D_1 = \frac{81 - 10(1.257)^2}{81} = 0.805$$

Under normality, the 10% significance level of D_1 is 0.490 (Tietjen and Moore, 1972, Table I-d). Since the calculated value of D_1 is not smaller than its significance level, we conclude that the observation 4.5747 is not an outlier. This explains the close agreement between the sample mean $\bar{x}$ and the MML10 estimate and their standard errors (see Example 5.8.1).

Some samples, however, display outliers both to the left and to the right. To test that a sample contains k outliers, irrespective of whether some are to the left or right, Tietjen and Moore (1972) defined the statistic

$$E_k = \frac{\Sigma_{i=1}^{n-k} (z_{(i)} - \bar{z}_{n-k})^2}{\Sigma_{i=1}^{n} (z_i - \bar{z})^2} \tag{7.2.5}$$

where $z_{(i)}$ are the order statistics of the observations $z_j = |x_j - \bar{x}|$

(j = 1, 2, ..., n) and $\bar{z}_{n-k} = \Sigma_{i=1}^{n-k} z_{(i)}/(n-k)$ with $k \leq [n/2]$. Under the assumption of normality, Tietjen and Moore have given the simulated percentage points of the null distributions of E_k for $k = 1(1)10$.

EXAMPLE 7.2.2 Consider the following 10 measurements, arranged in ascending order of magnitude, of percent elongation at break of certain material (Grubbs, 1969):

2.02 2.22 3.04 3.23 3.59 3.73 3.94 4.05 4.11 4.13

Here the outliers occur only on the left-hand side, if at all. We therefore use the statistic L_2^*:

$$n = 10 \quad \bar{x}_8 = \frac{\Sigma_{i=3}^{10} x_{(i)}}{8} = 3.728 \quad \sum_{i=3}^{10} (x_{(i)} - \bar{x}_8)^2 = 1.197$$

$$\sum_{i=1}^{10} (x_i - \bar{x})^2 = 5.351$$

Thus

$$L_2^* = \frac{1.197}{5.351} = 0.224$$

Under normality, the 10% significance level of L_2^* is 0.287 (Tietjen and Moore, 1972, Table I-d). Since the calculated value is smaller than the significance level, we conclude that the two smallest observations are outliers.

One might want to compare the classical (based on all 10 observations) and the MML (based on the largest eight observations) estimates of the location parameter. The classical estimate is

$\bar{x} = 3.41$, with standard error $\pm \frac{s}{\sqrt{n}} = \pm 0.24$

the MML estimate given by (2.7.11) is

$\hat{\mu}_c = 3.64$, with approximate standard error $\pm(\sqrt{\nu_{11}})_{\sigma=\hat{\sigma}_c} = \pm 0.15$ obtained from (2.7.17); this is also equal to $\pm\hat{\sigma}_c/\sqrt{m} = \pm 0.15$, since the element J_{12} is very small as compared to J_{11} and J_{22}

The standard error of the MML estimate is considerably smaller than that of $\bar{x}$, as expected. Since in calculating $\hat{\mu}_c$ and its standard error, one assumes normality or near-normality of a type II censored sample, one might also like to verify whether the observations above, other than the two smallest, can be regarded as a type II censored sample from a normal distribution

$N(\mu, \sigma)$. For calculating the goodness-of-fit statistic Z_c^*, given by (6.4.26), we have the following; the values of E_i are taken from Table 6.4.2:

i:	3	4	5	6	7	8	9	10
$x_{(i)}$	3.04	3.23	3.59	3.73	3.94	4.05	4.11	4.13
E_i	0.280	0.253	0.245	0.253	0.280	0.345	0.537	—
G_i	0.678	1.422	0.570	0.830	0.392	0.174	0.037	—

Therefore,

$$Z_c^* = \frac{2\ \Sigma_{i=3}^{8}\ (9 - i)G_i}{6\ \Sigma_{i=3}^{9}\ G_i}$$

$$= \frac{2(16.906)}{6(4.103)}$$

$$= 1.37$$

Now, for a complete sample of size $n = 10$ the upper 5% point of Z^* is $1 + 1.64\sqrt{0.0324} = 1.30$ (Table 6.4.2). Since the variance of Z_c^* based on a censored sample of size $n - r$ is surely larger than that of Z^* based on a complete sample of size n (Tiku, 1980b), the upper 5% point of Z_c^* is surely greater than 1.30. The calculated value of Z_c^* being close to 1.30, the normality assumption above is quite reasonable at the 10% significance level. This perhaps accounts for the smaller standard error of the MML estimate compared to the classical estimate $\bar{x}$, the latter apparently being inflated by the two outliers in the sample.

Another way of detecting multiple outliers, say to the right, under Dixon's outlier model would be to proceed sequentially; for example, one first calculates a certain statistic, suitable for detecting a single outlier, from the entire sample of size n. If it turns out to be significant, censor the largest order statistic $x_{(n)}$ and repeat this procedure with the reduced sample of size $n - 1$, adjusting the percentage points to the new sample size, and so on. However, this sequential procedure might turn out to be ineffective, due to the masking effect [the phenomenon of some observations being closer to each other than they are close to the bulk of observations (Pearson and Chandra and Sekar, 1936)]. In Example 7.2.2, for instance, let us proceed sequentially. At the first stage, we test whether the smallest observation is an outlier. The statistic

$$L_1 = \frac{\Sigma_{i=2}^{10} (x_{(i)} - \bar{x}_9)^2}{\Sigma_{i=1}^{10} (x_i - \bar{x})^2}$$

$$= \frac{3.217}{5.351}$$

$$= 0.601$$

Since this value is not smaller than the 10% significance level 0.490 (Tietjen and Moore, 1972, Table I-d), we conclude that the smallest observation 2.02 is not an outlier. We have, however, established that the two smallest observations are outliers. The sequential procedure does not even get started due to the masking effect.

To detect up to k outliers (k is known in advance) under Dixon's outlier model, Rosner (1975) has studied a somewhat different and interesting sequential procedure for $k = 2$. His procedure calculates repeatedly a certain statistic, suitable for detecting a single outlier. At the first stage, the statistic is calculated from the entire sample of size n. The largest (smallest) order statistic is censored whether or not the first-stage test yields a significant value. At the second stage, the statistic is calculated from the reduced sample of size $n - 1$. Using Monte Carlo methods, Rosner (1975) has found the statistic $\max|x_i - \bar{x}|/s$ slightly the superior one among several statistics designed to detect up to two outliers. Extension of Rosner's work to detect up to k outliers is due to Rosner (1977) for $k = 4$ and Hawkins (1979) for general k; Prescott (1979) and Jain (1981) have also studied such sequential procedures.

Tiku Approach

Let $x_1, x_2, \ldots, x_n$ be a random sample from any distribution of the type $(1/\sigma)f_0((x - \mu)/\sigma)$; f_0 is specified but μ and σ are not known. To test that r_1 smallest and r_2 largest observations are outliers, Tiku (1975c) defined the statistic

$$T(r_1, r_2) = \frac{h\tilde{\sigma}_c}{\tilde{\sigma}} \qquad h = \frac{E(\tilde{\sigma})}{E(\tilde{\sigma}_c)} \tag{7.2.6}$$

$\tilde{\sigma}_c$ is the ML or MML estimator of σ calculated from the type II censored sample

$$x_{(r_1+1)}, x_{(r_1+2)}, \ldots, x_{(n-r_2)} \tag{7.2.7}$$

and $\tilde{\sigma}$ is the ML estimator of σ calculated from the complete sample. In fact, $\tilde{\sigma}$ is the estimator $\tilde{\sigma}_c$ with $r_1 = r_2 = 0$. Note that the statistic T is location and scale invariant.

Exponential Distribution. For the exponential $E(\theta, \sigma)$, the statistic $T(r_1, r_2)$ is given by

$$T_E(r_1, r_2) = \frac{(n-1)[\Sigma_{i=r_1+1}^{n-r_2} x_{(i)} + r_2 x_{(n-r_2)} - (n-r_1)x_{(r_1+1)}]}{(n - r_1 - r_2 - 1)\,\Sigma_{i=1}^{n} (x_i - x_{(1)})} \qquad (7.2.8)$$

Small and large values indicate outliers, r_1 to the left and r_2 to the right.

THEOREM 7.2.1 Under exponentiality, the null distribution of $y = [(n - r_1 - r_2 - 1)/(n - 1)]\, T_E(r_1, r_2)$ is the beta distribution beta$(n - r_1 - r_2 - 1, r_1 + r_2)$.

Proof: It is easy to show that

$$y = \frac{\Sigma_{i=r_1+1}^{n-r_2-1} D_i}{\Sigma_{i=1}^{n-1} D_i}$$

where D_i $(i = 1, 2, \ldots, n - 1)$ are the exponential sample spacings (6.3.18). The result then follows exactly on the same lines as Theorem 6.3.2.

Uniform Distribution. For the uniform $U(\theta_1, \theta_2)$, the statistic T is given by

$$T_U(r_1, r_2) = \frac{(n-1)(x_{(n-r_2)} - x_{(r_1+1)})}{(n - r_1 - r_2 - 1)(x_{(n)} - x_{(1)})} \qquad (7.2.9)$$

Small and large values indicate outliers.

THEOREM 7.2.2 Under uniformity, the null distribution of $u = [(n - r_1 - r_2 - 1)/(n - 1)]\, T_U(r_1, r_2)$ is the beta distribution beta$(n - r_1 - r_2 - 1, r_1 + r_2)$.

Proof: It is easy to see that

$$u = \frac{\Sigma_{i=r_1+1}^{n-r_2-1} u_i}{\Sigma_{i=1}^{n-1} u_i}$$

where $u_i = (n + 1)(x_{(i+1)} - x_{(i)})$ $(i = 1, 2, \ldots, n - 1)$ are the uniform sample spacings. The result then follows on the same lines as Theorem 6.3.3.

Normal Distribution. For the normal distribution $N(\mu, \sigma)$,

$$T(r_1, r_2) = \frac{(1 - 1/n)\tilde{\sigma}_c}{(1 - 1/A)\tilde{\sigma}} \quad (A = n - r_1 - r_2) \tag{7.2.10}$$

where $\tilde{\sigma}_c = (B + \sqrt{B^2 + 4AC})/2A$ and $\tilde{\sigma} = \sqrt{\Sigma_{i=1}^n (x_i - \bar{x})^2/n}$ are the MML and ML estimators of σ, respectively; the expressions for B and C are given by (2.7.9). Small values indicate outliers, r_1 to the left and r_2 to the right.

It is difficult to work out the exact distribution of $T(r_1, r_2)$, but Tiku (1975c) has shown that the $100\alpha^*$ percentage points of the null distribution of $T(r_1, r_2)$ are closely approximated by

$$\frac{n - 1}{n - r_1 - r_2 - 1} u_{\alpha^*} + \frac{1}{5n}\left(1 + \frac{1}{n - 2r_2 + 1}\right) \tag{7.2.11}$$

where u_{α^*} are the $100\alpha^*$ percentage points of the beta distribution beta$(n - r_1 - r_2 - 1, r_1 + r_2)$ (see Tiku, 1975c, Table II). Extensive tables of the percentage points of beta distribution are available (Pearson and Hartley, 1956, Table 16).

EXAMPLE 7.2.3 Grubbs (1950) presented a classical set of data which have often been used as an illustration of outliers. The sample consists of 15 coded observations on the vertical semidiameters of Venus made in 1846 by Lt. Herdon. Arranged in ascending order of magnitude, the observations are

-1.40 -0.44 -0.30 -0.24 -0.22 -0.13 -0.05 0.06 0.10 0.18

0.20 0.39 0.48 0.63 1.01

The data are supposed to come from a normal distribution other than the two extreme observations -1.40 and 1.01, which are suspect. Here

$n = 15 \quad r_1 = r_2 = 1 \quad \alpha_1 = \alpha_2 = 0.637 \quad \beta_1 = \beta_2 = 0.880$ (Table B.1)

$$\tilde{\sigma}_c = \frac{0.6860 + \sqrt{0.4703 + 90.63}}{26} = 0.3935 \qquad \tilde{\sigma} = 0.5325$$

$$T(1,1) = \frac{(1 - 1/15)(0.3935)}{(1 - 1/13)(0.5325)}$$

$$= 0.747$$

The approximate lower 10% point of T(1, 1) is

$$\frac{14}{12}(0.732) + \frac{1.0714}{75} = 0.87$$

0.732 being the lower 10% point of the beta distribution beta(12, 2). Since the calculated value of T(1, 1) is smaller than its 10% significance level, we conclude that the two extreme observations, -1.40 and 1.01, are in fact outliers. This agrees with the conclusion reached by Tietjen and Moore (1972), who used the L_2 procedure.

Now, to compare the classical estimate (based on all the 15 observations) and the MML estimate (based on the middle 13 observations) of μ, we have the following:

$$\bar{x} = 0.018 \text{ with standard error } \pm \frac{s}{\sqrt{n}} = \pm 0.142$$

$$\hat{\mu}_c = \frac{0.66 + 0.1655}{14.722} = 0.056 \text{ with standard error } \pm \frac{\hat{\sigma}_c}{\sqrt{m}} = \pm 0.107$$

It is clear that $\bar{x}$ ceases to be an efficient estimator if the sample contains outliers.

To illustrate the usefulness of the statistic $T(r_1, r_2)$ for testing multiple outliers to the left as well as to the right, we consider the following data.

EXAMPLE 7.2.4 Consider the data of Example 2.7.1. Arranged in order of magnitude, the observations are

-67 -48 6 8 14 16 23 24 28 29 41 49 56 60 75

One would like to test whether the two smallest and the two largest observations are outliers.

From the middle 11 observations, we calculate $\tilde{\sigma}_c$:

$$\tilde{\sigma}_c = 24.311 \quad \text{and} \quad \tilde{\sigma} = 36.444$$

Now

$$T(2,2) = \frac{(1 - 1/15)(24.311)}{(1 - 1/11)(36.444)} = 0.685$$

The approximate 10% significance level of T(2, 2) is given by [equation (7.2.11)]

$$\frac{14}{10}(0.556) + \frac{1.0833}{75} = 0.79$$

0.556 being the lower 10% point of the beta distribution beta (10, 4). Since the calculated value of T(2, 2) is smaller than its 10% level, we conclude that the two smallest and the two largest observations are outliers. This perhaps explains why the same mean $\bar{x}$ has standard error considerably larger than the MML estimate (see Example 2.7.1).

Power Function

Various measures of performance of a test for outliers have been proposed, but the most useful one perhaps is that of power (David, 1981, p. 234; Hawkins, 1977). For a prespecified significance level α^*, the power of an outlier test V is the probability

$$1 - \beta^* = P(V < \nu_{\alpha^*} | H_1) \tag{7.2.12}$$

H_1 denotes the phenomenon that the sample contains a specified number r of outliers. Note here that small values of V indicate outliers, and ν_{α^*} is given by

$$\alpha^* = P(V < \nu_{\alpha^*} | H_0)$$

where H_0 denotes the phenomenon that the sample contains no outliers.

It is difficult to determine the power function of the L_r and T tests analytically. Using Monte Carlo methods, Tiku (1977) determined the values of the power of the L_r and $T(0, r)$ tests for testing r outliers to the right under both Dixon's and Tiku's location shift models, for samples of size $n = 10$, 20, and 40 and $r = 1$, 2, 3, and 4; $\alpha^* = 0.05$. To determine the power empirically, one generates n random observations from a normal distribution $N(\mu, \sigma)$, and for $r = 1$ adds $\xi\sigma$ $(\xi > 0)$ to an arbitrary x_i under Dixon's model and adds $\delta\sigma$ $(\delta > 0)$ to the largest order statistic $x_{(n)}$ under Tiku's model; for $r = 2$ one adds $\xi_1\sigma$ and $\xi_2\sigma$ (ξ_1 and ξ_2 both greater than zero) to two arbitrary observations x_i and x_j under Dixon's model and adds $\delta_1\sigma$ and $\delta_2\sigma$ (δ_1 and δ_2 both greater than zero) to $x_{(n-1)}$ and $x_{(n)}$ under Tiku's model, and similarly for larger values of r. In particular, ξ_1 and ξ_2 could both be equal and equal to ξ and δ_1 and δ_2 could both be equal and equal to δ. The empirical values of the power of $T(0, r)$ and L_r are given by a_1/N and a_2/N, respectively, where a_1 and a_2 are the number of samples, out of N (= 10,000) samples generated, which yielded significant values of $T(0, r)$ and L_r. The conclusion reached was that under Dixon's model, L_r can have somewhat higher power than $T(0, r)$; this was to be expected since L_r is known to be optimal under Dixon's model, at any rate for $r = 1$ (David, 1981; Hawkins, 1977). Under Tiku's model, however, $T(0, r)$ can have substantially higher power than L_r. To have an idea about the relative powers of $T(0, r)$ and L_r tests, we reproduce in Table 7.2.1 some of the values of the power given by Tiku (1977).

The asymptotic power functions of $T_E(r_1, r_2)$ and $T_U(r_1, r_2)$ have expressions exactly similar to (6.3.21); Tiku (1975c) has shown that these tests also have high power [see also Mann (1982), who has studied T_E and similar statistics and discussed their optimality].

Choice of r_1 and r_2

To use the $T(r_1, r_2)$ statistic and similar other statistics, one has to know in advance the number of possible outliers r_1 and r_2. In certain situations,

Table 7.2.1 Values of the Power for the 5% Significance Level; $L_r = L$, $T(0, r) = T$

		Tiku's model					Dixon's model			
		r = 1		r = 2			r = 1		r = 2	
n	$\delta\sigma$	T	L	T	L	$\xi\sigma$	T	L	T	L
10	0	0.05	0.05	0.05	0.05	0	0.05	0.05	0.05	0.05
	1	0.35	0.35	0.37	0.37	2	0.16	0.16	0.13	0.14
	2	0.88	0.84	0.85	0.82	4	0.66	0.67	0.59	0.61
	3	1.00	0.99	0.99	0.98	8	1.00	1.00	1.00	1.00
20	0	0.05	0.05	0.05	0.05	0	0.05	0.05	0.05	0.05
	1	0.50	0.49	0.67	0.61	3	0.39	0.42	0.44	0.50
	2	1.00	0.97	1.00	0.99	5	0.91	0.93	0.96	0.98
40	0	0.05	0.05	0.05	0.05	0	0.05	0.05	0.05	0.05
	1	0.68	0.56	0.89	0.74	3	0.39	0.42	0.51	0.57
	2	1.00	0.99	1.00	1.00	5	0.93	0.95	0.98	0.99

Table 7.2.2 Values of the Power of $T(0, r^*)$ Test Under Dixon's Location-Shift Model, r Being the Actual Number of Outliers to the Right

		r = 1		r = 2		
n	$\xi\sigma$	$r^* = 1$	$r^* = 2$	$r^* = 2$	$r^* = 3$	$r^* = 4$
10	0	0.05	0.05	0.05	0.05	0.05
	2	0.16	0.13	0.13	0.12	0.12
	4	0.66	0.54	0.59	0.54	0.45
	8	1.00	0.99	1.00	0.99	0.97
20	0	0.05	0.05	0.05	0.05	0.05
	3	0.39	0.37	0.44	0.46	0.46
	5	0.91	0.90	0.96	0.96	0.96

however, one might have no idea about r_1 and r_2. In such situations, Tiku (1977) recommended the use of $T(r_1^*, r_2^*)$, where $r_1^* \geq r_1$ and $r_2^* \geq r_2$ are the guessed values of the number of outliers to the left and right, respectively. This $T(r_1^*, r_2^*)$ test has remarkably high power. For example, Tiku (1977) has given empirical values (based on 10,000 Monte Carlo runs) of the power of this $T(r_1^*, r_2^*)$ test under normality (Table 7.2.2) and these may be compared with the values given in Table 7.2.1.

EXAMPLE 7.2.5 A set of eight mass spectrometer measurements on a particular isotope of uranium is arranged in ascending order of magnitude as (Tietjen and Moore, 1972, Example 3)

199.31 199.53 200.19 200.82 201.92 201.95 202.18 245.57

We calculate the statistic $T(0, r^*)$ with $r^* = 2$. Censoring the two largest observations we obtain the following:

$$\tilde{\sigma}_c = \frac{1.6351 + 14.7197}{12} = 1.3629 \qquad \tilde{\sigma} = 14.8252$$

$$T(0, 2) = \frac{(1 - 1/8)(1.3629)}{(1 - 1/6)(14.8252)}$$

$$= 0.096$$

Since the calculated value of $T(0, r^*)$ is smaller than even its 1% significance level $(7/5)(0.294) + (1.2/40) = 0.442$, we conclude that the observations 245.57 and 202.18 (particularly 245.57) may be outliers.

To compare the MML estimate of the location parameter μ obtained by censoring both the observations 202.18 and 245.57 with the MML estimate obtained by censoring only the largest observation 245.57, we have the following estimates, worked out exactly on the same lines as in Example 7.2.3:

With one largest observation censored:

$$\hat{\mu}_c = 201.113 \text{ with standard error } \pm\frac{\hat{\sigma}_c}{\sqrt{m}} = \pm\frac{1.419}{\sqrt{7.828}} = \pm 0.507$$

With two largest observations censored:

$$\hat{\mu}_{1c} = 201.197 \text{ with standard error } \pm\frac{\hat{\sigma}_{1c}}{\sqrt{m}} = \pm\frac{1.493}{\sqrt{7.536}} = \pm 0.544$$

The two estimates and their standard errors are in very close agreement. However, since $\hat{\mu}_c$ has smaller standard error than that of $\hat{\mu}_{1c}$, it indicates that the observation 202.18 is perhaps not an outlier.

The classical estimate (MML0) estimate) of μ is

$$\bar{x} = 206.434 \text{ with standard error } \pm\frac{s}{\sqrt{n}} = \pm\frac{15.849}{\sqrt{8}} = \pm 5.603$$

$\bar{x}$ is highly inefficient. This again illustrates the point we made earlier: for example, to obtain an efficient estimator of μ through the MML approach, it is very crucial to censor all the outliers in the sample; censoring a small proportion of "good" observations does not, however, render the MML estimator too inefficient.

7.3 ESTIMATION OF r_1 AND r_2

In situations where the actual number of outliers r_1 and r_2 are not known in advance, instead of guessing their values, one may like to estimate them and adopt a two-stage procedure (Tietjen and Moore, 1972; Tiku, 1975c). At the first stage, one estimates the values of r_1 and r_2; denote these estimates by $\tilde{r}_1$ and $\tilde{r}_2$. At the second stage, one calculates the statistic $T(\tilde{r}_1, \tilde{r}_2)$, or any other similar statistic, and compares it with the appropriate percentage points of the null distribution; the distribution theory is, however, very difficult. For estimating r_1 and r_2, the following proposals have been made.

Tietjen and Moore (1972): Calculate the gaps to the right of the sample median [e.g., $L_i = (x_{(i+1)} - x_{(i)})$ $(i = [n/2] + 1, \ldots, n - 1)$]. If L_j is the largest gap, then $\tilde{r}_2 = n - j$, and similarly for $\tilde{r}_1$.

Tiku (1975c): Calculate the sample spacings [given by (6.3.17)] to the right of the sample median. If G_j is the largest spacing, then $\tilde{r}_2 = n - j$, and similarly for $\tilde{r}_1$.

Jain and Pingel (1981): Calculate

$$RDM_i = \frac{G_i}{(x_{(i)} - \text{median})/\mu_{i:n}} \tag{7.3.1}$$

to the right of the median. If RDM_j is the largest RDM_i, then $\tilde{r}_2 = n - j$, and similarly for $\tilde{r}_1$.

Under normality, Jain and Pingel (1981) have found their proposal most satisfactory; for example, the RDM_j estimates are mostly larger than, if not equal to, the actual number of outliers in the sample; all other proposals, particularly the Tietjen-Moore proposal, produce estimates which are often smaller than the actual number of outliers in the sample and are therefore not satisfactory in view of the fact that most of the subsequent tests would be based on statistics of the type

$$\frac{f(x_{(\tilde{r}_1+1)}, \ldots, x_{(n-\tilde{r}_2)})}{f(x_1, x_2, \ldots, x_n)}$$

and for the power to be high it is very crucial that $\tilde{r}_1$ and $\tilde{r}_2$ be larger than, if not equal to, the actual number of outliers in the sample (Jain and Pingel, 1981). Jain et al. (1982) have studied a few such two-stage outlier detection procedures and found them quite effective.

EXAMPLE 7.3.1 The following data, arranged in order of magnitude, represent the values of 31 contrasts in a 2^5 factorial experiment (Daniel, 1959) rescaled by Dempster and Rosner (1971):

-3.143 -2.666 -1.305 -0.8980 -0.8138 -0.8138 -0.7577 -0.7437

-0.4771 -0.3087 -0.2526 -0.0982 -0.0842 -0.0561 0.000 0.0281

0.1263 0.1684 0.1964 0.2245 0.2947 0.3929 0.4069 0.4209 0.4350

0.4630 0.5472 0.6595 0.7437 1.080 2.147

For applying the statistic $T(r_1^*, r_2^*)$, clearly

$$r_1^* = 3 \quad \text{and} \quad r_2^* = 2$$

To calculate the estimate $\tilde{\sigma}_c$, we have the following:

$$n = 31 \quad q_1 = \frac{r_1^*}{n} = 0.097 \quad q_2 = \frac{r_2^*}{n} = 0.064 \quad \alpha_1 = 0.688 \quad \beta_1 = 0.832$$

$$\alpha_2 = 0.660 \quad \beta_2 = 0.852$$

$$\tilde{\sigma}_c = \frac{2.806 + \sqrt{7.876 + 934.304}}{52} = 0.644 \quad \tilde{\sigma} = \sqrt{\frac{30.001}{31}} = 0.984$$

$$T(3,2) = \frac{(1 - 1/31)(0.644)}{(1 - 1/26)(0.984)}$$

$$= 0.659$$

The 10% significance level of $T(3,2)$ is approximately [equation (7.2.11)]

$$\frac{30}{25}(0.736) + \frac{1.0357}{155} = 0.89$$

We conclude that as many as five observations in the data above, three to the left and two to the right, may be outliers.

To compare the classical estimate with the MML estimate (based on the data above) with three smallest and two largest observations censored), we have the following:

$$\text{Classical: } \bar{x} = -0.132 \text{ with standard error } \pm\frac{s}{\sqrt{n}} = \pm\frac{1}{\sqrt{31}} = \pm 0.180$$

$$\text{MML: } \hat{\mu}_c = -0.054 \text{ with standard error } \pm\frac{\hat{\sigma}_c}{\sqrt{m}} = \pm\frac{0.650}{\sqrt{30.2}} = \pm 0.118$$

As suggested in Chapter 4, suppose that we would have routinely calculated the MML10 estimate of μ; for calculating this estimate from (2.7.18), we would have the following:

$$n = 31 \quad r = [0.5 + 0.1(31)] = 3 \quad q = \frac{r}{n} = 0.097 \quad \alpha = 0.688 \quad \beta = 0.832$$

$$\hat{\mu}_{1c} = -\frac{1.535}{29.992} = -0.051 \qquad \hat{\sigma}_{1c} = \frac{3.215 + \sqrt{10.334 + 854.053}}{48.99} = 0.666$$

Therefore, the MML10 (with three smallest and three largest observations censored) estimate of μ is -0.051 with standard error $\pm\hat{\sigma}_{1c}/\sqrt{m} = \pm 0.666/\sqrt{29.992} = \pm 0.122$.

The MML estimates above and their standard errors are in close agreement, but the sample mean $\bar{x}$ is inefficient.

Assuming that the number of outliers in the data above is less than or equal to 4, Rosner (1977) applied his one-at-a-time sequential procedure, based on the statistic

$$\frac{\max |x_i - \tilde{\mu}_T|}{\tilde{\sigma}_T} \tag{7.3.2}$$

where $\tilde{\mu}_T = \Sigma_{i=k+1}^{n-k} x_{(i)}/(n - 2k)$ is the trimmed sample mean and $\tilde{\sigma}_T^2$ is the matching variance, and concluded that the two smallest and one largest observations are outliers. The MML estimate of μ, with these two smallest and one largest observations censored, is given by

$$\hat{\mu}_{2c} = -0.073 \text{ with standard error } \pm\frac{\hat{\sigma}_{2c}}{\sqrt{m}} = \pm\frac{0.650}{\sqrt{30.58}} = \pm 0.135$$

The fact that the standard error of this estimate is larger than that of $\hat{\mu}_c$ clearly indicates that the data contain more than the three outliers detected by Rosner's (1977) procedure. Furthermore, the fact that $\hat{\mu}_c$ has standard error not larger than that of $\hat{\mu}_{1c}$ indicates that the third largest observation 0.7437 is not an outlier.

It may be noted that for Rosner's one-at-a-time sequential procedure to be effective, it is very crucial to choose k larger than, if not equal to, the actual number of outliers in the sample; underspecification of k may render this procedure completely ineffective (Rosner, 1977, p. 311).

For two-stage procedures mentioned earlier, we have the following estimates of r_1 and r_2 for the data above:

Tietjen and Moore (1972): The leaps L_i ($i = 1, 2, \ldots, 30$) are 0.477, 1.361, 0.407, ..., 0.084, 0.336, 1.067, respectively; therefore, $\tilde{r}_1 = 2$ and $\tilde{r}_2 = 1$.

Tiku (1975c): Assuming normality, the spacings G_i ($i = 1, 2, \ldots, 30$) are 1.123, 5.466, 2.204, 0.558, ..., 0.744, 0.455, 1.350, 2.511, respectively; therefore, $\tilde{r}_1 = 2$ and $\tilde{r}_2 = 1$.

Jain and Pingel (1981): Assuming normality, the values of RDM_i ($i = 1, 2, \ldots, 30$) are 0.68, 5.67, 2.85, ..., 0.87, 2.61, 3.92, respectively; therefore, $\tilde{r}_1 = 2$ and $\tilde{r}_2 = 1$.

The expected values $\mu_{i:n}$ of normal order statistics which are needed to calculate G_i and RDM_i are given by Pearson and Hartley (1972, Table 9) for samples of size $n = 2(1)100(25)200$.

The three methods give exactly the same estimates $\tilde{r}_1 = 2$ and $\tilde{r}_2 = 1$ for these data.

As a result of a second-stage outlier detection procedure based on these first-stage estimates, Jain et al. (1982) have found the two smallest and one largest observations to be outliers.

If, however, one examines the spacings G_i carefully (realize that the expected values of G_i are all equal and equal to σ if the sample has no outliers), it seems more reasonable that $r_1 = 3$ and $r_2 = 2$, that is, three outliers to the left and two outliers to the right; this was also accomplished as a result of the $T(r_1^*, r_2^*)$ test.

7.4 TESTING THE SAMPLE FOR OUTLIERS

In certain situations, one simply wants to test whether a sample contains outliers irrespective of how many (Ferguson, 1961). If the statistic employed for this purpose turns out to be significant, the sample is rejected. The use of the statistic $\sqrt{b_1}$ [given by (6.3.1)] has been recommended if outliers occur only on one side of the sample; the use of the statistic b_2 [given by (6.3.3)] has been recommended if outliers occur on both sides of the sample (Ferguson, 1961). The distribution theory of $\sqrt{b_1}$ and b_2 has already been discussed in Chapter 6. As an alternative to $\sqrt{b_1}$, Tiku (1977) has recommended the use of $T(r_1^*, r_2^*)$; r_1^* and r_2^* are guessed values and are greater than, if not equal to, the actual number of outliers r_1 and r_2 to the left and right of the sample, respectively. He investigated the power properties of the $\sqrt{b_1}$ and $T(0, r^*)$ tests and concluded that even if r^* is chosen to be twice as large as the actual number of outliers r to the right of the sample, the $T(0, r^*)$ test is considerably more power than the $\sqrt{b_1}$ test. Under Dixon's location-shift outlier model, for example, Tiku (1977) has given the empirical values (based on 10,000 Monte Carlo runs) of the power, under normality (Table 7.4.1). Note that under Dixon's location-shift model, $\sqrt{b_1}$ is known to be locally optimal (Ferguson, 1961; Hawkins, 1977).

EXAMPLE 7.4.1 Consider the following 10 observations on breaking strength (in pounds) of 0.104-in. hard-drawn copper wire (Grubbs, 1969, Example 1):

568 570 570 570 572 572 572 578 584 596

Does the sample contain outliers?

If at all, the outliers clearly occur only to the right of the sample. For the $T(0, r^*)$ test, it clearly suffices to take $r^* = 2$. Now

Table 7.4.1 Values of the Power of $\sqrt{b_1}$ and $T(0, r^*)$ for Rejecting the Sample, r Being the Actual Number of Outliers to the Right

		r = 1			r = 2			
			T			T		
n	$\xi\sigma$	$\sqrt{b_1}$	r* = 2	r* = 3	$\sqrt{b_1}$	r* = 3	r* = 4	r* = 6
10	0	0.05	0.05	0.05	0.05	0.05	0.05	0.05
	1	0.06	0.06	0.05	0.05	0.06	0.06	0.06
	3	0.26	0.30	0.23	0.13	0.29	0.25	0.16
	5	0.75	0.76	0.62	0.39	0.77	0.66	0.37
20	0	0.05	0.05	0.05	0.05	0.05	0.05	0.05
	1	0.05	0.06	0.06	0.06	0.06	0.06	0.06
	3	0.29	0.37	0.34	0.35	0.46	0.46	0.42
	5	0.83	0.90	0.85	0.92	0.96	0.96	0.92

$$\tilde{\sigma}_c = 4.3815 \qquad \tilde{\sigma} = 8.2559$$

$$T(0,2) = \frac{(1 - 1/10)(4.3815)}{(1 - 1/8)(8.2559)} = 0.546$$

The 10% significance level of $T(0,2)$ is given by

$$\frac{9}{7}(0.594) + \frac{1.1429}{50} = 0.787$$

0.594 being the lower 10% point of the beta (7, 2). Since the calculated value of $T(0,2)$ is smaller than its 10% significance level, we conclude that the sample does contain outliers.

For the $\sqrt{b_1}$ test, we have

$$\sqrt{b_1} = \frac{(10)(9702.336)}{26,702.469} = 1.149$$

Large values indicate outliers. The upper 10% point of $\sqrt{b_1}$ is 0.950 (D'Agostino and Tietjen, 1973, Table 2). Since the calculated value is larger than 0.950, we conclude that the sample does contain outliers.

To compare the classical and the MML estimates of μ, we have the following:

$$\bar{x} = 575.20 \text{ with standard error } \pm \frac{s}{\sqrt{n}} = \pm \frac{10.101}{\sqrt{10}} = \pm 3.19$$

$$\hat{\mu}_c = 573.306 \text{ with standard error } \pm \frac{\hat{\sigma}_c}{\sqrt{m}} = \pm \frac{4.684}{\sqrt{9.5828}} = \pm 1.513$$

The superiority of the MML estimate over the classical estimate is obvious.

In conclusion it may be stated that the statistic $T(r_1, r_2)$ is versatile: (1) it can be used for testing outliers in samples from any distribution of the form $(1/\sigma)f((x - \mu)/\sigma)$, (2) it can be used for testing any specified number of outliers on either side of the sample, and (3) it can be used for testing whether the sample contains outliers, irrespective of how many. Besides, the exact null distributions of $T(r_1, r_2)$ are beta $(n - r_1 - r_2 - 1, r_1 + r_2)$ for the uniform $U(\theta_1, \theta_2)$ and exponential $E(\theta, \sigma)$ populations; for the normal $N(\mu, \sigma)$, the beta approximation (7.2.11) provides accurate values for the percentage points of the null distribution of $T(r_1, r_2)$.

7.5 MULTISAMPLE GENERALIZATION

The statistic $T(r_1, r_2)$ readily generalizes to multisample situations. For testing that r_{1i} and r_{2i} outliers exist to the left and right, respectively, of the ith independent random sample

$$x_{i1}, x_{i2}, \ldots, x_{in_i} \quad (i = 1, 2, \ldots, k) \tag{7.5.1}$$

the generalized statistic is given by

$$T^* = \frac{1}{k} \sum_{i=1}^{k} T_i(r_{1i}, r_{2i}) \tag{7.5.2}$$

$T_i(r_{1i}, r_{2i})$ being the statistic $T(r_1, r_2)$ calculated from the ith sample (7.5.1). The distributional properties of T* are similar to those of the generalized statistics given in Chapter 6 (see also Tiku, 1975c).

The statistic $T(r_1, r_2)$ also generalizes to multivariate populations [see Tiku and Singh, 1981c, for details]. For a Bayesian approach to outlier detection, see Hawkins (1980, Chap. 9).

7.6 CONCLUSIONS AND SUMMARY

Five to 10 percent wrong values in a data set seem to be the rule rather than the exception (Hampel, 1973). These wrong values may occur as outliers (e.g., observations that are too large or too small compared to the bulk of the data). In this chapter we have given procedures for testing whether certain extreme observations are in fact outliers. The power properties of these procedures are studied under two outlier models, the Dixon model and the Tiku model. Under the Dixon model, outliers are generated by shifting randomly chosen observations away from their natural positions. Under the Tiku model, outliers are generated by shifting the smallest and/or largest order statistics away from their natural positions. A statistic that is the ratio of the MML estimator of the scale parameter σ calculated from a suitably censored sample to the ML estimator of σ calculated from the complete sample is shown to have good power properties under the Dixon model as well as the Tiku model. Besides, this statistic is shown to be very versatile: (1) it can be used for testing outliers in a sample from any distribution of the type $(1/\sigma)f((x - \mu)/\sigma)$, (2) it can be used to test any specified number of outliers on either side of the sample, and (3) it can be used to test whether the sample contains outliers, irrespective of how many. Numerical examples are given to illustrate the usefulness of the outlier detection procedures.

EXERCISES

7.1 The observations (Tiku and Singh, 1981c)

-0.038 -0.211 0.604 -1.59 -0.597 4.424 1.67 0.669 0.062

1.368 0.220 1.449 -0.841 1.714 0.932 -0.290 -0.395 0.786

-0.573 -1.003

are supposed to constitute a random sample from a normal distribution. The largest observation, 4.424, is, however, a suspect.

(a) Test whether this observation is in fact an outlier.
(b) Censor this observation and calculate the MMLE of the population mean from the remaining observations. Show that this estimate has considerably smaller standard error than the sample mean.
(c) Calculate the Huber estimates w24 and BS82 and their standard errors. How do they compare with the MMLE?

7.2 The following are the order statistics of a random sample (Hawkins, 1980, p. 117) supposed to come from an exponential distribution $E(0, \sigma)$:

0.13 0.27 0.43 0.63 0.88 1.22 1.72 2.72 6.0 7.0

(a) Test whether the two largest observations are outliers.
(b) Censor these two observations and calculate the MMLE from the remaining observations from (3.4.15). Calculate the standard error of this estimate and compare it with the standard error of the traditional estimate, the sample mean. Is the MMLE more efficient?

Special Appendix C
Graphical Display of Data

We have already given examples in Chapters 4 and 5 to illustrate the usefulness of robust procedures. We have shown that these procedures are particularly useful if the underlying distribution has long tails or the sample contains outliers. For using these procedures, therefore, one might like to examine a given data set for an indication of long-tailedness or outliers. The formal procedures developed in Chapters 6 and 7 can of course be used for that purpose. These procedures, however, might be construed as being too time consuming in certain practical situations. For a quick overall assessment of a data set, however, plotting techniques can be used, for example, a notched box-and-whisker plot (Kafadar, 1985, pp. 367-370), which is a graphical display of univariate data and imparts a quick overall assessment of the distribution (symmetry, skewness, outliers). However, to assess normality vis-à-vis long-tailedness or presence of outliers, a simpler method is to plot the order statistics of a sample against the expected values of normal order statistics. This plot is usually called a Q-Q plot. The expected values of normal order statistics are readily available (Owen, 1962; Pearson and Hartley, 1972). Under normality, one would expect the data points to lie on a straight line. A noticeable departure of the end points from a straight line is an indication of long-tailedness or outliers. To illustrate this, we give a few examples.

EXAMPLE C.1 Consider the data of Example 2.7.1; for example,

x: 49 -67 8 16 6 23 28 41 14 29 56 24 75 60 -48

Here the sample size n = 15. The order statistics of this sample are

X: -67 -48 6 8 14 16 23 24 28 29 41 49 56 60 75

The expected values of the order statistics of a random sample of size n = 15 from a normal distribution N(0, 1) are (Pearson and Hartley, 1972, Table 9)

Z: -1.74 -1.25 -1.95 -0.71 -0.52 -0.34 -0.17 0.0 0.17 0.34

0.52 0.71 0.95 1.25 1.74

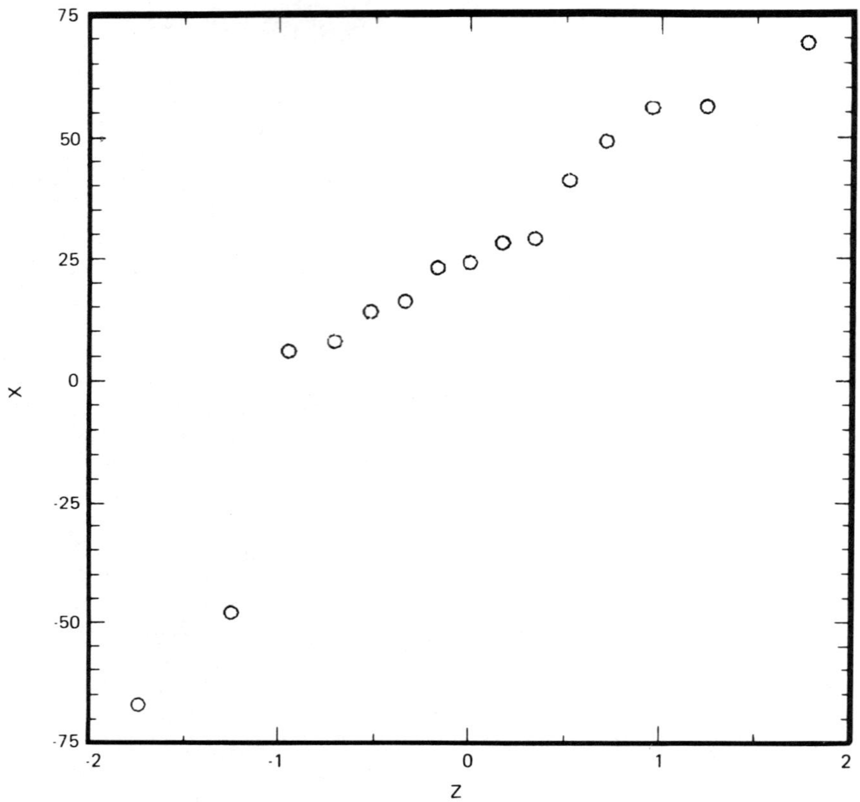

Figure C.1

Figure C.1 illustrates a Q-Q plot constructed from the data above. It is clear that all the points, other than the two extreme points on the left and perhaps one extreme point on the right, form almost a straight line. This indicates the presence of outliers. In such situations, robust procedures have a clear advantage over classical procedures based on sample mean and variance. For example, the classical and the MML10 estimates of the population mean are:

$$\text{Classical:}\quad \bar{x} = 20.933 \text{ with standard error } \pm \frac{s}{\sqrt{n}} = \pm \frac{37.744}{\sqrt{15}} = \pm 9.745$$

$$\text{MML10:}\quad \hat{\mu} = 27.713 \text{ with standard error } \pm \frac{\hat{\sigma}}{\sqrt{m}} = \pm \frac{25.498}{\sqrt{14.3}} = \pm 6.743$$

The sample mean has considerably larger standard error than the robust estimator $\hat{\mu}$.

EXAMPLE C.2 The following data represent the examination scores of 43 students in an introductory statistics course (Lindeman et al., 1980, pp. 4-5):

y: 32 24 34 40 37 24 12 31 28 38 27 33 33 36 13 39 20 38
32 24 28 21 28 38 23 30 23 30 19 33 24 34 27 32 21 38
36 30 29 24 33 21 29

Here the sample size n = 43. The order statistics of this sample are

Y: 12 13 19 20 21 21 21 23 23 24 24 24 24 24 27 27 28 28
28 29 29 30 30 30 31 32 32 32 33 33 33 33 34 34 36 36
37 38 38 38 38 30 40

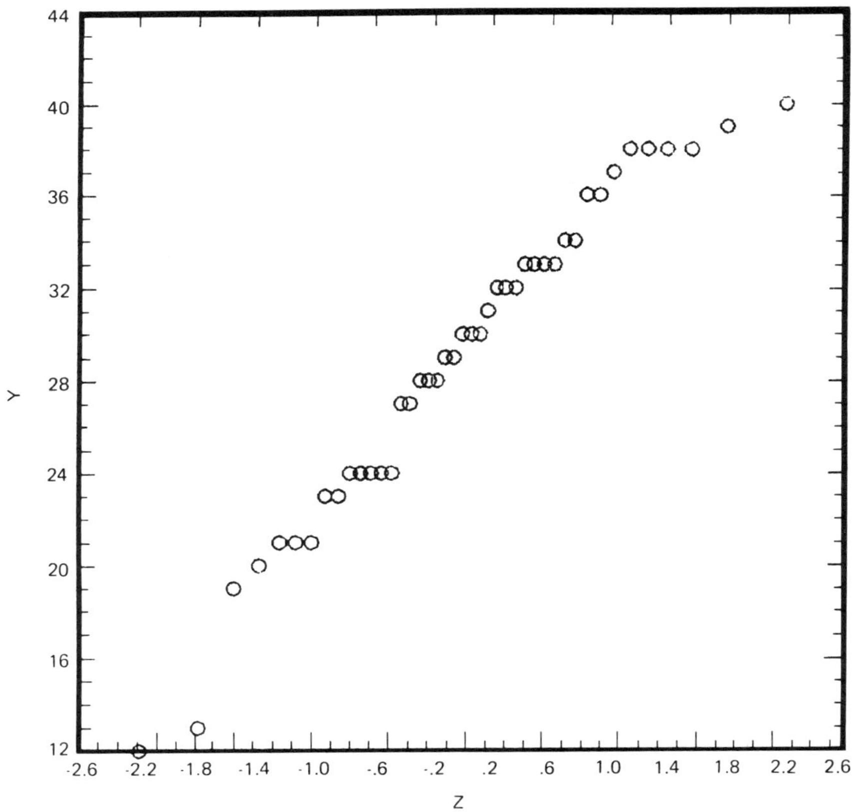

Figure C.2

The expected values of order statistics of a random sample of size n = 43 from a normal distribution N(0, 1) are (Pearson and Hartley, 1972, Table 9)

Z: -2.19 -1.79 -1.55 -1.38 -1.25 -1.13 -1.02 -0.93 -0.84
-0.76 -0.69 -0.62 -0.55 -0.48 -0.42 -0.36 -0.29 -0.23
-0.17 -0.12 -0.06 0.00 0.06 0.12 0.17 0.23 0.29 0.36
0.42 0.48 0.55 0.62 0.69 0.76 0.84 0.93 1.02 1.13 1.25
1.38 1.55 1.79 2.19

Figure C.2 illustrates a Q-Q plot constructed from the data above. No serious departure from normality is indicated since all the data points lie close to a straight line. In such situations, the use of a classical procedure is preferred to a robust procedure; the latter is expected to be slightly less efficient. For example, the classical and the MML10 estimates of the population mean are:

Classical: $\bar{y} = 28.98$ with standard error $\pm \frac{s}{\sqrt{n}} = \pm \frac{6.868}{\sqrt{43}} = \pm 1.047$

MML10: $\hat{\mu} = 29.37$ with standard error $\pm \frac{\hat{\sigma}}{\sqrt{m}} = \pm \frac{7.017}{\sqrt{41.7104}} = \pm 1.086$

The robust estimator $\hat{\mu}$ has a slightly larger standard error than the sample mean $\bar{y}$, as expected.

EXAMPLE C.3 Consider the data of Example 5.7.1, for example, the pollution levels of water samples from two lakes at randomly chosen locations:

x_1: -1.48 1.25 -0.51 0.46 0.60 -4.27 0.63 -0.14 -0.38 1.28
0.93 0.51 1.11 -0.17 -0.79 -1.02 -0.91 0.10 0.41 1.11

x_2: 1.32 1.81 -0.54 2.68 2.27 2.70 0.78 -4.62 1.88 0.86
2.86 0.47 -0.42 0.16 0.69 0.78 1.72 1.57 2.14 1.62

Suppose that one wants to test whether the average pollution levels of the two lakes are equal. Here the sample size n = 20. The expected values of order statistics of a random sample of size n = 20 from a normal distribution N(0, 1) are

Z: -1.87 -1.41 -1.13 -0.92 -0.74 -0.59 -0.45 -0.31 -0.19
-0.06 0.06 0.19 0.31 0.45 0.59 0.74 0.92 1.13 1.41 1.87

Figure C.3 illustrates a Q-Q plot constructed from the data above; the ordered x_1 points are represented by circles and the ordered x_2 points by triangles. The indication is that both the samples contain outliers, and therefore the use of a robust method is recommended in preference to a classical method based on sample means and variances. For the MML10 method

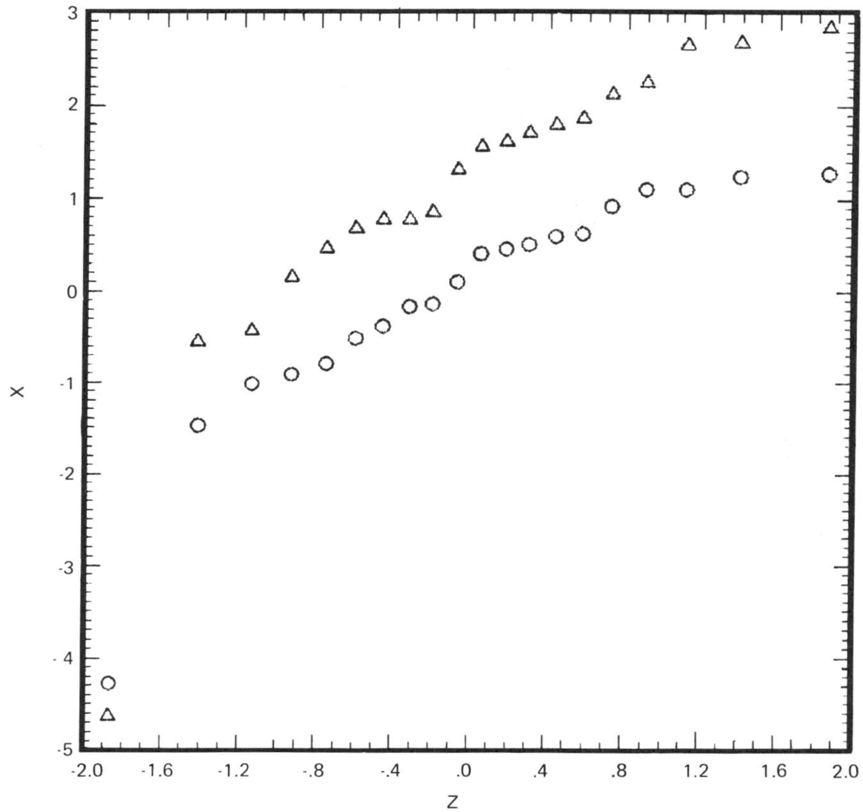

Figure C.3

$$\hat{\mu}_1 = 0.1080 \quad \hat{\sigma}_1 = 0.9664 \quad \hat{\mu}_2 = 1.2462 \quad \hat{\sigma}_2 = 1.2613$$

$$\hat{\sigma} = \sqrt{\frac{15(0.9664)^2 + 15(1.2613)^2}{30}} = 1.1236 \text{ (pooled estimate)}$$

$$|T| = \left| \frac{0.1080 - 1.2462}{1.1236\sqrt{2/19.3604}} \right| = 3.15 \text{ [equation (4.3.1)]}$$

Since the upper 0.5% point of Student's t distribution with 30 degrees of freedom is 2.75, the |T| value of 3.15 above is significantly large at the 1% significance level. One would therefore conclude that the two lakes have different pollution levels.

If one uses the classical method, the t value is

$$|t| = \left|\frac{\bar{x}_1 - \bar{x}_2}{s\sqrt{2/n}}\right| = \frac{0.971}{1.4798\sqrt{2/20}} = 2.07$$

Since the upper 0.5% point of Student's t distribution with 38 degrees of freedom is 2.72, the $|t|$ value of 2.07 above is not significant at the 1% significance level. One would therefore conclude that the two lakes have the same average pollution level. This conclusion is clearly erroneous (see Figure C.3).

EXAMPLE C.4 Consider the following data, which are similar to the data given by Kambo and Awad (1985, Example 1). The ordered data are:

Sample 1: 0.020 0.100 0.142 0.156 0.157 0.158 0.160 0.292
0.315 0.633 0.784 0.820 0.903 0.959 1.001 1.016
1.206 1.329 2.988 3.098

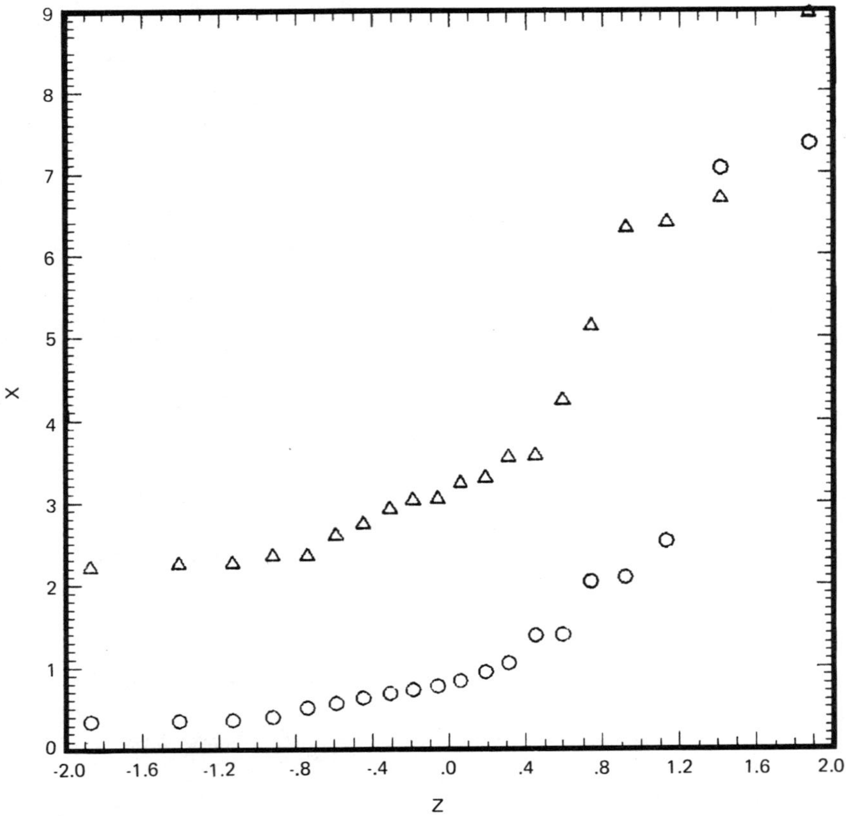

Figure C.4

Sample 2:	2.089	2.148	2.162	2.221	2.352	2.437	2.451	2.522
	2.569	2.631	2.836	2.928	3.074	3.364	3.536	3.715
	3.716	3.955	4.999	6.239				

Figure C.4 illustrates a Q-Q plot constructed from the data above; the data points in the first sample are represented by circles and the data points in the second sample by triangles. It is clear that in both samples quite a few data points lie away from a straight line only on the right-hand side. This indicates positively skew distributions of the extreme type (e.g., exponential, log-normal, etc.). For testing equality of location parameters of the two distributions, the robust method developed in Section 4.4 is therefore recommended in preference to the Student two-sample t test based on sample means and variances.

References

Abe, S. (1971). Simplified linear unbiased estimators for location and scale parameters in doubly censored samples. Rep. Statist. Appl. Res. JUSE 18, 43-55, 83-96.

Aho, M., Bain, L. J., and Engelhardt, M. (1983). Goodness-of-fit tests for the Weibull distribution with unknown parameters and censored sampling. J. Statist. Comp. Simul. 18, 59-69.

Aitchison, J., and Brown, J. A. C. (1957). The Lognormal Distribution. Cambridge: Cambridge University Press.

Ali, M. M., and Chan, L. K. (1964). On Gupta's estimates of the parameters of the normal distribution. Biometrika 51, 498-501.

Anderson, T. W. (1951). Classification by multivariate analysis. Psychometrika 16, 31-50.

Anderson, T. W. (1966). Some nonparametric multivariate procedures based on statistically equivalent blocks. Proc. International Symposium on Multivariate Analysis. New York: Academic

Anderson, T. W., and Darling, D. A. (1952). Asymptotic theory of certain goodness-of-fit criteria based on stochastic processes. Ann. Math. Statist. 23, 193-212.

Andrews, D. F. (1974). A robust method for multiple linear regression. Technometrics 16, 523-531.

Andrews, D., Bickel, P. J., Hampel, F., Huber, P., Rogers, W., and Tukey, J. W. (1972). Robust Estimates of Location: Survey and Advances. Princeton, N.J.: Princeton University Press.

Ariyawansa, K. A., and Templeton, J. G. C. (1984). Structural inference on the parameter of the Rayleigh distribution from doubly censored samples. Statist. Hefte 25, 181-199.

Atkinson, A. C. (1969). A test for discriminating between models. Biometrika 56, 337-347.

Atkinson, A. C. (1970). A method for discriminating between models (with discussion). J. R. Statist. Soc. B32, 323-353.

Atkinson, A. C. (1973). Testing transformations to normality. J. R. Statist. Soc. B35, 473-479.

Atkinson, A. C. (1982). Robust and diagnostic regression analyses. Commun. Statist.—Theor. Meth. 11(22), 2559-2571.

Atkinson, A. C., and Cox, D. R. (1977). Robust regression via discriminant analysis. Biometrika 64, 15-19.

Bain, L. J. (1972). Inferences based on censored sampling from the Weibull or extreme value distribution. Technometrics 14, 693-702.

Bain, L. J. (1978). Statistical Analysis of Reliability and Life-Testing Models. New York: Marcel Dekker.

Bain, L. J., and Engelhardt, M. (1980). Probability of correct selection of Weibull versus gamma based on likelihood ratio. Commun. Statist., A9, 375-381.

Bain, L. J., and Weeks, D. L. (1964). A note on the truncated exponential distribution. Ann. Math. Statist. 35, 1366-1367.

Balakrishnan, N. (1983). Empirical power study of a multi-sample test of exponentiality based on spacings. J. Statist. Comp. Simul. 18, 265-271.

Balakrishnan, N. (1984). Order statistics from the half logistic distribution. J. Statist. Comp. Simul. 20, 287-309.

Balakrishnan, N., and Joshi, P. C. (1981). Moments of order statistics from doubly truncated power function distribution. Aligarh J. Statist. 1, 98-105.

Balakrishnan, N., and Joshi, P. C. (1982). Moments of order statistics from doubly truncated Pareto distribution. J. Ind. Statist. Ass. 20, 109-117.

Balakrishnan, N., and Joshi, P. C. (1984). Product moments of order statistics from the doubly truncated exponential distribution. Naval Res. Logist. Quart. 31, 27-31.

Balakrishnan, N., and Kocherlakota, S. (1985a). On the double Weibull distribution: order statistics and estimation. Sankhya B. (in press).

Balakrishnan, N., and Kocherlakota, S. (1985b). On the moments of order statistics from doubly truncated logistic distribution. J. Statist. Plann. Inf., (1986), 13, 117-129.

Balakrishnan, N., and Malik, H. J. (1985a). Order statistics from the log-logistic distribution. Technical Report, Dept. of Mathematics, Univ. of Guelph.

Balakrishnan, N., and Malik, H. J. (1985b). Some general identities involving order statistics. Commun. Statist.—Theor. Meth. 14(2), 333-339.

Balakrishnan, N., Tiku, M. L., and Shaarawi, A. H. El. (1985). Robust univariate two-way classification. Biom. J. 27, 123-138.

Barlow, R. E., and Proschan, F. (1975). Statistical Theory of Reliability and Life Testing. New York: Holt, Rinehart and Winston.

Barnett, F. C., Mullen, K., and Saw, J. G. (1967). Linear estimates of a population scale parameter. Biometrika 54, 551-554.

Barnett, V. D. (1966). Order statistics estimators of the location of the Cauchy distribution. J. Amer. Statist. Ass. 61, 1205-1218. Correction 63, 383-385.

Barnett, V. (1975). Probability plotting methods and order statistics. Appl. Statist. 24, 95-108.

Barnett, V. (1976). Convenient probability plotting positions for the normal distribution. Appl. Statist. 25, 47-50.

Barnett, V. (1978). The study of outliers: purpose and model. Appl. Statist. 27, 242-250.

Barnett, V., and Lewis, T. (1978). Outliers in Statistical Data. Chichester, England: Wiley.

Barr, D. R., and Davidson, T. (1973). A Kolmogorov-Smirnov test for censored samples. Technometrics 15, 739-757.

Bartholomew, D. J. (1957). A problem in life testing. J. Amer. Statist. Ass. 52, 350-355.

Bartholomew, D. J. (1963). The sampling distribution of an estimate arising in life testing. Technometrics 5, 361-374.

Bartlett, M. S. (1953). Approximate confidence intervals. Biometrika 40, 12-19.

Barton, D. E., and Dennis, K. E. (1952). The condition under which Gram-Charlier and Edgeworth curves are positive definite and unimodal. Biometrika 39, 425-427.

Barton, D. E., David, F. N., and Merrington, M. (1966). Table for the solution of the exponential equation, exp (b) - b/(1 - p) = 1. Biometrika 50, 169-176.

Basu, A. P. (1965). On some tests of hypotheses relating to the exponential distribution when some outliers are present. J. Amer. Statist. Ass. 60, 548-559.

Basu, A. P., and Ghosh, J. K. (1980). Asymptotic properties of a solution to the likelihood equation with life-testing applications. J. Amer. Statist. Ass. 75, 410-414.

Beaton, A. E., and Tukey, J. W. (1974). The fitting of power series, meaning polynomials, illustrated on band-spectroscopic data. Technometrics 16, 147-186.

Belsley, D. A., Kuh, E., and Welsch, R. E. (1980). Regression Diagnostics: Identifying Influential Data and Sources of Collinearity. New York: Wiley.

Benjamini, Y. (1983). Is the t test really conservative when parent population is long-tailed? J. Amer. Statist. Ass. 78, 645-654.

Bennett, C. A. (1952). Asymptotic properties of ideal linear estimators. Doctoral dissertation, University of Michigan.

Berkson, J. (1951). Why I prefer logits to probits. Biometrics 7, 327-339.

Beyer, J. N., Moore, A. H., and Harter, H. L. (1976). Estimation of the standard deviation of the logistic distribution by the use of selected order statistics. Technometrics 18, 313-332.

Bickel, P. J. (1973). Analogues of linear combinations of order statistics in the general linear model. Ann. Statist. 1, 597-616.

Bickel, P. J. (1975). One-step Huber estimates in the linear model. J. Amer. Statist. Ass. 70, 428-434.

Bickel, P. J. (1976). Another look at robustness: a review of reviews and some new developments. Scand. J. Statist. 3, 145-168.

Bickel, P. J. (1978). Using residuals robustly: I. Tests for heteroscedasticity, nonlinearity. Ann. Statist. 6, 266-291.

Bickel, P. J., and Doksum, K. A. (1981). An analysis of transformation revisited. J. Amer. Statist. Ass. 76, 296-311.

Billman, B., Antle, C., and Bain, L. J. (1972). Statistical inferences from censored Weibull samples. Technometrics 14, 831-840.

Binkley, D. A., and Birch, J. B. (1984). Estimation of regression coefficients under model misspecification and non-normality of errors. J. Statist. Comp. Simul. 19, 155-170.

Birch, J. B., and Myers, R. H. (1982). Robust analysis of covariance. Biometrics 38, 699-713.

Birnbaum, A., and Laska, E. (1967). Optimal robustness: a general method, with applications to linear estimators of location. J. Amer. Statist. Ass. 62, 1230-1240.

Birnbaum, A., and Mike, V. (1970). Asymptotically robust estimators of location. J. Amer. Statist. Ass. 65, 1265-1282.

Birnbaum, A., Laska, E., and Meisner, M. (1971). Optimally robust linear estimators of location. J. Amer. Statist. Ass. 66, 302-310.

Bliss, C. I. (1934). The method of probits. Science 79, 38-39.

Blom, G. (1958). Statistical Estimates and Transformed Beta Variables. New York: Wiley.

Blom, G. (1962). Nearly best linear estimates of location and scale parameters. In Contributions to Order Statistics (A. E. Sarhan and B. G. Greenberg, eds.), 34-46. New York: Wiley.

Bondesson, L. (1976). When is the sample mean BLUE? Scand. J. Statist. 3, 116-120.

Boos, D. D. (1980). A new method for constructing approximate confidence intervals from M-estimates. J. Amer. Statist. Ass. 75, 142-145.

Bowman, K. O., and Shenton, L. R. (1973a). Notes on the distribution of b in sampling from Pearson distributions. Biometrika 60, 155-167.

Bowman, K. O., and Shenton, L. R. (1973b). Space of solutions for a normal mixture. Biometrika 60, 629-636.

Bowman, K. O., and Shenton, L. R. (1975). Omnibus test contours for departures from normality based on b_1 and b_2. Biometrika 62, 243-250.

Box, G. E. P. (1953). Non-normality and tests on variances. Biometrika 40, 318-335.

Box, G. E. P., and Cox, D. R. (1964). An analysis of transformations (with discussion). J. R. Statist. Soc. B26, 211-252.

Box, G. E. P., and Cox, D. R. (1982). An analysis of transformations revisited, rebutted. J. Amer. Statist. Ass. 77, 209-210.

Box, G. E. P., and Tiao, G. C. (1964a). A Bayesian approach to the importance of assumptions applied to the comparison of variances. Biometrika 51, 153-167.

Box, G. E. P., and Tiao, G. C. (1964b). A note on criterion robustness and inference robustness. Biometrika 51, 169-173.

Box, G. E. P., and Tiao, G. C. (1973). Bayesian Inference in Statistical Analysis. Reading, Mass.: Addison-Wesley.

Bradley, J. V. (1980). Nonrobustness in Z, t and F tests at large sample sizes. Bull. Psychonomic Soc. 16, 333-336.

Brown, G. R., and Forsythe, A. D. (1974). Robust test for the equality of variances. J. Amer. Statist. Ass. 69, 364-367.

Brown, M. B. (1975). Exploring interaction effects in the ANOVA. J. R. Statist. Soc. C24, 288-298.

Brownlee, K. A. (1965). Statistical Theory and Methodology in Science and Engineering, 2nd ed. New York: Wiley.

Canfield, R. V., and Borgman, L. E. (1975). Some distributions of time to to failure for reliability applications. Technometrics 17, 263-268.

Carroll, R. J. (1979). On estimating variance of robust estimators when the errors are asymmetric. J. Amer. Statist. Ass. 74, 674-679.

Carroll, R. J. (1980). A robust method for testing transformations to achieve approximate normality. J. R. Statist. Soc. B42, 71-78.

Carroll, R. J. (1982). Tests for regression parameters in power transformation models. Scand. J. Statist. 9, 217-222.

Carroll, R. J., and Gallo, P. P. (1982). Some aspects of robustness in the functional error-in-variables regression model. Commun. Statist.—Theor. Meth. 11(22), 2573-2585.

Chan, L. K. (1967a). Linear estimation of the location and scale parameters from type II censored samples from symmetric unimodal distributions. Naval Res. Logist. Quart. 14, 135-145.

Chan, L. K. (1967b). Remark on the linearized maximum likelihood estimate. Ann. Math. Statist. 38, 1876-1881.

Chan, L. K. (1969). Linear quantile estimates of the location and scale parameters of the logistic distribution. Statist. Hefte 10, 277-282.

Chan, L. K. (1970a). On the best linear estimates of the mean or standard deviation of the normal distribution based on order statistics. J. Statist. Res. 4, 71-92.

Chan, L. K. (1970b). Linear estimation of the location and scale parameters of the Cauchy distribution based on sample quantiles. J. Amer. Statist. Ass. 65, 851-859.

Chan, L. K. (1971). Some asymptotic properties of the linearized maximum likelihood estimate and best linear unbiased estimate. Ann. Inst. Statist. Math. 23, 225-232.

Chan, L. K. (1974). On the asymptotic normality of some unbiased estimates. Scand. Actuar. J. 1974, 151-156.

Chan, L. K., Chan, N. N., and Mead, E. R. (1971). Best linear unbiased estimates of the parameters of the logistic distribution based on selected order statistics. J. Amer. Statist. Ass. 66, 889-892.

Chernoff, H., and Gastwirth, J. L. (1967). Asymptotic distribution of linear combinations of functions of order statistics with applications to estimation. Ann. Math. Statist. 38, 52-72.

Chernoff, H., and Lieberman, G. J. (1954). Use of normal probability paper. J. Amer. Statist. Ass. 49, 778-785.

Chernoff, H., and Lieberman, G. J. (1956). The use of generalized probability paper for continuous distributions. Ann. Math. Statist. 27, 806-818.

Cleveland, W. S. (1984). Graphical methods for data presentation. Amer. Statist. 38, 270-280.

Cochran, W. G. (1938). Some difficulties in the statistical analysis of replicated experiments. Empire J. Experimental Agri. 6, 157-163.

Cochran, W. G. (1946). Relative accuracy of systematic and stratified random samples for a certain class of populations. Ann. Math. Statist. 17, 164-177.

Cochran, W. G. (1968). Commentary on estimation of error rates in discriminant analysis. Technometrics 10, 204-205.

Cohen, A. C. (1951). Estimating parameters of log-normal distribution by maximum likelihood. J. Amer. Statist. Ass. 46, 206-212.

Cohen, A. C. (1955). Censored samples from truncated normal distributions. Biometrika 42, 516-519.

Cohen, A. C. (1957). On the solution of estimating equations from truncated and censored samples from normal populations. Biometrika 44, 225-236.

Cohen, A. C. (1959). Simplified estimators for the normal distribution when samples are singly censored or truncated. Technometrics 1, 217-237.

Cohen, A. C. (1961). Tables for maximum likelihood estimates; singly censored samples. Technometrics 3, 535-541.

Cohen, A. C. (1963). Progressively censored samples in life testing. Technometrics 5, 327-339.

Cohen, A. C., and Whitten, B. (1980). Estimation in the three-parameter log-normal distribution. J. Amer. Statist. Ass. 75, 399-404.

Cohen, A. C., and Whitten, B. (1982). Modified maximum likelihood and modified moment estimators for the three-parameter Weibull distribution. Commun. Statist.—Theor. Meth. 11(23), 2631-2656.

Cohen, A. C., and Woodward, J. (1953). Tables of Pearson-Lee-Fisher functions of singly truncated normal distributions. Biometrics 9, 489-497.

Conover, W. J., Johnson, M. E., and Johnson, M. M. (1981). A comparative study of tests for homogeneity of variances, with applications to the outer continental shelf bidding data. Technometrics 23, 351-362.

Cook, D. R., and Weisberg, S. (1982). Residuals and Influence in Regression. London: Chapman & Hall.

Cox, D. R. (1959). The analysis of exponentially distributed lifetimes with two types of failure. J. R. Statist. Soc. B21, 411-421.

Cox, D. R. (1961). Tests of separate families of hypotheses. Proc. Fourth Berkeley Symp. 1, 105-123.

Cox, D. R. (1962). Further results on tests of separate families of hypotheses. J. R. Statist. Soc. B24, 406-423.

Cox, D. R., and Lewis, P. A. W. (1966). The Statistical Analysis of Series of Events. London: Methuen.

Cressie, N. A. C. (1980a). Relaxing assumptions in the one-sample t-test. Austral. J. Statist. 22, 143-153.

Cressie, N. A. C. (1980b). M-estimation in the presence of unequal scale. Statist. Neerland. 34, 19-32.

Csörgö, M., Seshadri, V., and Yalovsky, M. (1973). Some exact tests for normality in the presence of unknown parameters. J. R. Statist. Soc. B35, 507-522.

Csörgö, M., Seshadri, V., and Yalovsky, M. (1975). Applications of characterizations in the area of goodness-of-fit. In Model Building and Model Selection, Vol. 2: Statistical Distributions in Scientific Work (G. P. Patil, S. Kotz, and J. K. Ord, eds.), 79-90. Boston: D. Reidel.

Cushney, A. R., and Peebles, A. R. (1905). The action of optimal isomers. II. Hyocines. J. Physiology 32, 501-510.

D'Agostino, R. B. (1970). Transformations to normality of the null distribution of g_1. Biometrika 57, 679-680.

D'Agostino, R. B. (1971a). Linear estimation of the Weibull parameters. Technometrics 13, 171-182.

D'Agostino, R. B. (1971b). An omnibus test of normality for moderate and large sample sizes. Biometrika 58, 341-348.

D'Agostino, R. B., and Austin, F. S. (1977). Robustness of location estimators under changes of population kurtosis. J. Amer. Statist. Ass. 72, 393-396.

D'Agostino, R. B., and Cureton, E. E. (1973). A class of simple linear estimators of the standard deviation of the normal distribution. J. Amer. Statist. Ass. 68, 207-210.

D'Agostino, R. B., and Lee, A. F. S. (1976). Linear estimation of the logistic parameters for complete or tail-censored samples. J. Amer. Statist. Ass. 71, 462-464.

D'Agostino, R. B., and Pearson, E. S. (1973). Tests for departure from normality. Empirical results for the distributions of b_2 and b_1. Biometrika 60, 613-622.

D'Agostino, R. B., and Rosman, B. (1974). The power of Geary's test of normality. Biometrika 61, 181-184.

D'Agostino, R. B., and Tietjen, G. L. (1971). Simulation probability points of b_2 for small samples. Biometrika 58, 669-672.

D'Agostino, R. B., and Tietjen, G. L. (1973). Approaches to the null distribution of b_1. Biometrika 60, 169-173.

Daniel, C., and Wood, F. S. (1971). Fitting Equations to Data. New York: Wiley.

Das Gupta, S. (1964). Nonparametric classification rules. Sankhya A26, 25-30.

David, F. N., and Johnson, N. L. (1951). The effect of non-normality on the power function of the F-test in the analysis of variance. Biometrika 38, 43-57.

David, F. N., and Johnson, N. L. (1954). Statistical treatment of censored data, I. Fundamental formulae. Biometrika 41, 228-240.

David, H. A. (1981). Order Statistics, 2nd ed. New York: Wiley.

David, H. A., Hartley, H. O., and Pearson, E. S. (1954). The distribution of the ratio, in a single normal sample, of range to standard deviation. Biometrika 41, 482-493.

David, H. A., Kennedy, W. J., and Knight, R. D. (1977). Means, variances, and covariance of normal order statistics in the presence of an outlier. In Selected Tables in Mathematical Statistics, Vol. 5, 75-204. Providence, R.I.: American Mathematical Society.

Davis, A. W. (1976). Statistical distributions in univariate and multivariate Edgeworth populations. Biometrika 63, 661-670.

Davis, D. J. (1952). An analysis of some failure data. J. Amer. Statist. Ass. 47, 113-150.

Deemer, W. L., and Votaw, D. F. (1955). Estimation of parameters of truncated or censored experimental distributions. Ann. Math. Statist. 26, 498-504.

Dempster, A. P., and Rosner, B. (1971). Detection of outliers. In Statistical Decision Theory and Related Topics. New York: Academic Press.

Dixon, W. J. (1950). Analysis of extreme values. Ann. Math. Statist. 21, 488-506.

Dixon, W. J. (1953). Processing data for outliers. Biometrics 9, 74-89.

Dixon, W. J. (1960). Simplified estimation from censored normal samples. Ann. Math. Statist. 31, 385-391.

Dixon, W. J., and Tukey, J. W. (1968). Approximate behavior of the distribution of Winsorized t (trimming/Winsorization 2). Technometrics 10, 83-98.

Doksum, K. A., and Wong, C. W. (1983). Statistical tests based on transformed data. J. Amer. Statist. Ass. 78, 411-417.

Donaldson, T. S. (1968). Robustness of the F-test to errors of both kinds and the correlation between the numerator and denominator of the F-ratio. J. Amer. Statist. Ass. 63, 660-667.

Downton, F. (1953). A note on ordered least-squares estimation. Biometrika 40, 457-458.

Downton, F. (1954). Least square estimates using ordered observations. Ann. Math. Statist. 25, 303-316.

Downton, F. (1966a). Linear estimates of the parameters in the extreme value distribution. Technometrics 8, 3-17.

Downton, F. (1966b). Linear estimates with polynomial coefficients. Biometrika 53, 129-141.

Draper, N. R., and Smith, H. (1966). Applied Regression Analysis. New York: Wiley.

Dumonceaux, R., and Antle, C. E. (1973). Discrimination between the lognormal and the Weibull distributions. Technometrics 15, 923-926.

Dumonceaux, R., Antle, C. E., and Haas, G. (1973). Likelihood ratio test for discrimination between two models with unknown location and scale parameters. Technometrics 15, 19-27.

Dunnett, C. W. (1982). Robust multiple comparisons. Commun. Statist.—Theor. Meth. 11(22), 2611-2629.

Durbin, J. (1961). Some methods of constructing exact tests. Biometrika 48, 41-55.

Durbin, J. (1973). Distribution theory for tests based on the sample distribution function. Regional Conference Series Appl. Math. No. 9.

Durbin, J. (1975). Kolmogorov-Smirnov tests when parameters are estimated with applications to tests of exponentiality and tests on spacings. Biometrika 62, 5-22.

Durbin, J., Knott, M., and Taylor, C. C. (1975). Components of Cramer-von Mises statistics: II. J. R. Statist. Soc. B37, 216-237.

Dyer, A. R. (1973). Discrimination procedures for separate families of hypotheses. J. Amer. Statist. Ass. 68, 970-974.

Dyer, A. R. (1974). Hypothesis testing procedures for separate families of hypotheses. J. Amer. Statist. Ass. 69, 140-145.

Dyer, D., and Harbin, M. S. (1981). An empirical power study of multi-sample tests for exponentiality. J. Statist. Comp. Simul. 12, 277-291.

Dyer, D. D., and Whisenand, C. W. (1973a). Best linear unbiased estimator of the parameter of the Rayleigh distribution: I. Small sample theory for censored order statistics. IEEE Trans. Reliab. R-22, 27-34.

Dyer, D. D., and Whisenand, C. W. (1973b). Best linear unbiased estimator of the parameter of the Rayleigh distribution: II. Optimum theory for selected order statistics. IEEE Trans. Reliab. R-22, 229-231.

Edgeworth, F. Y. (1887). The choice of means. Philosophical Magazine 24, 268-271.

El-Sayyad, G. M. (1967). Estimation of the parameter of an exponential distribution. J. R. Statist. Soc. B29, 525-532.

Engelhardt, M., and Bain, L. J. (1974). Some results on point estimation for the two parameter Weibull or extreme-value distribution. Technometrics 16, 49-56.

Engelhardt, M., and Bain, L. J. (1979). Prediction limits and two-sample problems with complete or censored Weibull data. Technometrics 21, 233-237.

Epstein, B. (1956). Simple estimators of the parameters of exponential distributions when samples are censored. Ann. Inst. Statist. Math. 8, 15-26.

Epstein, B. (1958). Exponential distribution and its role in life testing. Industrial Quality Control 15, 4-9.

Epstein, B. (1960a). Tests for the validity of the assumption that the underlying distribution of life is exponential: Parts I and II. Technometrics 2, 83-101, 167-183.

Epstein, B. (1960b). Estimation of the parameters of two exponential distributions from censored samples. Technometrics 2, 403-406.

Epstein, B. (1960c). Estimation from life test data. Technometrics 2, 447-454.

Epstein, B., and Sobel, M. (1953). Life-testing. J. Amer. Statist. Ass. 48, 486-502.

Epstein, B., and Sobel, M. (1954). Some theorems relevant to life testing from an exponential distribution. Ann. Math. Statist. 25, 373-381.

Epstein, B., and Tsao, C. K. (1953). Some tests based on ordered observations from two exponential populations. Ann. Math. Statist. 24, 456-466.

Ferguson, T. S. (1961). On the rejection of outliers. Proc. Fourth Berkeley Symp. I, 253-287.

Filliben, J. J. (1975). The probability plot correlation coefficient test for normality. Technometrics 17, 111-117.

Finney, D. J. (1941). On the distribution of a variate whose logarithm is normally distributed. J. R. Statist. Soc. B7, 155-161.

Fisher, R. A. (1930). Moments and product moments of sampling distributions. Proc. Lond. Math. Soc., Series 2 30, 199-238.

Fisher, R. A. (1931). The truncated normal distribution. British Association for the Advancement of Science, Math. Tables 5, 33-34, London.

Fisher, R. A. (1966). The Design of Experiments, 8th ed. Edinburgh: Oliver & Boyd.

Fisher, R. A. (1970). Statistical Methods for Research Workers, 14th ed. Edinburgh: Oliver & Boyd.

Fisk, P. R. (1961). The graduation of income distributions. Econometrica 29, 171-185.

Fix, E., and Hodges, J. L., Jr. (1951). Discriminatory analysis: Parametric discrimination; consistency properties. USAF School of Aviation Medicine, Report No. 4.

Fleming, T. R., and Harrington, D. P. (1980). A class of hypothesis tests for one and two sample censored survival data. School of Engineering and Applied Sciences, University of Virginia. DAMACS Tech. Rep. No. 80-9.

Fligner, M. A., and Killeen, T. J. (1976). Distribution-free two-sample tests for scale. J. Amer. Statist. Ass. 71, 210-213.

Forsythe, A. B. (1972). Robust estimation of straight line regression coefficients by minimizing the p-th power deviations. Technometrics 14, 159-166.

Fraser, D. A. S. (1976). Necessary analysis and adaptive inference (with discussion). J. Amer. Statist. Ass. 71, 99-113.

Fuller, W. A. (1970). Simple estimators for the mean of skewed populations. Tech. Rep.: Iowa State University.

Gaddum, J. H. (1933). Reports on biological standards: II. Methods on biological assay depending on a quantal response. Special Report Series, Medical Research Council, No. 183, London.

Gail, M. H., and Gastwirth, J. L. (1978). A scale-free goodness of fit test for the exponential distribution based on the Gini statistic. J. R. Statist. Soc. B40, 350-357.

Gartside, P. S. (1972). A study of methods for comparing several variances. J. Amer. Statist. Ass. 67, 342-346.

Gayen, A. K. (1949). The distribution of Student's t in random samples of any size drawn from non-normal universes. Biometrika 36, 353-369.

Gayen, A. K. (1950). The distribution of the variance-ratio on random samples of any size drawn from non-normal universes. Biometrika 37, 236-255.

Geary, R. C. (1947). Testing for normality. Biometrika 34, 209-220.

Gehan, E. A., and Thomas, D. G. (1969). The performance of some two-sample tests in small samples with and without censoring. Biometrika 56, 127-132.

Gessaman, M. P. (1970). A consistent nonparametric multivariate density estimator based on statistically equivalent blocks. Ann. Math. Statist. 41, 1344-1346.

Gibbons, J. D., and Freund, R. J. (1980). Organizations for statistical consultancy at colleges and universities. Amer. Statist. 34, 140-145.

Gibrat, R. (1930). Une loi des repartitions economiques: l'effet proportionnel. Bull. Statistique General, France 19, 469ff.

Godambe, V. P. (1955). A unified theory of sampling from finite populations. J. R. Statist. Soc. B17, 269-278.

Goldberger, A. S. (1962). Best linear unbiased prediction in the generalized linear regression model. J. Amer. Statist. Ass. 57, 369-375.

Golub, G. H., and Reinsch, C. H. (1970). Singular value decomposition and least squares solution. Numer. Math. 14, 402-420.

Gore, S. M., Jones, I. G., and Rytter, E. F. (1977). Misuse of statistical methods: a critical assessment of articles in BMJ from January to March, 1976. British Med. J. 1, 85-87.

Govindarajulu, Z. (1963). On moments of order statistics and quasi-ranges from normal populations. Ann. Math. Statist. 34, 633-651.

Govindarajulu, Z. (1964). A supplement to Mendenhall's bibliography on life testing and related topics. J. Amer. Statist. Ass. 59, 1231-1291.

Govindarajulu, Z. (1966). Best linear estimates under symmetric censoring of the parameters of a double exponential population. J. Amer. Statist. Ass. 61, 248-258.

Govindarajulu, Z. (1968). Certain general properties of unbiased estimates of location and scale parameters based on ordered observations. SIAM J. Appl. Math. 16, 533-551.

Govindarajulu, Z., and Eisenstat, S. (1965). Best estimates of location and scale parameters of a chi (1 d.f.) distribution using ordered observations. Rep. Statist. Appl. Res. JUSE 12, 149-164.

Govindarajulu, Z., and Joshi, M. (1968). Best linear unbiased estimation of the location and scale parameters of Weibull distribution using ordered observations. Rep. Statist. Appl. Res. JUSE 15, 57-70.

Gross, A. M. (1976). Confidence interval robustness with long-tailed symmetric distributions. J. Amer. Statist. Ass. 71, 409-416.

Gross, A. M. (1977). Confidence intervals for bisquare regression estimates. J. Amer. Statist. Ass. 72, 341-354.

Grubbs, F. E. (1950). Sample criteria for testing outlying observations. Ann. Math. Statist. 21, 27-58.

Grubbs, F. E. (1969). Procedures for detecting outlying observations. Technometrics 11, 1-21.

Grundy, P. M. (1951). The expected frequencies in a sample of an animal population in which the abundances of species are log-normally distributed, Part I. Biometrika 38, 427-434.

Gupta, A. K. (1952). Estimation of the mean and standard deviation of a normal population from a censored sample. Biometrika 39, 260-273.

Gupta, A. K., and Kim, B. K. (1978). On a distribution-free discriminant analysis. Biom. J. 20, 729-736.

Gupta, S. S., Qureishi, A. S., and Shah, B. K. (1967). Best linear unbiased estimators of the parameters of the logistic distribution using order statistics. Technometrics 9, 43-56.

Hajek, J., and Sidak, Z. (1967). Theory of Rank Tests. New York: Academic Press.

Hall, D. L. (1980). On the limiting behaviour of the mode and median of a sum of independent random variables. Ann. Prob. 8, 419-430.

Halperin, M. (1952). Maximum likelihood estimation in truncated samples. Ann. Math. Statist. 23, 226-238.

Hamouda, E. M., and Leone, F. C. (1974). The O-BLUE estimators for complete and censored samples in linear regression. Technometrics 16, 441-446.

Hamouda, E. M., and Leone, F. C. (1977). The robustness of efficiency of adjusted trimmed estimators in linear regression. Technometrics 19, 19-34.

Hampel, F. R. (1971). A general qualitative definition of robustness. Ann. Math. Statist. 42, 1887-1896.

Hampel, F. R. (1973). Robust estimation: a condensed partial survey. Z. Wahr. verw. Gebiete 27, 87-104.

Hampel, F. R. (1974). The influence curve and its role in robust estimation. J. Amer. Statist. Ass. 62, 1179-1186.

Harter, H. L. (1964). Exact confidence bounds, based on one order statistic, for the parameter of an exponential population. Technometrics 6, 301-317.

Harter, H. L. (1969). Order Statistics and Their Use in Testing and Estimation. Washington, D.C.: U.S. Government Printing Office.

Harter, H. L., and Moore, A. H. (1966). Local maximum likelihood estimation of the parameters of three-parameter log-normal populations from complete and censored samples. J. Amer. Statist. Ass. 61, 842-851.

Harter, H. L., and Moore, A. H. (1967). Maximum likelihood estimation from censored samples of the parameters of a logistic distribution. J. Amer. Statist. Ass. 62, 675-684.

Harter, H. L., and Moore, A. H. (1968). Maximum likelihood estimation, from doubly censored samples, of the parameters of the first asymptotic distribution of extreme value. J. Amer. Statist. Ass. 63, 889-901.

Harter, H. L., and Moore, A. H. (1976). An evaluation of exponential and Weibull test plans. IEEE Trans. Reliab. R25, 100-104.

Harter, H. L., Moore, A. H., and Curry, T. F. (1979). Adaptive robust estimation of location and scale parameters of symmetric populations. Commun. Statist.—Theor. Math. A8(15), 1473-1491.

Hatch, T. (1933). Determination of average particle size from the screen-analysis of non-uniform particles. J. Franklin Inst. 207, 369-388.

Hatch, T., and Choute, S. P. (1929). Statistical description of the size properties of non-uniform particles. J. Franklin Inst.

Hawkins, D. M. (1977). Comment on "A new statistic for testing suspected outliers." Commun. Statist. A6, 435-438.

Hawkins, D. M. (1979). Fractiles of an extended multiple outlier test. J. Statist. Comp. Simul. 8, 227-236.

Hawkins, D. M. (1980). Identification of Outliers. London: Chapman & Hall.

Hernandez, F., and Johnson, R. A. (1980). The large sample behaviour of transformations to normality. J. Amer. Statist. Ass. 75, 855-861.

Hill, B. M. (1963). The three parameter lognormal distribution and Bayesian analysis of a point source epidemic. J. Amer. Statist. Ass. 58, 72-84.

Hinkley, D. V. (1975). On power transformations to symmetry. Biometrika 62, 101-112.

Hinkley, D. V., and Runger, G. (1984). The analysis of transformed data (with comments). J. Amer. Statist. Ass. 79, 302-320.

Hoaglin, D. C., Mosteller, F., and Tukey, J. W. (1983). Understanding Robust and Exploratory Data Analysis. New York: Wiley.

Hodges, J. L., Jr., and Lehmann, E. L. (1963). Estimates of location based on rank tests. Ann. Math. Statist. 34, 598-611.

Hogg, R. V. (1967). Some observations on robust estimation. J. Amer. Statist. Ass. 62, 1179-1186.

Hogg, R. V. (1972). More light on the kurtosis and related statistics. J. Amer. Statist. Ass. 67, 422-424.

Hogg, R. V. (1974). Adaptive robust procedures: a partial review and some suggestions for future applications and theory. J. Amer. Statist. Ass. 69, 909-927.

Hogg, R. V. (1979). Statistical robustness: our view of its use in applications today. Amer. Statist. 33, 108-115.

Hogg, R. V. (1982). On adaptive statistical inferences. Commun. Statist.—Theor. Meth. 11(22), 2531-2542.

Hogg, R. V., and Craig, A. T. (1978). Introduction to Mathematical Statistics, 4th ed. New York: Macmillan.

Hogg, R. V., Fisher, D. M., and Randles, R. H. (1975). A two-sample adaptive distribution-free test. J. Amer. Statist. Ass. 70, 656-661.

Hogg, R. V., Horn, P. S., and Lenth, R. V. (1984). On adaptive estimation. J. Statist. Plann. Inf. 9, 333-343.

Hogg, R. V., Uthoff, V. A., Randles, R. H., and Davenport, A. S. (1972). On the selection of the underlying distribution and adaptive estimation. J. Amer. Statist. Ass. 67, 597-600.

Holland, P. W., and Welsch, R. E. (1977). Robust regression using iteratively reweighted least-squares. Commun. Statist. A6(9), 813-827.

Huang, J. S. (1975). A note on order statistics from Pareto distribution. Scand. Actuar. J. 1975, 187-190.

Huber, P. J. (1964). Robust estimation of a location parameter. Ann. Math. Statist. 35, 73-101.

Huber, P. J. (1968). Robust estimation. Mathematical Centre Tracts, Selected Statistical Papers 27, 3-25. Amsterdam: Mathematisch Centrum.

Huber, P. J. (1970). Studentizing robust estimates In Nonparametric Techniques in Statistical Inference (M. L. Puri, ed.). Cambridge: Cambridge University Press.

Huber, P. J. (1972). Robust statistics: a review. Ann. Math. Statist. 43, 1041-1067.

Huber, P. J. (1973). Robust regression: asymptotics, conjectures and Monte Carlo. Ann. Statist. 1, 799-821.

Huber, P. J. (1977). Robust Statistical Procedures. Regional Conference Series in Applied Mathematics No. 27, Soc. Industr. Appl. Math., Philadelphia.

Huber, P. J. (1981). Robust Statistics. New York: Wiley.

Hudimoto, H. (1964). On a distribution-free two-way classification. Ann. Inst. Statist. Math. 16, 247-253.

Irwin, J. O. (1942). The distribution of the logarithm of survival times when the true law is exponential. J. Hyg. 42, 328-333.

Ito, K. (1980). Robustness of ANOVA and MANOVA test procedures. In Handbook of Statistics, Vol. 1 (P. R. Krishnaiah, ed.). Amsterdam: North Holland.

Jaeckel, L. A. (1971a). Robust estimates of location: symmetry and asymmetric contamination. Ann. Math. Statist. 42, 1020-1034.

Jaeckel, L. A. (1971b). Some flexible estimates of location. Ann. Math. Statist. 42, 1540-1552.

Jaeckel, L. A. (1972). Estimating regression coefficients by minimizing the dispersion of the residuals. Ann. Math. Statist. 43, 1449-1458.

Jain, R. B. (1981). Percentage points of many-outlier detection procedures. Technometrics 23, 71-75.

Jain, R. B., and Pingel, L. A. (1981). A procedure for estimating the number of outliers. Commun. Statist.—Theor. Meth. A10(10), 1029-1041.

Jain, R. B., Pingel, L. A., and Davidson, J. L. (1982). A unified approach for estimation and detection of outliers. Commun. Statist.—Theor. Meth. A11(25), 2953-2976.

Jeffreys, H. (1961). Theory of Probability. Oxford: Clarendon Press.

Johnson, N. L. (1949). Systems of frequency curves generated by methods of translation. Biometrika 36, 149-176.

Johnson, N. L. (1965). Tables to facilitate fitting S_U frequency curves. Biometrika 52, 547-558.

Johnson, N. L., and Kotz, S. (1970). Continuous Univariate Distributions, Vol. 1. Boston: Houghton Mifflin.

Johnson, N. L., Nixon, E., Amos, D. E., and Pearson, E. S. (1963). Tables of percentage points of Pearson curves. Biometrika 50, 459-498.

Johnson, R. A. (1974). Some optimality results for one or two sample procedures based on the smallest r order statistics. In Reliability and Biometry (F. Proschan and R. J. Serfling, eds.), Philadelphia: SIAM.

Joiner, B. L., and Hall, D. L. (1983). The uniquitous role of f'/f in efficient estimation of location. Amer. Statist. 37, 128-133.

Joshi, P. C. (1971). Recurrence relations for the mixed moments of order statistics. Ann. Math. Statist. 42, 1096-1098.

Joshi, P. C. (1978). Recurrence relations between moments of order statistics from exponential and truncated exponential distributions. Sankhya B39, 362-371.

Joshi, P. C. (1982). A note on the mixed moments of order statistics from exponential and truncated exponential distributions. J. Statist. Plann. Inf. 6, 13-16.

Joshi, P. C., and Balakrishnan, N. (1981). An identity for the moments of normal order statistics with applications. Scand. Actuar. J. 1981, 203-213.

Joshi, P. C., and Balakrishnan, N. (1982). Recurrence relations and identities for the product moments of order statistics. Sankhya B44, 39-49.

Jung, J. (1955). On linear estimates defined by a continuous weight function. Ark. Math. 3, 199-209.

Jung, J. (1962). Approximation of least squares estimates of location and scale parameters. In Contributions to Order Statistics (A. E. Sarhan and B. G. Greenberg, eds.), 28-33. New York: Wiley.

Jureckova, J. (1971). Nonparametric estimates of regression coefficients. Ann. Math. Statist. 42, 1328-1338.

Kabir, A. B. M. L., and Ahsanullah, M. (1974). Estimation of the location and scale parameters of a power-function distribution by linear functions of order statistics. Commun. Statist. 3, 463-467.

Kafadar, K. (1985). Notched box-and-whisker plot. In Encyclopedia of Statistical Sciences, Vol. 6 (S. Kotz and N. L. Johnson, eds.). New York: Wiley.

Kambo, N. S. (1978). Maximum likelihood estimators of the location and scale parameters of the exponential distribution from a censored sample. Commun. Statist.—Theor. Meth. A7(12), 1129-1132.

Kambo, N. S., and Awad, A. M. (1985). Testing equality of location parameters of exponential distributions. Commun. Statist.—Theor. Meth. 14(3), 567-583.

Karlin, S. (1966). A First Course in Stochastic Processes. New York: Academic Press.

Kendall, M. G. (1945). Advanced Theory of Statistics, Vol. 1. London: Charles Griffin.

Kendall, M. G., and Stuart, A. (1969). The Advanced Theory of Statistics, Vol. 1, 3rd ed. London: Charles Griffin.

Kendall, M. G., and Stuart, A. (1973). The Advanced Theory of Statistics, Vol. 2. New York: Hafner Press.

Kim, P. J., and Jennrich, R. J. (1973). Tables of the exact sampling distributions of two-sample Kolmogorov-Smirnov criterion D_{mn} ($m \leq n$). In Selected Tables in Mathematical Statistics, Vol. 1 (H. L. Harter and D. B. Owen, eds.), 79-170. Providence, R.I.: American Mathematical Society.

Klotz, J. (1964). On the normal scores two-sample test. J. Amer. Statist. Ass. 59, 652-664.

Kolmogorov, A. N. (1941). Uber das logarithmisch normale verteilungsgesetz der Dimensionen der Teilchen bei zerstuckelung. Doklady Akademii Nauk SSSR 31, 99-101.

Kottler, F. (1950). The distribution of particle sizes: I. J. Franklin Inst. 250, 339-356.

Kotz, S. (1973). Normality vs. lognormality with applications. Commun. Statist. 1, 113-132.

Krumbein, W. C. (1936). Application of logarithmic moments of size frequency distributions of sediments. J. Sedimentary Petrology 6, 35-47.

Kulldorff, G., and Vannman, K. (1973). Estimating location and scale parameters of a Pareto distribution by linear functions of order statistics. J. Amer. Statist. Ass. 68, 218-227.

Lachenbruch, P. A., and Mickey, M. R. (1968). Estimation of error rates in discriminant analysis. Technometrics 10, 1-11.

Lagakos, S. W. (1979). General right censoring and its impact on the analysis of survival data. Biometrics 35, 139-156.

Lawless, J. F. (1971). A prediction problem concerning samples from the exponential distribution, with applications in life testing. Technometrics 13, 725-730.

Lawless, J. F. (1975). Construction of tolerance bounds for the extreme-value and Weibull distributions. Technometrics 17, 255-261.

Lawless, J. F. (1977). Prediction intervals for the two-parameter exponential distribution. Technometrics 19, 469-472.

Lawless, J. F. (1979). Statistical inference for type I censored data from the exponential and Weibull distributions. Unpublished manuscript.

Lawless, J. F. (1982). Statistical Models & Methods for Life Data. New York: Wiley.
Lawless, J. F., and Singhal, K. (1980). Analysis of data from life-test experiments under exponential model. Naval Res. Logist. Quart. 27, 323-334.
Layard, M. W. J. (1973). Robust large-sample tests for homogeneity of variance. J. Amer. Statist. Ass. 68, 195-198.
Lee, A. F. S., and D'Agostino, R. B. (1976). Levels of significance of some two-sample tests when observations are from compound normal distributions. Commun. Statist. A5(4), 342-352.
Lee, A. F. S., and Gurland, J. (1977). One-sample t-test when sampling from a mixture of normal distributions. Ann. Statist. 5, 803-807.
Lee, K. R., Kapadia, C. H., and Dwight, B. B. (1980). On estimating the scale parameter of the Rayleigh distribution from doubly censored samples. Statist. Hefte 21, 14-29.
Lehmann, E. L. (1959). Testing Statistical Hypotheses. New York: Wiley.
Lehmann, E. L. (1975). Nonparametrics: Statistical Methods Based on Ranks. New York: McGraw-Hill.
Leone, F. C., and Hamouda, E. M. (1973). Relative efficiencies of "O-BLUE" estimators in simple linear regression. J. Amer. Statist. Ass. 68, 953-959.
Levene, H. (1960). Robust tests for equality of variances. In Contributions to Probability and Statistics (I. Olkin, ed.), 278-292. Stanford, Calif.: Stanford University Press.
Lewis, P. A. W. (1965). Some results on tests for Poisson processes. Biometrika 52, 67-77.
Lilliefors, H. W. (1967). On the Kolmogorov-Smirnov test for normality with mean and variance unknown. J. Amer. Statist. Ass. 62, 399-402.
Lindeman, R. H., Merenda, P. F., and Gold, R. Z. (1980). Introduction to Bivariate and Multivariate Analysis. Glenview, Ill.: Scott, Foresman.
Littell, C. R., McClave, J. R., and Offen, W. W. (1979). Goodness-of-fit tests for two parameter Weibull distribution. Commun. Statist.—Simul. Comp. B8, 257-269.
Lloyd, E. H. (1952). Least squares estimation of location and scale parameters using order statistics. Biometrika 39, 88-95.
Locke, C., and Spurrier, J. D. (1976). The use of U-statistics for testing normality against nonsymmetric alternatives. Biometrika 63, 143-147.
Locke, C., and Spurrier, J. D. (1977). The use of U-statistics for testing normality against alternatives with both tails heavy or both tails light. Biometrika 64, 638-640.
Loftsgaarden, D. O., and Quesenberry, C. P. (1965). A nonparametric estimate of the multivariate density function. Ann. Math. Statist. 36, 1049-1051.
Low, B. B. (1959). Mathematics. Edinburgh: Neill and Co.

Lwin, T. (1976). Optimal linear estimators of location and scale parameters using order statistics and related empirical Bayes estimation. Scand. Actuar. J. 1976, 79-91.

Malik, H. J. (1967). Exact moments of order statistics from a power-function distribution. Skand. Aktuarietidskr. 1967, 144-157.

Mann, H. B., and Whitney, D. R. (1947). On a test of whether one of two random samples is stochastically larger than the other. Ann. Math. Statist. 18, 50-60.

Mann, N. R. (1967). Tables for obtaining the best linear invariant estimates of parameters of the Weibull distribution. Technometrics 9, 629-645.

Mann, N. R. (1968). Point and interval estimation procedures for the two-parameter Weibull and extreme-value distributions. Technometrics 10, 231-256.

Mann, N. R. (1969). Optimum estimators for linear function of location and scale parameters. Ann. Math. Statist. 40, 2149-2155.

Mann, N. R. (1971). Best linear invariant estimation for Weibull parameters under progressive censoring. Technometrics 13, 521-534.

Mann, N. R. (1972). Design of over-stress life-test experiments when failure times have the two-parameter Weibull distribution. Technometrics 14, 437-451.

Mann, N. R. (1982). Optimal outlier tests for a Weibull model—to identify process changes or to predict failure times. TIMS/Studies in the Management Sciences 19, 261-279 (Amsterdam: North-Holland).

Mann, N. R., and Fertig, K. W. (1973). Tables for obtaining Weibull confidence bounds and tolerance bounds based on best linear invariant estimates of parameters of the extreme-value distribution. Technometrics 15, 87-102.

Mann, N. R., Schafer, R. E., and Singpurwalla, N. D. (1974). In Methods for Statistical Analysis of Reliability and Lifetime Data (N. R. Mann, R. E. Schafer, and N. D. Singpurwalla, eds.). New York: Wiley.

Mann, N. R., Scheuer, E. M., and Fertig, K. W. (1973). A new goodness-of-fit test for the two-parameter Weibull or extreme-value distribution. Commun. Statist. 2, 383-400.

Mantel, N., and Myers, M. (1971). Problem of convergence of maximum likelihood iterative procedures in multiparameter situation. J. Amer. Statist. Ass. 66, 484-491.

Mardia, K. V. (1980). Tests of univariate and multivariate normality. In Handbook of Statistics, Vol. 1 (P. R. Krishnaiah, ed.), 279-320. Amsterdam: North-Holland.

Margolin, B. H., and Maurer, W. (1976). Tests of the Kolmogorov-Smirnov type for exponential data with unknown scale and related problems. Biometrika 63, 149-160.

Marquardt, D. W. (1963). An algorithm for least-squares estimation of non-linear parameters. J. Soc. Ind. Appl. Math. 11, 431-441.

Marshall, A. W. (1958). The small sample distribution of nw_n^2. Ann. Math. Statist. 29, 307-309.

McCool, J. I. (1965). The construction of good linear unbiased estimates from the best linear estimates for a smaller sample size. Technometrics 7, 543-552.

Mehrotra, K. G., and Nanda, P. (1974). Unbiased estimation of parameters by order statistics in the case of censored samples. Biometrika 61, 601-606.

Mendenhall, W., and Hader, R. J. (1958). Estimation of parameters of mixed exponential distributed failure times from censored life test data. Biometrika 45, 504-520.

Menon, M. V. (1963). Estimation of the shape and scale parameters of the Weibull distribution. Technometrics 5, 175-182.

Miller, R. G. (1968). Jackknifing variances. Ann. Math. Statist. 39, 567-582.

Milton, R. C. (1970). Rank Order Probabilities. New York: Wiley.

Montfort, M. A. J. V. (1970). On testing that the distribution of extremes is Type I when Type II is the alternative. J. Hydrology 11, 421-427.

Mulholland, H. P. (1977). On the distribution of b_1 for samples of size at most 25, with tables. Biometrika 64, 401-409.

Nair, K. R. (1950). Efficiencies of certain linear systematic statistics for estimating dispersion from normal samples. Biometrika 37, 182-183.

Nelson, W. B. (1970). Statistical methods for accelerated lifetest data—the inverse power law model. General Electric Co. Tech. Rep. 71-C011. Schenectady, New York.

Nelson, W. B. (1972). Theory and applications of hazard plotting for censored failure data. Technometrics 14, 945-965.

Nelson, W. B. (1982). Applied Life Data Analysis. New York: Wiley.

Owen, D. B. (1962). Handbook of Statistical Tables. Reading, Mass.: Addison-Wesley.

Parzen, E. (1962). On estimation of a probability density and mode. Ann. Math. Statist. 33, 1065-1076.

Patil, G. P. (1963). A characterization of the exponential-type distribution. Biometrika 50, 205-207.

Patnaik, P. B. (1949). The noncentral χ^2 and F distributions and their applications. Biometrika 36, 202-232.

Pearl, R. (1940). Medical Biometry and Statistics. Philadelphia: W. B. Saunders.

Pearson, E. S. (1932). The analysis of variance in cases of nonnormal variation. Biometrika 23, 114-133.

Pearson, E. S. (1963). Some problems arising in approximating to probability distributions using moments. Biometrika 50, 95-112.

Pearson, E. S., and Adyanthaya, N. K. (1929). The distribution of frequency constants in small samples from non-normal symmetrical and skew populations. Biometrika 21, 259-286.

Pearson, E. S., and Chandra Sekar, C. (1936). The efficiency of statistical tools and a criterion for the rejection of outlying observations. Biometrika 28, 308-320.
Pearson, E. S., and Hartley, H. O. (1956). Biometrika Tables for Statisticians, Vol. 1. Cambridge: Cambridge University Press.
Pearson, E. S., and Hartley, H. O. (1972). Biometrika Tables for Statisticians, Vol. 2. Cambridge: Cambridge University Press.
Pearson, K. (1934). Tables of the Incomplete Beta Function. Cambridge: Cambridge University Press.
Pearson, K., and Lee, A. (1908). On the generalized probable error in multiple normal correlation. Biometrika 6, 69-68.
Pearson, E. S., and Stephens, M. A. (1962). The goodness-of-fit tests based on W_N^2 and U_N^2. Biometrika 49, 397-402.
Pearson, E. S., and Tiku, M. L. (1970). Some notes on the relationship between the distributions of central and noncentral F. Biometrika 57, 175-179.
Perks, W. F. (1932). On some experiments in the graduation of mortality statistics. J. Inst. Actuar. 58, 12-57.
Persson, T., and Rootzen, H. (1977). Simple and highly efficient estimators for a Type I censored normal sample. Biometrika 64, 123-128.
Pettitt, A. N. (1976). Cramér-von Mises statistics for testing normality with censored samples. Biometrika 63, 475-481.
Pettitt, A. N. (1977). Tests for the exponential distribution with censored data using Cramér-von Mises statistics. Biometrika 64, 629-632.
Pettitt, A. N., and Stephens, M. A. (1976). Modified Cramér-von Mises statistics for censored data. Biometrika 63, 291-298.
Plackett, R. L. (1958). Linear estimation from censored data. Ann. Math. Statist. 29, 131-142.
Posten, H. O. (1978). The robustness of the two-sample t-test over the Pearson system. J. Statist. Comp. Simul. 6, 295-311.
Posten, H. O. (1982). Two-sample Wilcoxon power over the Pearson system and comparison with the t-test. J. Statist. Comp. Simul. 16, 1-18.
Posten, H. O., Yeh, H. C., and Owen, D. B. (1982). Robustness of the two-sample t-test under violations of the homogeneity of variance assumption. Commun. Statist.—Theor. Meth. A11(2), 109-126.
Prescott, P. (1974). Variances and covariances of order statistics from the gamma distribution. Biometrika 61, 607-613.
Prescott, P. (1976a). On a test for normality based on sample entropy. J. R. Statist. Soc. B38, 254-256.
Prescott, P. (1976b). Comparison of tests for normality using stylized sensitivity surfaces. Biometrika 63, 285-290.
Prescott, P. (1979). Critical values for a sequential test for many outliers. Appl. Statist. 28, 36-39.
Proschan, F. (1963). Theoretical explanation of observed decreasing failure rates. Technometrics 5, 375-383.

Puri, M. L., and Rao, C. R. (1976). Augmenting Shapiro-Wilk test for normality. In Contributions to Applied Statistics, 129-139. Stuttgart, West Germany: Birkhauser (Grossohaus).

Pyke, R. (1965). Spacings (with discussion). J. R. Statist. Soc. B27, 395-439.

Rao, C. R. (1975). Linear Statistical Inference and Its Applications. New York: Wiley.

Rao, J. N. K. (1975). On the foundations of survey sampling. In A Survey of Statistical Designs and Linear Models (J. N. Srivastava, ed.). Amsterdam: North-Holland.

Rasch, D. (1980). Probleme der angewandten Statistik, Robusttheit I: Fz. fur Tierproduktion, Dummerstorf-Rostock, Heft 4.

Rasch, D. (1983). Robustness of sequential tests. In Robustness of Statistical Methods and Nonparametric Statistics (D. Rasch and M. L. Tiku, eds.). East Berlin: VEB Deutscher Verlag der Wissenschaften.

Reed, L. J., and Berkson, J. (1929). The application of the logistic function to experimental data. J. Physical Chemistry 33, 760-779.

Reider, H. (1978). A robust asymptotic testing model. Ann. Statist. 6, 1080-1094.

Reider, H. (1979). Robustness of one and two sample rank tests against gross errors. Ann. Statist. 9, 245-265.

Relles, D. A., and Rogers, W. H. (1977). Statisticians are fairly robust estimators of location. J. Amer. Statist. Ass. 72, 107-111.

Rey, W. J. J. (1983). Introduction to Robust and Quasi-robust Statistical Methods. New York: Springer-Verlag.

Ringland, J. T. (1983). Robust multiple comparison. J. Amer. Statist. Ass. 78, 145-151.

Rocke, D. M., Downs, G. W., and Rocke, A. J. (1982). Are robust estimators really necessary? Technometrics 24, 95-101.

Rosner, B. (1975). On the detection of many outliers. Technometrics 17, 221-227.

Rosner, B. (1977). Percentage points for the RST many outlier procedure. Technometrics 19, 307-312.

Roy, J., and Tiku, M. L. (1962). A Laguerre series approximation to the sampling distribution of the variance. Sankhya 24, 181-184.

Ruppert, D., and Carroll, R. J. (1980). Trimmed least squares estimation in the linear model. J. Amer. Statist. Ass. 75, 828-838.

Sacks, J., and Ylvisaker, D. (1972). A note on Huber's robust estimation of a location parameter. Ann. Math. Statist. 43, 1068-1075.

Sacks, J., and Ylvisaker, D. (1978). Linear estimation for approximately linear models. Ann. Statist. 6, 1122-1137.

Saleh, A. K. M. E. (1966). Estimation of the parameters of the exponential distribution based on order statistics in censored samples. Ann. Math. Statist. 37, 1717-1735.

Samford, M. R., and Taylor, J. (1959). Censored observations in randomized block experiments. J. R. Statist. Soc. B21, 214-237.
Sansing, R. C., and Owen, D. B. (1974). The density of the t-statistic for non-normal distributions. Commun. Statist. 3(2), 139-155.
Sarhan, A. E. (1955). Estimation of the mean and standard deviation by order statistics: Part III. Ann. Math. Statist. 26, 576-592.
Sarhan, A. E., and Greenberg, B. G. (1955). Estimation of location and scale parameters by order statistics from singly and doubly censored samples: Part I. The normal distribution up to samples of size 10. Ann. Math. Statist. 27, 427-451; correction 40, 325.
Sarhan, A. E., and Greenberg, B. G. (1957). Tables of the best linear estimates by order statistics of the parameters of single exponential distributions from singly and doubly censored samples. J. Amer. Statist. Ass. 52, 58-87.
Sarhan, A. E., and Greenberg, B. G., eds. (1962). Contributions to Order Statistics. New York: Wiley.
Sarhan, A. E., Greenberg, B. G., and Ogawa, J. (1963). Simplified estimates for the exponential distribution. Ann. Math. Statist. 34, 102-116.
Sarkadi, K. (1975). The consistency of the Shapiro-Francia test. Biometrika 62, 445-450.
Schafer, R. E., Finkelstein, J. M., and Collins, J. (1972). On a goodness-of-fit test for the exponential distribution with mean unknown. Biometrika 59, 222-224.
Scheffé, H. (1959). The Analysis of Variance. New York: Wiley.
Schrader, R. M., and McKean, J. W. (1977). Robust analysis of variance. Commun. Statist.—Theor. Meth. A6(9), 879-894.
Seshadri, V., Csörgö, M., and Stephens, M. A. (1969). Tests for the exponential distribution using Kolmogorov-type statistics. J. R. Statist. Soc. B31, 499-509.
Shah, B. K. (1966). On the bivariate moments of order statistics from a logistic distribution. Ann. Math. Statist. 37, 1002-1010.
Shah, B. K. (1970). Note on moments of logistic order statistics. Ann. Math. Statist. 41, 2151-2152.
Shapiro, S. S., and Brain, C. W. (1981). A review of distributional testing procedures and development of a censored sample distributional test. In Statistical Distributions in Scientific Work, Vol. 5 (C. Taillie, G. P. Patil, and B. A. Baldessari, eds.), 1-24. Boston: D. Reidel.
Shapiro, S. S., and Francia, R. S. (1972). An approximate analysis of variance test for normality. J. Amer. Statist. Ass. 67, 215-216.
Shapiro, S. S., and Wilk, M. B. (1965). An analysis of variance test for normality (complete samples). Biometrika 52, 591-611.
Shapiro, S. S., and Wilk, M. B. (1972). An analysis of variance test for exponential distribution. Technometrics 14, 355-370.
Shapiro, S. S., Wilk, M. B., and Chen, J. (1968). A comparative study of various tests of normality. J. Amer. Statist. Ass. 63, 1343-1372.

Shenton, L. R., and Bowman, K. O. (1977). A bivariate model for the distribution of b_1 and b_2. J. Amer. Statist. Ass. 72, 206-211.

Shorack, G. R. (1972). Functions of order statistics. Ann. Math. Statist. 43, 412-427.

Shorack, G. R. (1974). Random means. Ann. Statist. 2, 661-675.

Siddiqui, M. M. (1962). Some problems connected with Rayleigh distributions. J. Res. Nat. Bur. Stand. 660, 167-174.

Siddiqui, M. M. (1963). Optimum estimators of the parameters of negative exponential distributions from one or two order statistics. Ann. Math. Statist. 34, 117-121.

Singh, D., Singh, P., and Kumar, P. (1982). Handbook on Sampling Methods. New Delhi: Indian Agricultural Statistics Research Institute (I.C.A.R.).

Sinha, S. K. (1973). Distributions of order statistics and estimation of mean life when an outlier may be present. Canad. J. Statist. 1, 119-121.

Smith, R. M., and Bain, L. J. (1976). Correlation type goodness-of-fit statistics with censored sampling. Commun. Statist. A5(2), 119-132.

Smith, W. B., Zeis, C. D., and Syler, G. W. (1973). Three parameter lognormal estimation from censored data. J. Ind. Statist. Ass. 11, 15-31.

Spiegelhalter, D. J. (1977). A test for normality against symmetric alternatives. Biometrika 64, 415-418.

Spjøtvoll, E., and Aastveit, A. H. (1980). Comparison of robust estimators on data from field experiments. Scand. J. Statist. 7, 1-13.

Sprott, D. A. (1978). Robustness and non-parametric procedures are not the only nor the safe alternatives to normality. Canad. J. Psychol. 32, 180-185.

Sprott, D. A. (1982). Robustness and maximum likelihood estimation. Commun. Statist.—Theor. Meth. 11(22), 2513-2529.

Srinivasan, R. (1970). An approach to testing the goodness-of-fit of incompletely specified distributions. Biometrika 57, 605-611.

Srivastava, A. B. L. (1958). Effect of non-normality on the power function of t-test. Biometrika 45, 421-429.

Srivastava, A. B. L. (1959). Effect of non-normality on the power of the analysis of variance test. Biometrika 46, 114-122.

Stein, C. (1956). Efficient nonparametric testing and estimation. Proc. Third Berkeley Symp. on Mathematical Statistics and Probability, Vol. 1. Berkeley: University of California Press.

Stephens, M. A. (1963). The distribution of goodness-of-fit statistics, U_N^2: I. Biometrika 50, 303-313.

Stephens, M. A. (1964). The distribution of goodness-of-fit statistics, U_N^2: II. Biometrika 51, 393-397.

Stephens, M. A. (1974). EDF statistics for goodness of fit and some comparisons. J. Amer. Statist. Ass. 69, 730-737.

Stephens, M. A. (1976). Asymptotic results for goodness of fit statistics with unknown parameters. Ann. Statist. 4, 357-369.

Stephens, M. A. (1977). Goodness-of-fit for the extreme value distribution. Biometrika 64, 583-588.

Stigler, S. M. (1977). Do robust estimators work with real data? Ann. Statist. 5, 1005-1098.

Stoller, D. C. (1954). Univariate two population distribution-free discrimination. J. Amer. Statist. Ass. 49, 770-775.

Stone, C. J. (1975). Adaptive maximum likelihood estimators of a location parameter. Ann. Statist. 3, 267-284.

Subrahmaniam, K., Subrahmaniam, K., and Messeri, J. Y. (1975). On the Robustness of some tests of significance in sampling from a compound normal population. J. Amer. Statist. Ass. 70, 435-438.

Sukhatme, P. V. (1937). Tests of significance for samples of the chi-square population with two degrees of freedom. Ann. Eugen. 8, 52-56.

Susara, V., and Pathala, K. S. (1965). A probability distribution for the time of first birth. J. Scientific Res., Banaras Hindu Univ., 16, 59-62.

Syler, G. W. (1968). Estimation of parameters in the lognormal distribution with censored samples. Unpublished thesis, Texas A & M University.

Szegö, G. (1959). Orthogonal Polynomials. Colloquium Publications, No. 23. New York: American Math. Soc.

Takeuchi, K. (1971). A uniformly asymptotically efficient estimator of a location parameter. J. Amer. Statist. Ass. 66, 292-301.

Tallis, G. M., and Light, R. (1968). The use of fractional moments for estimating the parameters of a mixed exponential distribution. Technometrics 10, 161-175.

Tan, W. Y. (1971). Note on an extension of Gauss-Markov theorem to multivariate regression models. SIAM Jour. Appl. Math. 20, 24-29.

Tan, W. Y. (1977). On the distributions of sums of squares in balanced random effect models. Statistica 37, 439-462.

Tan, W. Y. (1979). On the quadratic estimation of covariance matrix in multivariate linear models. J. Multivariate Anal. 9, 452-459.

Tan, W. Y. (1982). Sampling distributions and robustness of t, F and variance-ratio in two samples and ANOVA models with respect to departure from normality. Commun. Statist.—Theor. Meth. 11(22), 2485-2511.

Tan, W. Y. (1985a). On Tiku's robust procedure—a Bayesian insight. J. Statist. Plann. Inf. 11, 329-340.

Tan, W. Y. (1985b). Bayesian approach to robust procedures for ANOVA models under heteroscedasticity and non-normality. Sankhya B 47, 33-46.

Tan, W. Y., and Cheng, S. S. (1981). Cumulants of bilinear forms and quadratic forms and its applications. Commun. Statist.—Theor. Meth. 10(A), 283-298.

Tan, W. Y., and Cheng, S. S. (1984). On testing variance components in three stages unbalanced random effects model. Sankhya B 46, 188-200.

Tan, W. Y., and Tabatabai, M. A. (1986). Some robust ANOVA procedures under heteroscedasticity and nonnormality. Commun. Statist. 14(B), 1007-1026.

Tan, W. Y., and Wong, S. P. (1977). On the Roy-Tiku approximation to the distributions of sample variances from nonnormal universe. J. Amer. Statist. Ass. 72, 875-881.

Tan, W. Y., and Wong, S. P. (1978). On the effects of departure from normality on the F-ratios in a nested random effect model by mixture of normals. Biomet. Zeit. 20, 441-451.

Tan, W. Y., and Wong, S. P. (1980). On approximating the null and non-null distributions of the F-ratio in unbalanced random effect models from non-normal universe. J. Amer. Statist. Ass. 75, 652-665.

Taylor, J. (1973). The analysis of design experiments with censored observations. Biometrics 29, 35-43.

Thomas, D. R., Bain, L. J., and Antle, C. E. (1969). Inferences on the parameters of the Weibull distribution. Technometrics 11, 445-460.

Tiao, G. C., and Ali, M. M. (1971). Effects of non-normality on inferences about variance components. Technometrics 13, 635-650.

Tiao, G. C., and Lund, D. (1970). The use of OLUMV estimators in inference robust studies of the location parameter of a class of symmetric distribution. J. Amer. Statist. Ass. 65, 370-386.

Tietjen, G. L., and Moore, R. H. (1972). Some Grubbs-type statistics for the detection of several outliers. Technometrics 14, 583-597.

Tietjen, G. L., Kahaner, D. K., and Beckman, R. J. (1977). Variances and covariances of the normal order statistics for sample sizes 2 to 50. Selected Table in Mathematical Statistics, Vol. 5, 1-73. Providence, R. I.: American Mathematical Society.

Tiku, M. L. (1964). Approximating the general non-normal variance-ratio sampling distributions. Biometrika 51, 83-95.

Tiku, M. L. (1965a). Chi-square approximations for the distributions of goodness-of-fit statistics U_N^2 and W_N^2. Biometrika 52, 630-633.

Tiku, M. L. (1965b). Laguerre series forms of noncentral χ^2 and F-distributions. Biometrika 52, 415-427.

Tiku, M. L. (1967). Estimating the mean and standard deviation from a censored normal sample. Biometrika 54, 155-165.

Tiku, M. L. (1968a). Estimating the parameters of log-normal distribution from censored samples. J. Amer. Statist. Ass. 63, 134-140.

Tiku, M. L. (1968b). Estimating the parameters of normal and logistic distributions from censored samples. Austral. J. Statist. 10, 64-74.

Tiku, M. L. (1968c). Estimating the mean and standard deviation from progressively censored normal samples. J. Ind. Soc. Agri. Statist. 20, 20-25.

Tiku, M. L. (1970). Monte Carlo study of some simple estimators in censored normal samples. Biometrika 57, 207-211.

Tiku, M. L. (1971). Student's t-distribution under non-normal situations. Austral. J. Statist. 13, 142-148.

Tiku, M. L. (1972a). Power function of the F-test under non-normal situations. J. Amer. Statist. Ass. 66, 913-916.

Tiku, M. L. (1972b). A note on the distribution of the doubly non-central F distribution. Austral. J. Statist. 14, 37-40.

Tiku, M. L. (1973). Testing group effects from Type II censored normal samples in experimental design. Biometrics 29, 25-33.

Tiku, M. L. (1974a). A new statistic for testing for normality. Commun. Statist. 3(3), 223-232.

Tiku, M. L. (1974b). Testing normality and exponentiality in multi-sample situations. Commun. Statist. 3(8), 777-794.

Tiku, M. L. (1975a). Laguerre series forms of the distributions of classical test-statistics and their robustness in non-normal situations. In Applied Statistics (R. P. Gupta, ed.). New York: American Elsevier.

Tiku, M. L. (1975b). A new statistic for testing an assumed distribution. In Statistical Distributions in Scientific Work (G. P. Patil, S. Kotz, and J. K. Ord, eds.). Boston: D. Reidel.

Tiku, M. L. (1975c). A new statistic for testing suspected outliers. Commun. Statist. A4, 737-752.

Tiku, M. L. (1977). Rejoinder: "Comment on 'A new statistic for testing suspected outliers'." Commun. Statist. A6, 1417-1422.

Tiku, M. L. (1980a). Robustness of MML estimators based on censored samples and robust test statistics. J. Statist. Plann. Inf. 4, 123-143.

Tiku, M. L. (1980b). Goodness of fit statistics based on the spacings of complete or censored samples. Austral. J. Statist. 22, 260-275.

Tiku, M. L. (1981a). Robust two-sample test and robust regression and analysis-of-variance via MML estimators. In Statistics and Related Topics (M. Csörgö, D. A. Dawson, J. N. K. Rao, and A. K. M. E. Saleh, eds.), 297-314. Amsterdam: North-Holland.

Tiku, M. L. (1981b). A goodness-of-fit statistic based on the sample spacings for testing a symmetric distribution against symmetric alternatives. Austral. J. Statist. 23, 149-158.

Tiku, M. L. (1982a). Testing linear contrasts of means in experimental design without assuming normality and homogeneity of variances. Biom. J. 24, 613-627.

Tiku, M. L. (1982b). Robust statistics for testing equality of means or variances. Commun. Statist.—Theor. Meth. 11(22), 2543-2558.

Tiku, M. L. (1983a). Robust location-tests and classification procedures. In Robustness of Statistical Methods and Nonparametric Statistics (D. Rasch and M. L. Tiku, eds.). East Berlin: VEB Deutscher Verlag der Wissenschaften.

Tiku, M. L. (1983b). Efficiencies of some robust estimators in sample survey. Commun. Statist.—Theor. Meth. 12(17), 2043-2051.

Tiku, M. L., and Balakrishnan, N. (1984a). Testing equality of population variances the robust way. Commun. Statist.—Theor. Meth. 13(17), 2143-2159.

Tiku, M. L., and Balakrishnan, N. (1984b). Robust multivariate classification procedures based on the MML estimators. Commun. Statist.—Theor. Meth. 13(8), 967-986.

Tiku, M. L., and Balakrishnan, N. (1985). Testing the equality of variance-covariance matrices the robust way. Commun. Statist.—Theor. Meth. 14(12), 3033-3051.

Tiku, M. L., and Bhasin, P. (1982). Usefulness of robust estimators in sample survey. Commun. Statist.—Theor. Meth. 11(22), 2597-2610.

Tiku, M. L., and Jones, P. W. (1971). Best linear unbiased estimators for a distribution similar to the logistic. In Proc. "Statistics 71 Canada" (C. S. Carter and T. D. Dwivedi, eds.), 412-419.

Tiku, M. L., and Kumra, S. (1981). Expected values and variances and covariances of order statistics for a family of symmetric distributions (Student's t). In Selected Tables in Mathematical Statistics, Vol. 8. Providence, R.I.: American Mathematical Society; 1985, 141-270.

Tiku, M. L., and Singh, M. (1981a). Robust test for means when population variances are unequal. Commun. Statist.—Theor. Meth. A10(20), 2057-2071.

Tiku, M. L., and Singh, M. (1981b). Testing the two-parameter Weibull distribution. Commun. Statist.—Theor. Meth. A10, 907-918.

Tiku, M. L., and Singh, M. (1981c). Testing outliers in multivariate data. In Statistical Distributions in Scientific Work, Vol. 5 (C. Taillie, G. P. Patil, and B. A. Baldessari, eds.). Boston: D. Reidel.

Tiku, M. L., and Singh, M. (1982). Robust statistics for testing mean vectors of multivariate distributions. Commun. Statist.—Theor. Meth. A11(9), 985-1001.

Tiku, M. L., and Stewart, D. E. (1977). Estimating and testing group effects from Type I censored normal samples in experimental design. Commun. Statist. A6(15), 1485-1501.

Tiku, M. L., and Tamhankar, M. V. (1980). Improved tests of exponentiality in single- and multi-sample situations. J. Ind. Soc. Agri. Statist. 22, 40-50.

Tiku, M. L., Rai, K., and Mead, E. (1974). A new statistic for testing exponentiality. Commun. Statist. 3(5), 485-493.

Tukey, J. W. (1958). A problem of Berkson, and minimum variance orderly estimators. Ann. Math. Statist. 29, 588-592.

Tukey, J. W. (1960). A survey of sampling from contamination distributions. In Contributions to Probability and Statistics (I. Olkin, ed.), 448-485. Stanford, Calif.: Stanford University Press.

Tukey, J. W. (1970). Exploratory Data Analysis, mimeographed preliminary edition. (1977) Reading, Mass.: Addison-Wesley.

Uthoff, V. A. (1970). An optimum test property of two well-known statistics. J. Amer. Statist. Ass. 65, 1597-1600.

Uthoff, V. A. (1973). The most powerful scale and location invariant test of the normal versus the double exponential. Ann. Statist. 1, 170-174.

Varde, S. D. (1969). Life testing and reliability estimation for the two parameter exponential distribution. J. Amer. Statist. Ass. 64, 621-631.

Wang, Y. H., and Chang, S. A. (1977). A new approach to the nonparametric tests of exponential distribution with unknown parameters. In The Theory and Applications of Reliability, Vol. 1 (C. P. Tsokos and I. N. Shimi, eds.). New York: Academic Press.

Watson, G. S. (1961). Goodness-of-fit tests on a circle. Biometrika 48, 109-114.

Watterson, G. A. (1959). Linear estimation in censored samples from multivariate normal populations. Ann. Math. Statist. 30, 814-824.

Wegman, E. J., and Carroll, R. J. (1977). A Monte Carlo study of robust estimators of location. Commun. Statist. A6, 795-812.

Weisberg, S., and Bingham, C. (1975). An approximate analysis of variance test for non-normality suitable for machine calculation. Technometrics 17, 133-134.

Weissman, I. (1978). Estimation of parameters and large quantiles based on the k largest observations. J. Amer. Statist. Ass. 73, 812-815.

Welch, B. L. (1937). The significance of the difference between two means when the population variances are unequal. Biometrika 29, 2057-2071.

White, J. S. (1964). Least squares unbiased censored linear estimation for the log-Weibull (extreme value) distribution. Indust. Math. 14, 21-60.

Wilcoxon, F., Katti, S. K., and Wilcox, R. A. (1973). Critical values and probability levels for the Wilcoxon signed rank test. In Selected Tables in Mathematical Statistics, Vol. 1 (H. L. Harter and D. B. Owen, eds.). Providence, R.I.: American Mathematical Society.

Wilkinson, G. N. (1970). A general recursive procedure for analysis of variance. Biometrika 57, 19-46.

Wilks, S. S. (1962). Mathematical Statistics. New York: Wiley.

Williams, C. B. (1937). The use of logarithms in the interpretation of certain entomological problems. Ann. Appl. Biol. 24, 404-414.

Wise, M. E. (1952). The geometry of log-normal and related distributions and an application to tracer-dilution curves. Statist. Neerland. 20, 119-142.

Wolynetz, M. S. (1974). Analysis of type I censored normally distributed data. Unpublished Ph.D. thesis, University of Waterloo, Canada.

Wright, F. T., Engelhardt, M., and Bain, L. J. (1978). Inferences for the two-parameter exponential distribution under type I censored sampling. J. Amer. Statist. Ass. 73, 650-655.

Yohai, V. J., and Maronna, R. A. (1979). Asymptotic behavior of M-estimators for the linear model. Ann. Statist. 7, 258-268.

Yuen, K. K. (1974). The two sample trimmed t for unequal population variances. Biometrika 61, 165-170.

Zacks, S. (1971). The Theory of Statistical Inference. New York: Wiley.

Zelen, M., and Dannemiller, M. C. (1961). The robustness of life testing procedures derived from the exponential distribution. Technometrics 3, 29-49.

Zellner, A. (1971). An Introduction to Bayesian Inference in Econometrics. New York: Wiley.

Addendum

The following are some of the recent references that illustrate some additional properties and applications of the MML estimators and the robust procedures based on them.

Balakrishnan, N., and Kocherlakota, S. (1985). Robust two sided tolerance limits, Commun. Statist.—Theor. Meth. 14, 175-184.

Balakrishnan, N., and Tiku, M. L. (1986). Robust classification procedures based on continuous and dichotomous variables, Technical Report: Department of Mathematics and Statistics, McMaster University, Hamilton, Canada.

Bhattacharyya, G. K. (1985). The asymptotics of maximum likelihood and related estimators based on type II censored data, J. Amer. Statist. Assoc. 80, 398-404.

Fung, K. Y., and Paul, S. R. (1985). Comparisons of outlier detection procedures in Weibull or extreme-value distribution, Commun. Statist.—Simula. Computa. 14, 895-917.

Kocherlakota, S., and Balakrishnan, N. (1984). Two sided acceptance sampling plans based on MML estimators, Commun. Statist.—Theor. Meth. 13, 3123-3131.

McLeish, D. L. (1984). Estimation for aggregate models: The aggregate Markov chain, The Canadian J. Statist. 12, 265-282.

Singh, M. (1982). A robust procedure for estimating components of means from plant breeding data, Proc. Indian Statistical Institute Golden Jubilee International Conference, Calcutta, Sept. 27-Oct. 1, 1982 (Ed., S. K. Roy).

Tan, W. Y. (1986). Inferences on regression coefficients in a regression model under heteroscedasticity and robustness with respect to departure from normality, Commun. Statist.—Simula. Computa. 15, 35-60.

Tan, W. Y., and Tabatabai, M. A. (1985). Some ANOVA procedures under heteroscedasticity and non-normality, Commun. Statist.—Simula. Computa. 14, 1007-1026.

Tan, W. Y., and Tabatabai, M. A. (1986). Some robust procedures for comparing several straight lines under heteroscedasticity and nonnormality, J. Statist. Comput. Simul. 24 (to appear).

Tan, W. Y., and Balakrishnan, N. (1986). Bayesian insight into Tiku's robust procedures based on asymmetric censored samples, J. Statist. Comput. Simul. 24 (to appear).

Tabatabai, M. A., and Tan, W. Y. (1985). Some comparative studies on testing parallelism of several straight lines under heteroscedastic variances, Commun. Statist.—Simula. Computa. 14, 837-844.

Tiku, M. L., and Balakrishnan, N. (1985). Robust classification procedures based on the MML estimators, Proc. International Conference on Advances in Multivariate Statistical Analysis: Indian Statistical Institute, Calcutta, Dec. 16-20, 1985 (Ed., S. Das Gupta), (to appear).

Tiku, M. L., and Balakrishnan, N. (1986). A robust test for testing the correlation coefficient, Technical Report: Department of Mathematics and Statistics, McMaster University, Hamilton, Canada.

Tiku, M. L., and Balakrishnan, N. (1986). A robust Hotelling-type T^2 statistic, Technical Report: Department of Mathematics and Statistics, McMaster University, Hamilton, Canada.

Index